GLOBAL HEALTH IMPACTS OF NANOTECHNOLOGY LAW

GLOBAL HEALTH IMPACTS OF NANOTECHNOLOGY LAW

A Tool for Stakeholder Engagement

Ilise L. Feitshans

PAN STANFORD PUBLISHING

Published by

Pan Stanford Publishing Pte. Ltd.
Penthouse Level, Suntec Tower 3
8 Temasek Boulevard
Singapore 038988

Email: editorial@panstanford.com
Web: www.panstanford.com

British Library Cataloguing-in-Publication Data
A catalogue record for this book is available from the British Library.

Global Health Impacts of Nanotechnology Law

ISBN 978-981-4774-84-0 (Hardcover)
ISBN 978-1-351-13447-7 (eBook)

Dedicated to my children,
Jay Levy Feitshans and Emalyn Levy Feitshans,
my nephews, Jason Levy and David Levy,
and my grandchildren and their grandchildren.
With special thanks to
my beloved new spouse, Dominique Charoy
And with undying gratitude to my parents,
Sylvia Feelus Levy and Jack Levy Esq.,
whose love death could not part.
Very special thanks to Dr. Mark Hoover, US NIOSH;
Mme. Evelyne Morali, CIG, Geneva, Switzerland;
and Ms. Archana Ziradkar, editor par excellence.

Contents

Books by the Same Author

1. *Designing an Effective OSHA Compliance Program*
2. *Bringing Health to Work*
3. *Walking Backwards to Undo Prejudice: Report of the US Capitol Conference Including Disabled Students; What Works What Doesn't*
4. *Genetic Destiny: Today's Science Tomorrow's Laws*

Foreword

The mission of the US National Institute for Occupational Safety and Health (NIOSH) is to generate new knowledge in the field of occupational safety and health and to transfer that knowledge into practice for the benefit of workers and their families. Emerging nanotechology applications promise great societal benefits. There is a need to be sure that both the commercial applications and risk implications of nanotechnology are communicated to consumers, workers, and those who value the environment.

While much has been published about the various applications and risk implications of nanotechnology from scientific, ethical, and legal perspectives, publications aimed at a nontechnical, nonlegal audience are less common and are usually oriented to a particular country's legal rules. This is why *Global Health Impacts of Nanotechnology Law: A Tool for Stakeholder Engagement* by Dr. Ilise L. Feitshans fills a void in addressing the legal, social, and policy implications of nanotechnology from a global governance perspective.

The book makes the nanotechnology applications of daily life accessible to readers who are curious about this new scientific and legal field but do not have a doctorate in either aerosol physics or administrative law. Dr. Feitshans wants her readers to feel comfortable moving across the fields of science, ethics, and law in order to obtain a fuller understanding of how nanotechnology can reshape both commerce and public health, producing social benefits globally.

The benefits and risks of nanotechnology will be felt broadly across industries and countries. Nanotechnology will usher in new and improved methods for information technology, homeland security, medicine, transportation, energy, food safety, environmental science, and advanced manufacturing. It is important for all of us to understand this emerging technology. I invite you to read *Global*

Health Impacts of Nanotechnology Law. I think you will enjoy reading this book.

John Howard

Director, US National Institute for Occupational Safety and Health

2018

Preface

Small things add up: products applying nanotechnology have been marketed to consumers for over a decade, and therefore nanotechnology applications represent a huge slice of daily economic life [1], whether people know it or not [2]. "The impact of nanotechnology on the health, wealth, and lives of people could be at least as significant as the combined influences of microelectronics, medical imaging, computer-aided engineering, and man-made polymers," according to the report to the president of the United States, which launched an international vision of new materials and new products for the 21st century. "Developments in these emerging fields are likely to change the way almost everything – from vaccines to computers to automobile tires to objects not yet imagined . . . Such new forms of materials and devices herald a revolutionary age for science and technology," according to that report [3]. Long-term nanoscale research and development. . . materials and manufacturing, nanoelectronics, medicine and health care, environment, energy, chemicals, biotechnology, agriculture, information technology, and national security . . . . *How can the benefits of nanotechnology be realized, while reducing risk to public health?*

The sheer economic importance of nanotechnology will change several antiquated systems regarding industrial processes, scientific understanding and categorization of chemical informatics, and, ultimately, the health care delivery systems that must use or correct the end products of these changes anyway. The global health package embraces nanotechnology: new economic frontiers with wide horizons, promising new medicines, strong packaging to protect goods from contamination, cheaper consumer products, and new medicines to fight cancer. Nanotechnology's revolution provides the perfect vehicle to fix old problems. Therefore, its arrival in commerce provides an unprecedented, excellent opportunity to change society for the better. This book examines the following:

- Definition of nanotechnology under law

- The role of stakeholders in setting society's standards for acceptable nanosafety and risk management
- Emerging laws of risk management for nanotechnology; to be harmonized in a unified risk governance framework that will have positive impacts on global health

The conclusions include draft text based on legal principles for harmonizing risk management, as applied to emerging nanosafety laws, with a goal to incubate new products, while having positive global health impacts.

Nanotechnology's Revolution for the Global Economy Can Also Revolutionize Public Health!

Law and science have partnered together in the recent past to solve major public health problems using *global health diplomacy* [4], woven from international scientific collaborations and emerging multinational efforts for risk prevention. Global health diplomacy embraces nanotechnology—a revolutionary approach to the long-established rules in science and nature about matter and the properties of key elements such as titanium, silver, and gold. Nano-enabled products manipulate these properties to improve the quality of life worldwide. Using international collaborative research and treaty-based tools of diplomacy for global commerce, nanotechnology, nano-enabled applications of new knowledge, and nanomedicine bring products from lab to market, which, in turn, impact global health concerns about food, clothing, shelter, transportation, and medical care. Nanotechnology products and the local, national, and international laws that may govern them therefore touch the life of everyone. "It is expected that nanotechnology will play the role electronics played in the 20th century and metallurgy played in the 19th. . . . Manufactured nanomaterials are expected to yield significant innovation . . . a new competitive edge to European industry and strong benefits . . . from medicine to agriculture, from biology to electronics" [5].

Nanotechnology is here. Not only is it true that "you can't put the toothpaste back into the tube," but the toothpaste you use already also has the latest nanotechnology inside: a quiet but important example of the daily application of nanotechnology to consumer products [6]. The greatest challenge for successfully applying nanotechnology's promises involves designing and implementing flexible laws that will both incubate new commerce and ensure global health protections. This beautiful, bouncing new arrival in commerce, however, is well planned and therefore should not surprise policymakers, consumers, or stakeholders.

Figure P.1 Examples of nanotechnology products already in commerce. Courtesy: Dr. John Howard, Director, US National Institute of Occupational Safety and Health (NIOSH), as presented during his guest lecture at the International Labour Organization (ILO), Geneva, Switzerland: "Nanotechnology: The Newest Slice of Global Economic Life," November 27, 2008. Source: US government.

Opinion leaders in science, law, and health policy have heralded nanotechnology as a revolution [7] destined to realize unprecedented economic growth by applying smart new scientific developments since the beginning of the 21st century. In 2000, presidential advisers in the United States proclaimed nanotechnology the "next Industrial Revolution" [3, 8]. "The impact of nanotechnology on the health, wealth, and lives of people could be at least as significant as the combined influences of microelectronics, medical imaging, computer-aided engineering, and man-made polymers developed in this century," according to the report to the president of the United

States at the outset of the 21st century from the National Science and Technology Council. "Compared to the physical properties and behavior of isolated molecules or bulk materials . . . exhibit important changes for which traditional models and theories cannot explain. Developments in these emerging fields are likely to change the way almost everything – from vaccines to computers to automobile tires to objects not yet imagined – is designed and made. . . . Such new forms of materials and devices herald a revolutionary age for science and technology, provided we can discover and fully utilize the underlying principles," according to the US government's report in 2000 [9].

That same report to the president of the United States also successfully advocated for nearly a quarter of a billion dollars for research and development in 2001, which rose to 23 billion dollars a decade and a half later. Anticipating the cross-cutting importance of nanotechnology in every facet of commerce and daily life, the Executive Office of the President of the United States created a network of federal agencies within its government, spanning health [10], homeland security, space exploration, food and drug regulation, environmental protection, household and consumer goods, the Department of Commerce, the Department of Justice, and branches of the government associated with military defense [11]. The network is called the National Nanotechnology Initiative (NNI). Similar programs exist around the world. The NNI is credited with having tagged nanotechnology as a "revolution" [12] for industry and commerce, in 2000 [3]. Under the 21st-century Nanotechnology Research and Development Act of 2003 [13], NNI agencies are required to develop an updated NNI strategic plan. The Office of Science and Technology Policy (OSTP) describes the NNI mission as "the vision of the NNI is a future in which the ability to understand and control matter at the nanoscale leads to a revolution in technology and industry that benefits society" [14].

Policymakers and scientists have consistently agreed. These sentiments have been echoed by the Woodrow Wilson Center for Scholars [15]. By 2013, this mantra was unchanged: the European Union (EU) NanoSafety Cluster praised nanotechnology as "one of the key technological drivers in building an innovation [sic] European Union based on smart, sustainable and inclusive growth."

Scientists in the NanoSafety Cluster use policy words, not scientific terms, describing "tremendous growth potential for a large number of industry [sic] sectors." The EU has also committed billions to baseline nanotechnology research and commercialization of nano-enabled products [17]. Basic research building on the foundation of knowledge that has been developed in areas such as precision medicine and precision materials has expanded the already large range of applications for nanomaterials and nano-enabled products. Recognizing this evolution, the focus of the NNI has broadened in the 2016 draft strategy to embrace the evolving importance of previously unmet needs.

The EU, in collaboration and sometimes competition with the NNI, has kept pace with every facet of the NNI strategy. The EU NanoReg2 program is tasked with charting a path for registering, licensing, and certifying the nanosafety of a wide variety of substances and end products. The clear role of science policy in new laws at the national and international levels demonstrates that technology and economic factors in the globalization of commerce have transformed laws governing risk. In the 20th century, the main focus of international law strategies and national public health law agendas concerned law balancing diversity by maintaining cultural differences without reinforcing cultural prejudices such as racism and sexism. The 21st century has overcome many of those challenges and confronts the needs for one world with new institutions for governance and a new role for the rule of law. Nanotechnology is a key component of the social forces shaping society in the wake of legislated prohibitions on discrimination. Nanotechnology therefore will have a major impact on government and the shape of risk governance as a social good under law. The social determinants of health will also be influenced by nanotechnolgies, offering researchers, policymakers, and stakeholders in greater society the possibility of removing the health disparities that were the result of culturally embedded prejudices. This crossroads in science influencing social policy raises new questions about how people will survive when applying nanotechnology across a gamut of medical, security, travel, housing, and nutrition venues. It is not surprising, therefore, that the perception that nanotechnology is a revolution remains successfully echoed in the literature ranging from clinical nanomedicine [17] to the popular press.

"Nanotechnology represents the possibility of revolutionizing many aspects of our lives," according to Dr. Varvara Karagkozaki, a cardiologist and a leading researcher in nanomedicine.

"Nano-architecture is a new architectural style of the 21st century that will revolutionize architecture in every aspect" because nano applications in architecture can change the architect's vision, design techniques and ideas, construction methods, structure technologies, statics, form and aesthetics of mechanical systems, lighting, topic of energy, maintenance and repair techniques, decoration and interior design" [18].

"Nanomedicine is expected to dramatically exceed what has occurred in the field thus far, and our belief is that it will revolutionize medicine," said Dr. Gangs Bao regarding the opening of the Center for Pediatric Nanomedicine (CPN) at Emory University, Pediatric Nanomedicine Center, in partnership with Georgia Tech and Morehouse School of Medicine in Atlanta in March 2011. "We plan to make this new pediatric nanomedicine center a leader in applying these unique discoveries to treating and curing children's diseases." And he further noted that "because nano-scale structures are compatible in size to biomolecules, nanomedicine provides unprecedented opportunities for achieving better control of biological processes and drastic improvements in disease detection, therapy and prevention" [19].

"From a fundamental point of view, it is very interesting to investigate nano systems like nanoparticles," said Alexander Fabian from Justus-Liebig University, Giessen, Germany. "Since they can be fabricated in a very controlled manner, they can also be studied in a systematic approach. Properties of the nanoparticles different from the bulk, or even new properties like superparamagnetism in nanoparticles, make them also interesting for fundamental research" [20].

Echoing both government voices and the boasts of scientific research academia, delegates at the World Economic Forum (WEF) in Davos, Switzerland, declared, "Technology is the fourth Industrial Revolution" embracing nanotechnology in 2016 [21]. Klaus Schwab, founder and executive chairman of the WEF, stated, "We feel we are not prepared sufficiently for this fourth industrial revolution which will come over us like a tsunami which will change whole systems."

Putting all these attributes together, in totality, the arrival of nanotechnology and nano-enabled products in commerce means improved security, while reducing the price of goods used in health care, thus positively impacting not merely the global economy but also the global quality of life. Therefore the nanotechnology revolution for scientific theory brings important health impacts with its new economic frontiers with wide horizons, promising new medicines, strong packaging to protect goods from contamination, cheaper consumer products and new commerce from their trade, and new approaches to governance for their control. Dr. John Howard, director of the US National Institute for Occupational Safety and Health (NIOSH) openly predicted the advent of nanotechnology as a game changer [22]. In 2004, Dr. Howard predicted that nanotechnology was at that time already like a train having left the station and the general public was like a commuter who was running after it to catch up [22]. Nevertheless, he noted that such new technologies may reshape our understanding of the interaction between human health and the context of work as well as the nature of work itself. "These are exciting times," he said. "This time, we may be on the train [of technologic change], not just running behind it" He urged the leaders to "leap on that train before it leaves the station—if it hasn't left already."

Figure P.2 Photo of Einstein thinking. Final slide from the guest lecture by Dr. John Howard, Director, US NIOSH, at the ILO, Geneva, Switzerland: "Nanotechnology: The Newest Slice of Global Economic Life," November 27, 2008. Source: US government.

The progress between the 2004 speech and his lecture at the International Labour Organisation (ILO) in Geneva, Switzerland, just four years later in 2008 was startling. By 2008 research for anticancer drug delivery systems was well underway and lists of products using nanotechnology filled an entire slide (which Dr. Howard later humbly apologized was outdated). As Dr. Howard concluded in his pathbreaking lecture [6] about nanotechnology in 2008:

> "In the long term, nanotechnology will demand a revolutionary re-thinking of . . . health and safety."

References

1. John Howard and Vladimir Murashov, National nanotechnology partnership to protect workers, *Journal of Nanoparticle Research*, July 2009.
2. Mihail C. Roco, Chad A. Mirkin, and Mark C. Hersam, *Nanotechnology Research Directions for Societal Needs in 2020 Retrospective and Outlook*, World Technology Evaluation Center, Arlington, Virginia, September 2010. "The market is doubling every three years as a result of successive introduction of new products," p. 39.
3. National Science and Technology Council Committee on Technology, Subcommittee on Nanoscale Science, Engineering and Technology, *National Nanotechnology Initiative: The Initiative and Its Implementation Plan*, July 2000, Washington, DC, report to the president of the United States.
4. Shinzo Abe, Japan's strategy for global health diplomacy: Why it matters, *Lancet*, **382**:915–916, 2013.
5. Nicolas Segebarth and Georgios Kagarianakis, Foreword, in Michael Riediker and Georgios Kagarianakis, eds., *NanoSafety Cluster "Compendium of Projects in the European Nanosafety Cluster*, 2011 edition, Brussels, Belgium.
6. John Howard slide from "Nanotechnology the Newest Slice of Global Economic Life," guest lecture at the International Labour Organization, Geneva, Switzerland, November 27, 2008.
7. Louis Theodore and Robert G. Kunz, *Nanotechnology: Environmental Implications and Solutions*, Wiley Interscience, Hoboken, New Jersey, 2005. "The authors believe that nanotechnology is the second coming

of the Industrial Revolution, . . . Industrial revolution II," preface on page xii.

8. *National Nanotechnology Initiative: The Initiative and Its Implementation Plan*. The report further states, "The development of a healthy global marketplace for nanotechnology products and ideas will require the establishment of consumer confidence, common approaches to nanotechnology environmental, health, and safety issues, efficient and effective regulatory schemes, and equitable trade practices for nanotechnology worldwide."
9. "The initiative will support long-term nanoscale research and development leading to potential breakthroughs in areas such as materials and manufacturing, nanoelectronics, medicine and healthcare, environment, energy, chemicals, biotechnology, agriculture, information technology, and national security. The effect of nanotechnology on the health, wealth, and lives of people could be at least as significant as the combined influences of microelectronics, medical imaging, computer-aided engineering, and man-made polymers."
10. Centers for Disease Control and Prevention (CDC), NIOSH Publications and Products, *Approaches to Safe Nanotechnology: Managing the Health and Safety Concerns Associated with Engineered Nanomaterials*, http://www.cdc.gov/niosh/docs/2009-125/.
11. www.nano.gov.
12. National Nanotechnology Initiative (NNI) webpage, "FAQs," October 7, 2013: "Nanotechnology has been recognized as a revolutionary field of science and technology, comparable to the introduction of electricity, biotechnology, and digital information revolutions." This repeats concepts from the first NNI strategy, www.nano.gov.
13. 21st-century Nanotechnology Research and Development Act (15 U.S.C. §7501(c)(4), P.L. 108-153); www.gpo.gov/fdsys/pkg/PLAW-108publ153/html/PLAW-108publ153.htm.
14. Executive Office of the President of the United States, Office of Science and Technology Policy, National Nanotechnology Initiative (NNI) 2016 Draft Strategy for Public Comment, pp. I, 3, Washington, DC, US Government Printing Office, 2016
15. Daniel Fiorino, *Voluntary Initiatives, Regulation, and Nanotechnology Oversight: Charting a Path*, Woodrow Wilson Center for Scholars and the Pew Charitable Trust, 2010. This report examines nanotechnology voluntary initiatives and oversight.

16. European Union, Seventh Framework for Scientific Research, "Nanosafety Strategic Research Agenda 2015–2020 Safe and Sustainable Development and Use of Nanomaterials."
17. Varvara Karagkiozaki and Stergios Logothetidis, eds. *Horizons in Clinical Nanomedicine*, Pan Stanford, Singapore, 2015.
18. F. H. S. Javad, R. Zeynali, F. Shahsavari, and Z. M. Alamouti, Study of Nanotechnology application in construction industry (case study: houses in north of Iran), Current World Environment, **10**(Special Issue May 2015), 2015, http://www.cwejournal.org/?p=11691.
19. Press release March 28, 2011, Pediatric Nanomedicine Center links health care and engineering. First-of-its kind research center includes physicians and scientists from Georgia Tech, Emory University, and Children's Healthcare of Atlanta.
20. Alexander Fabian, Elm, Hoffman Klar, Magnets built from nanoparticles take shape, *Journal of Applied Physics*, **3**, July 2017.
21. Jane Onyanga-Omara, Davos 2016 examines 4th Industrial Revolution, *USA Today*, Section B, page 2, January 18, 2016.
22. Ilise Feitshans, NIOSH Director Launches New Millennium from Mt. Sinai, ERC OEM Press, Spring 2004.

Introduction: Crossing Impermeable Borders (The Uncharted Frontier of Law and Science Governing Nanotechnology)

Nanotechnology's revolution for commerce can revolutionize global public health: the scientific revolution that began at the dawn of the 21st century has taken hold, surpassing 3 trillion dollars in 2015 [1]. Every nation has a nanotechnology science strategic program under law, and most universities have big-ticket programs for nanoresearch and training, including nuclear research, nanomedicine, and genetic research using nano-enabled products. International treaties and national laws from countries where nanotechnology is a rapidly growing part of their economy abound, impacting every facet of production, storage, transport, distribution, use, and end-of-life-cycle removal or waste from manufactured goods.

This book brings together law and science for people who wish to understand the amazing role of nanotechnology in the global economy. People who use nanotechnology every day in their homes, for personal medical needs, and in cars, buses, airplanes, trains, and workplaces, need basic science policy information in order to (i) make informed choices of their own use of nano-enabled products, (ii) understand the information that is disclosed on labels regarding nano-enabled products, and (iii) have their say about laws and regulations pertaining to nanosafety in order to make the best use of these new technologies.

Global health impacts provides a brilliant example of how a dynamic moment in history finds society at an unusual policy crossroads: the changes wrought by technology offer the opportunity to choose which old values will be kept by the new order and which values will be thrown away. People who ignore vital social issues raised by the implementation of nanotechnology applications in commerce, or who shy away from discourse with people who disagree with them, risk ignoring the importance of these revolutionary developments. On the sidelines as spectators to nanotechnology's revolution reshaping society, they will then be clueless when old inequitable prejudices are accidentally embedded into the matrix for new nanotechnology laws or when prejudiced old rules no longer apply. Exploring key questions about nanotechnology offers people a unique window of opportunity to understand the cultural matrix shaping the laws, regulations, policies, and use of nanotechnology and then to seize the opportunities brought by the nanotechnology revolution to institute meaningful, long-needed social change to improve public health without sacrificing economic benefits promised by nanomedicine and other nanotechnology applications. This book therefore offers basic information to inform policy and future research for lawmakers, politicians, business leaders, and citizens throughout the world.

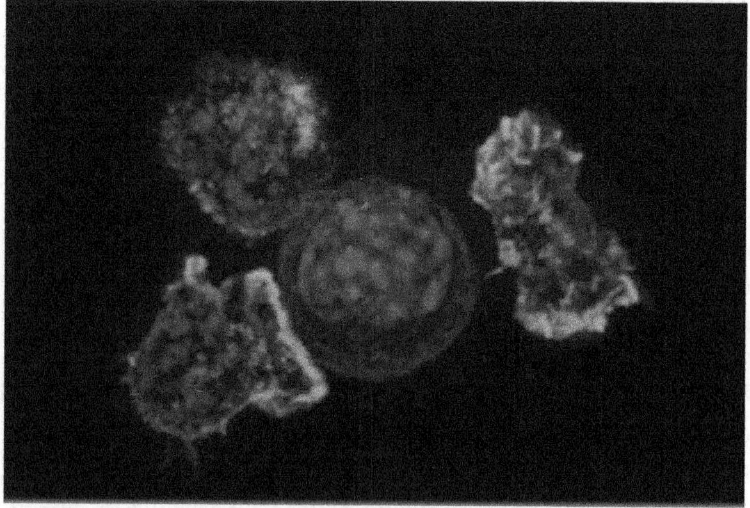

Figure I.1 Killer T-cells surround a cancer cell. Source: US National Institutes of Health (NIH).

Nanotechnology involves manipulating known chemicals at the molecular and atomic levels in order to create smaller, faster, stronger, lighter, reliable products [2]. The "nanoscale," which measures activities that apply nanotechnology, is very small. A human hair has a diameter of 100,000 nm. Within a few nanometers, nanoparticles can alter the biology of life. For example, natural mechanisms with complicated names such as "high-density lipoproteins" range from 8 to 10 nm, and ribosomes, the building blocks of DNA, are between 25 and 30 nm. This means that regulatory discussions about discoveries at the nanoscale under 100 nm might include many proteins that are ingredients for key genetic material. Additionally, there are new elements and substances with fancy names such as 1D and 2D matter and special properties for matter at the nanoscale: nanogold is combustible, nanosilver is antibacterial, and nanoscale titanium dioxide makes marvelous, frothy white foam for filling donuts, making whipped cream, and making shaving cream that billions of consumers use every day!

The prefix "nano-" is derived from the Greek word "dwarf" [3]. Yet nanotechnology is quite big! As predicted, the value of nanotechnology for commerce and industry rose to trillions of dollars by 2015 [4]. As nanotechnology's trajectory moves from experiments in laboratories into products in commerce, it becomes ubiquitous in daily life. Touching health care products, cosmetics [5], electronics, apparel, and automobiles, nanotechnology in the global economy impacts health. Miraculous-sounding developments that appear as if they are scenes from science fiction movies will enable patients to use nano-enabled medical products for new bones and organs from their own stem cells to defeat cancer or degenerative neurological diseases! And, engineered or manufactured nanoparticles are believed to traverse such historically mysterious places as the human placenta and the blood–brain barrier [6], thus offering new possibilities for treatment of illness or disease. Some of these treatments will use nanoparticle drug carriers to move between cells; others will enter cells such as those in tumors in order to alter them. Widespread application of nanotechnology in health care for drug delivery, and applications that regenerate tissue, teeth, and cells will redefine "health" and "disability" under law and in daily life by changing the nature of treatment for disability and redefining the population considered disabled culturally and under law.

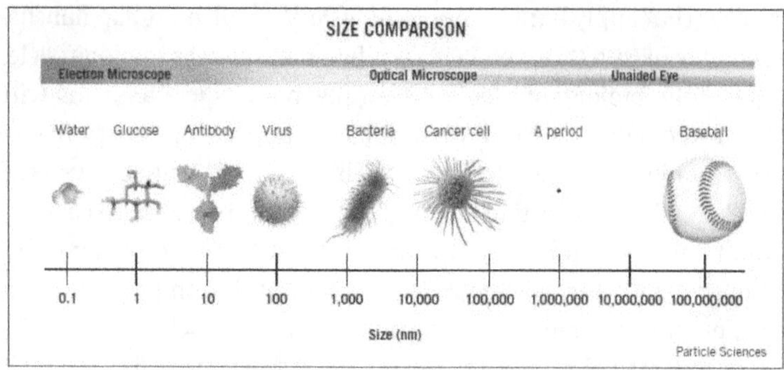

Figure I.2 Nanoparticles always existed. Photo source: Dr. John Howard, Director, US National Institute of Occupational Safety and Health, PowerPoint presentation at the United Nations ILO, Geneva, Switzerland, November 2008.

According to the White House Office of Science and Technology Policy (OSTP) created by the US Congress to advise the president of the United States about major issues in science, whole careers will someday be dedicated to advancing human understanding about the so-called *bio–nano interface* [7], then applying this understanding to medicine and commercial products, and then creating new methods to study its impact [8]. New uses for old materials, such as gold, silver, and diamonds, have already brought a breath of new life into old industries, where supply has diminished but not depleted, because industrial processes at the nanoscale require so much less in order to deliver their final product. And the same attributes of nanomaterials, although in different substances, are already in industry, making airplanes lighter and stronger, producing eyelashes longer and more alluring, and constructing houses and buildings that can withstand natural disasters.

Having new materials is very exciting! New materials and new methods for using those materials offers rejuvenating promising for commerce, for stock exchanges, for startup companies, for large-scale employers who use nano-enabled products, and for the future of humanity. Yet, the ubiquitous character of nano-enabled products may be a weakness as well as a strength. The very attractive feature of small nanoparticles that can traverse previously impermeable barriers also means that little is known about how to stop them from migrating, how to predict where they will go on their own despite human calculations, or how to predict which substances can interact as a trigger to make the nanoparticles cluster together or adhere to other substances.

Mystery surrounds the behavior of nanoparticles and the subsequent impacts. Estimating risk is problematic because so little is known about the emerging field of nanotoxicity. Therefore questions about so-called fate of nanoparticles loom important about controlling risk at the nanoscale. Thus, short-term benefits for the intended use of manufactured nanomaterials (MNMs) must be followed by vigilant attention to health and environmental concerns throughout the life cycle of the products in order to discover and understand results that cannot be predicted yet. Trying to stop nanoparticles from continuing to migrate elsewhere once the desired job is done, and then determining whether they cluster together or whether their collection in large clusters is useful, presents many questions. Some of these questions are already well understood about the same substances in a larger size, but the established rules of science do not apply to the same substance in the nanoscale.

Debating this dilemma of how to regulate unquantifiable risk in order to protect public health, while also protecting a culture of innovation, has been the hallmark of the first decade of juridical discussion regarding the application of nanotechnology to daily life [9]. There is an important but inescapable policy dilemma: whether unquantifiable risks can coexist with public health protections as the market expands for needed nano-enabled medicines and a wide variety of consumer products. Debate has not stopped the infusion of revolutionary nanotechnology into products across society. Nor has this debate stopped jurisdictions from writing laws about nanotechnology. Scientists and governments agree there are unknown risks. Examples include the Swiss Federation (Precautionary Matrix 2008) Royal Commission on Environmental Pollution (UK; 2008), the German Governmental Science Commission, public testimony sought by the US National Institute for Occupational Safety and Health (NIOSH; February 2011), the Organisation for Economic Co-operation and Development (OECD) working group (since 2007), the World Health Organization (WHO) working group (in process of formation), the International Organization for Standardization (ISO), the World Trade Organization (WTO), several industrial groups, and various nongovernmental organizations (NGOs). Legislators, therefore, have begun drafting laws despite the absence of clear and compelling information. Legislation is keeping surprisingly close pace with the rapid pace of emerging technology through global partnerships to draft treaties about nanosafety. The form of multinational regulation and its substance are reshaping government globally.

One of the fascinating features of nanotechnology is that it is not born of any specific discipline; the science is inherently interdisciplinary, and therefore the governance of social impacts must also be derived from a cluster of disciplines that have not previously worked closely together. This requires also a multidimensional approach to charting and measuring the social impacts. Lawyers can contribute information to this discourse [10]. Good legal training can inform every phase of this process. Creativity, however, is not random; cultivating innovations that save money and reduce duplication of efforts requires much forethought as well as new ideas. This means that policy documents and their regulatory content must be filled with more than compromise; it requires training outside one's own professional career path and then applying the lessons learned from that training outside of the academic bubble. Bringing together a diverse group of thoughtful people to create an accurate big picture is therefore one of the great challenges for people who wish to maximize the benefits of nanotechnology's revolutionary change. One former staffer in high-level science policy complained, "People in grad school end up being in their own little bubble."

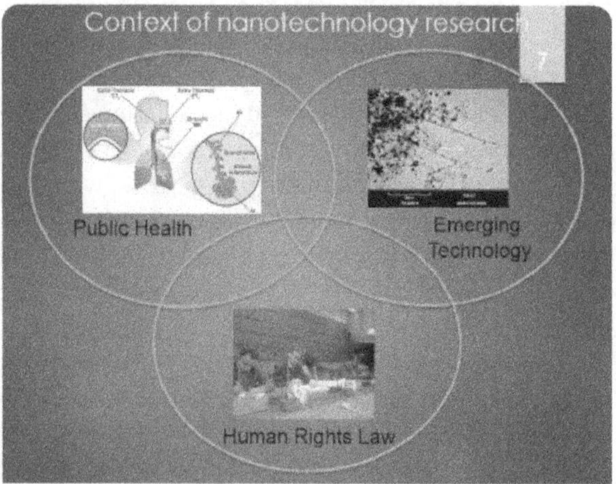

Figure I.3 The fascinating convergence of public health technology and law, prepared for the doctoral thesis "Forecasting Nano Law: Risk Management Protecting Public Health under International Law" by Dr. Ilise L. Feitshans, Geneva, Switzerland. Special thanks to the thesis advisor, Dr. Mark Hoover, NIOSH, USA.

Choice of Outcomes for Regulatory Policy

Key public health policy questions to explore regarding nanotechnology, nanomedicine, and their impact upon human health worldwide include:

- Rethinking the role of key illnesses and injuries in terms of the global disease burden when setting global health priorities
- Rethinking notions of informed consent and the right to refuse consent once presymptomatic testing diagnosis and treatment using nanomedicine become a reality, demanding new consent paradigms
- Rethinking the role of rehabilitation as a source of return to gainful work among aging populations who might not have considered working without the benefits of nanomedicine

This dilemma applies across many disciplines, including but not limited to, law, bench science, municipal bureaucracies, and academia. Science policy emerges with great difficulty when people from these various narrow tracks attempt to work together for the first time, as each discipline tries to surmount the learning curve for respecting the disciplines in other professions. International projects that require a delicate mix of law and science in global health diplomacy are therefore on the front lines, at the beginning stages of major bilateral and multilateral partnerships. At the highest levels, decisions can have lasting influence, but measuring the direct impacts is difficult when evaluating the long-term impact of research or diplomacy. Each facet of the new synthesis of disciplines for nanotechnology law and policy requires input from people who think logically about how to solve problems, how to work in teams, how to think across disciplines, and how to do so with respect for parallel professions. The seemingly impermeable barriers between disciplines in the life sciences, physics, and jurisprudence across professions must therefore also be crossed in order to create meaningful policies that are both workable and fair. This book attempts to prick a hole in the graduate school "bubble" that keeps people away from related disciplines, without destroying the sheltering structures that enable each discipline to flourish and gather stronger expertise in their own right.

References

1. Vladimir Murashov and John Howard, Essential features for proactive risk management, *Nature Nanotechnology*, **4**, 2009.
2. Royal Commission on Environmental Pollution, Chairman Sir John Lawton, *Twenty-Seventh Report: Novel Materials in the Environment; The Case Of Nanotechnology*, presented to Parliament by command of Her Majesty, November 2008.
3. European Commission, *Towards a European Strategy for Nanotechnology*, Luxembourg Office for Official Publications of the European Communities, 2004. "Originating from the Greek word meaning 'dwarf,' in science and technology the prefix nano signifies 10^{-9}, i.e., one billionth of a meter . . . ," p. 4.
4. John Howard and Vladimir Murashov, National nanotechnology partnership to protect workers, *Journal of Nanoparticle Research*, July 2009.
5. M. Crosera, M. Bovenzi, G. Maina, G. Adami, C. Zanette, C. Florio, and F. Filon Larese, Nanoparticle dermal absorption and toxicity: A review of the literature, *International Archives of Occupational and Environmental Health*, **82**(9):1043–1055, 2009, http://doi.org/ http://dx.doi.org/10.1007/s00420-009-0458-x.
6. Ilise Feitshans, *Final Report to the European Science Foundation*, European Science Foundation Epitome Grant Recipient, February–March 2012, Center for BioNano Interactions (CBNI), University College of Dublin, Ireland.
7. Executive Office of the President of the United States, Office of Science and Technology Policy, National Nanotechnology Initiative (NNI) 2016 Draft Strategy for Public Comment, p. i, Washington, DC, US Government Printing Office, 2016. As noted in each NNI strategy since 2003, "The Office of Science and Technology Policy (OSTP) was established by the National Science and Technology Policy, Organization, and Priorities Act of 1976. OSTP's responsibilities include advising the President in policy formulation and budget development on questions in which science and technology are important elements; articulating the President's science and technology policy and programs; and fostering strong partnerships among Federal, State, and local governments, and the scientific communities in industry and academia including space exploration and military defense."
8. From a legal standpoint, the term "bio–nano interface" has not been defined in the literature or in the background documents for discussion within the European Framework, where the term was coined.

Generally, the concept refers to the impact of nanoparticles upon the behavior, outcomes, and sustainability of ongoing biological processes. Use of the term "bio–nano interface" as an anchor for a sound legal definition remains limited, however, by the primitive state of the art regarding human understanding of these potential interactions. Despite the reality that some nanoparticles may have a biological life of their own that is not part of existing biological systems, setting forth the parameters of that interaction under law would be premature. A future legal definition of biological systems may be altered to reflect knowledge and discoveries that do not exist now but that are projected on the research horizon. For example, nanoparticle clusters that gather along the corona of proteins and the so-called fate of such nanoparticles may resemble existing biological processes that are not yet understood.

9. Ilise Feitshans, Ten years after: Ethical legal and social impacts of nanotechnology, *BAOJ Nanotechnology*, **1**, June 2016.
10. Ilise Feitshans, Continuing Legal Education (CLE) course in New York City, *Legal Implications of Nanotechnology within the Field of Public Health*, http://bit.ly/znlhip.

Chapter 1

Big Questions about Defining Little Nanoparticles

1.1 How Can Benefits of Nanotechnology Be Realized While Minimizing Risk?

Touching the economy globally, nanotechnology spans across almost every industry: food processing for retail markets, cosmetics, paintings and coatings, agriculture, equipment, and packaging. For every label that can be made lighter because of nanotechnology applied to thinner paper with less glue and less ink, there is reduced cost for transport, which is passed on to the consumer. For every bottle or can or container that is made thinner stronger and lighter, there is less chance of damage to the goods inside and reduced cost for storage, transport, insurance, and cleaning those goods before bringing them to market; and that cost saving is passed on to consumers, too. Now, imagine that each of those lighter papers is part of a greater packaging strategy: the paper around fast food, the plastic around boxes of fast food, even the ink on the paper and boxes, is lighter, stronger, cheaper to produce, rendering the benefit that it is easier to protect and insure the whole shipment compared to the same package a decade before. If so, any company that sells billions of hamburgers or sandwiches or liquid beverages per day can save money by simply reducing the weight of packaging

Global Health Impacts of Nanotechnology Law
Ilise L. Feitshans

ISBN 978-981-4774-84-0 (Hardcover), 978-1-351-13447-7 (eBook)
www.panstanford.com

by a few micrograms. Instant clothing is but one example of new and inexpensive applications of lighter, stronger textiles with less pigment for printing, flexible, and more resistant to destruction. And the airplane that has lighter, fluffier blankets for passengers with less chance of damage, with inflight movies whose video viewers go faster but weigh less because gold nanoparticles make their electronic systems more efficient, while being transported inside an airplane that nanotechnology has made stronger and lighter with less risk can also enjoy the benefit of cost savings. Now imagine how lighter paper and ink can impact hard-copy communication: books, magazines, and newspapers are less expensive to produce and transport, with the result that information can transfer quickly and freely. For this reason, nanotechnology applications can be found in every sector of the economy, even in places where one might not expect that anything is new or has been changed! Thus nanotechnology changes the calculus of risk and the methods for risk communication.

Nanotechnology involves the manipulation of matter at the level of a handful of molecules—the diameter of one human hair is 100,000 nm—and everything on earth is made up of nanoparticles. Some nanomaterials occur naturally: smoke, soot, dust, or sand. A sheet of paper is about 100,000 nm thick; a single gold atom is about a third of a nanometer in diameter. Dimensions between approximately 1 and 100 nm are known as the nanoscale. Unusual physical, chemical, and biological properties can emerge in materials at the nanoscale. These properties may differ in important ways from the properties of bulk materials and single atoms or molecules Matter at the so-called nanoscale exhibits unique characteristics compared to the same material in its more usual, regular bulk size. Some nanoparticles have been engineered or manufactured by humans, others occur naturally (e.g., volcanic ash), and yet another, mysterious group of nanoparticles seems to be generated as a by-product of events or industrial processes. The difference that may make ultrafine legal distinctions between the nanoparticles in each of these categories is blurry, and therefore the definitions for these categories may change with the use or context. Scientists may debate which types of nanoparticles fit into these categories, but regulators may have more concern about one category or another. And so, too, the task of drawing bright lines among types of nanoparticles in commerce is therefore very difficult. There may someday be definitions that will traverse several categories or bundle categories together on the basis of their use, context, and function. Significantly, the benefit or disadvantage of using nanoparticles of a specific substance may depend upon the intended use and its context.

Many people are just now discovering the beauty and pitfalls of extended use of nanotechnology in commerce, but this revolution created by nanotechnology is not really news. The US president's report called nanotechnology a "revolution" in 2000 because of the "unprecedented behavior by materials at the nanoscale that changes their chemical properties... usually between 1 and 100 nanometers." Nanotechnology is revolutionary because these discoveries challenge bedrock working assumptions in science. As demonstrated in Table 1.1, the presence of nano-enabled products is ubiquitous n society. This pervasive and new character of nanotechnology uses throughout the world means that nanotechnology also has an impact on cultural values, whether people realize it or not. Bringing new products to market, combined with the emerging social context for new rights for people with disabilities, will reshape society.

Table 1.1 Passive contact with nanotechnologies in daily life

- An *increased surface area* gives a promising future to solar energy. New uses for materials with a high surface area include catalysts, drug delivery molecules, energy storage, and popular cosmetics.
- *Electronics, organic electronics, and communication*: Interaction between nanostructures is *very* fast! Faster and energy-efficient systems enable communication at the speed of light.

Product	Application	Consumer contact
Food and agriculture	Production using nanosensors and creation of new foods, new materials, and new methods for using those materials offer rejuvenating promises for commerce, stock exchanges, startup companies, and large-scale employers who use nano-enabled products. **Packaging and storage** Many nanostructured materials can be harder yet less brittle than comparable bulk materials with the same composition, especially plastic wrapping!	Yes

Clothing 	New fabrics, rodenticides, and protection **Packaging and storage** Many nanostructured materials can be harder yet less brittle than comparable bulk materials with the same composition, especially plastic wrapping! This results in lighter, more secure transport with greatly reduced costs.			Yes
Shelter and lighting to protect homes    	Self-cleaning concrete and paints and coatings to protect buildings	Paints and coatings, interior and exterior Smart appliances and new electronics for security Refrigerator magnets and Christmas lights: electronic interactions influenced by activity at the nanometer scale, the so-called nanoscale Since nanoparticles of different sizes emit light at different frequencies, they can be used for different colors, like Christmas lights!	New building materials for larger, stronger buildings	Yes

Airplanes 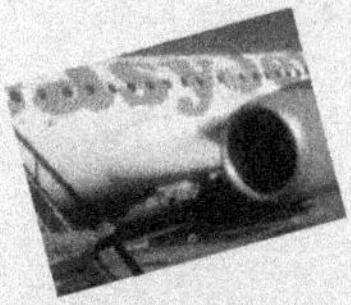	Lighter, stronger fuselage and wings	Interior fabrics and food service Cheaper construction costs translate into cheaper flights	Instrumentation	Yes
Cars and driverless cars	Tires	Paints and coatings	Instrumentation	Yes
Trains   	New systems for faster travel in Japan east hybrid train cars Self-cleaning concrete for clean stations	Tires, paints, coatings, and fabrics Electronics for rider passes and emergency communications	Instrumentation	Yes
Cargo transport and waste disposal 	Detectors for customs and duties for products in transit Hazardous goods and illegal goods can be detected and suspicious quantities noted.	Lighter, cheaper packaging that resists tampering and resists bugs. This means cheaper transfer of goods.	Lighter, cheaper, and strong vehicles mean cheaper transport of goods	No

Nanomedicine Killer T-cells surround a cancer cell. *Source*: US National Institutes of Health (NIH)	Biocompatible medical devices: Nanoscience enables health professionals to place artificial components and assemblies inside cells. These materials and components will be more biocompatible—less chance of rejection by the body and more chance of repair of the medical problem. Fighting cancer, Alzheimer's, Parkinson's, and bone loss; replacing teeth; and regenerating bones, neurological cells, and skin	Yes
Bioelectronics and electrical systems *Source*: World Health Organization	Nanoscale matter can sometimes be used to enhance specific fundamental properties of materials (such as magnetization, charge capacity, catalytic activity) without changing the chemical composition.	No
Cosmetics and personal care	Wow, lookin' good! Longer eyelashes, thanks to carbon nanotubes (CNTs) in mascara, longer-lasting color on the lips using less lipstick, new colors with sparkle and transparency, lighter cosmetics for foundation, perfumes, and a variety of skin creams . . .	Yes

Summary: Nanotechnology will, therefore, leave no corner of the global economy untouched, across industries: food processing, cosmetics, paintings, and coatings.

Some amazing manufactured nanomaterials (MNMs) are specially designed for their size and properties. According to BNA Bloomberg, "Between 2009 and 2016, revenue from nanotech products in the U.S. grew more than six-fold" [1]. Exploring the tremendous potential for nanotechnology to create nano-enabled products therefore opens up the possibility of new products and new durability for well-established products. Yet, the new interaction of nanoparticles that have been deliberately created with a view to altering biological systems or changing the durability of goods raises many questions compared to the use of the very same substance in its bulk form. This has implications for the cumulative dose from human exposure, for the synergy of the mixture of MNMs with other substances that might be harmless if encountered alone, and for the long-term disposal and recycling of nanomaterials.

"The impact of nanotechnology on the health, wealth, and lives of people could be at least as significant as the combined influences of microelectronics, medical imaging, computer-aided engineering, and man-made polymers developed in this century," according to the report to the president of the United States that kicked off nanotechnology research and development for the 21st century. "Developments in these emerging fields are likely to change the way almost everything –from vaccines to computers to automobile tires to objects not yet imagined . . . Such new forms of materials and devices herald a revolutionary age for science and technology, provided we can discover and fully utilize the underlying principles," according to that report [2].

National Nanotechnology Initiative (NNI) of the United States

The initiative and its implementation plan, July 2000, Washington, DC, report to the president of the United States, predicted: "Biotechnology and Agriculture. The molecular building blocks of life - proteins, nucleic acids, lipids, carbohydrates and their non-biological mimics - are examples of materials that possess unique properties determined by their size, folding and patterns at the nanoscale. Biosynthesis and bioprocessing offer fundamentally new ways to manufacture new chemicals and pharmaceutical products. Integration of biological building blocks into synthetic materials and devices will allow the combination of biological functions with other desirable materials properties. Imitation of biological systems provides a major area of research in several disciplines."

Roadmap for CNT Yarn Applications

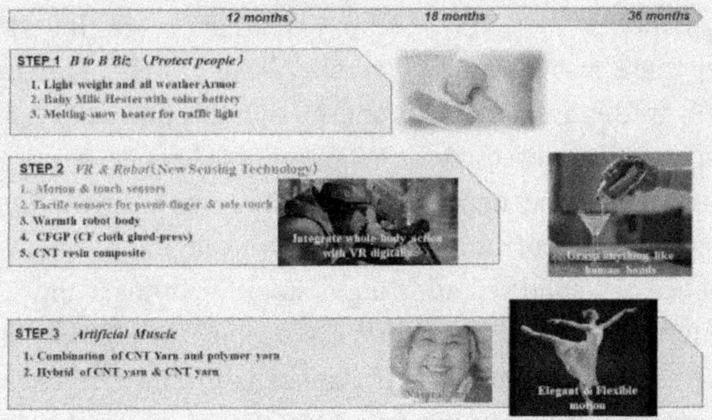

Figure 1.1 Carbon nanotube yarn for lighting and heating will have applications in armor, for warming food, in heaters for traffic control in snow, in motion and track sensors, in tactile sensors for robot skin, and eventually for artificial muscles. Science fiction or science fact? Photos courtesy of Siddarmark, LLC, Japan.

Nanoscience will contribute directly to advancements in agriculture in several key ways: molecular-engineered biodegradable chemicals for nourishing plants and protecting them against insects, genetic improvement for animals and plants, delivery of genes and drugs to animals, and nanoarray-based testing technologies for DNA testing. Plant scientists will be able to determine which genes are expressed in a plant when it is exposed to salt or drought stress.

The application of nanotechnology in agriculture has only begun to sprout the benefits to be reaped later. Nanosensors on harvesting equipment can detect which parts of the crop are ready to be cut and which parts of the crop should be left until a specified time later. Nanotechnology is also applied to packing and transport of foods. Additionally, nanotechnology is being studied to create new foods and new plants that will be viewed as commonplace in the later 21st century. Industries using nanotechnology traverse all the economic domains.

Yet, risks that may arise when applying such technologies emerge slowly compared to the many new vistas of prosperity and efficiency that nanotechnology promises to humanity throughout the world. These properties may differ in important ways from the properties of bulk materials and single atoms or molecules. For example, researchers have started to explore the health impact of exposure to ultrafine nanoparticles following combustion processes involving carbon nanotubes (CNTs). Some researchers are investigating whether there may be an accidental release of toxic nanoparticles following traffic accidents. No one knows the potential harms or benefits from these interactions. While opinion leaders in science, law and health policy may herald nanotechnology as a "revolution" with unprecedented opportunity for human development and growth, the absence of a solid scientific understanding of these interactions threatens to bring bad consequences despite an alluring mystery and fascinating new economic frontiers with new products. With forethought, however, the maximum benefits of nanotechnology can improve global health and the quality of life.

1.1.1 Simple Applications of Nanotechnology to Daily life

Thousands of products using nanotechnology are already on the market because the prospect of lighter, stronger materials for packaging and for reshaping industrial processes enables manufacturers to apply nanotechnology to trusted brands of established product lines. Nanosilver, for example, is an excellent antibacterial and rodenticide that is used to coat textiles in uniforms, furniture, rugs, infantwear, and designer clothing. Nanosilver is also used as a refrigerator lining for home use and in transport of goods and foods. CNTs have a wide variety of uses, but their ability to make stronger mascara and packaging is a likely point of contact for many consumers. Bottles, bags, and wrapping materials are excellent examples of CNTs in daily life. Fluffy products from donut fillings to shaving foam may also contain nanoscale quantities of titanium dioxide. Nanotechnology therefore creates employment and new products for industry, stimulating global commerce in every facet of the economy, including nuclear research and genetic research, while also raising questions about the cumulative impact of these new technologies on the environment, on cultural values shaping the quality of life and impacting world health.

Nanomedicine redefining health and offering an opportunity to end embedded health disparities (life cycle based on WHO points in the human life cycle):

1. Birth to 5 years
2. Early childhood
3. Adolescence
4. Reproductive years
5. Advanced aging

Source: Ref. [3].

1.1.2 An Illusion of Consensus: The "Nano" That "Everybody" Knows

Consumers enjoy applications of nanotechnology in paint coatings, refrigerator linings, sun tan lotions, and even a car called the "nano." In 2011, the chain store Migros in Switzerland started an ad campaign for "nano mania": a set of toys marketed nationally in the stores.

1.1.2.1 Are all of these nanoproducts made of pure nanotechnology that must be regulated?

At first, nanotechnology policy questions seem to be easy to answer. "Everybody" knows the true and clear answers to policy questions, or at least "everybody in the scientific community" knows. The European Parliament has called for the adoption of a "comprehensive science-based" definition of "nanomaterial" without hesitating to ask whether consumer-based definitions may embrace bulk substances that scientists would exclude. The European Union (EU) has also committed billions of eurodollars to baseline nanotechnology research and commercializing nano-enabled products. Thus, there may appear to be an immediately clear difference regarding a "truly scientific" discussion of nanotechnology and nanoparticles, compared to use in commerce when the term "nano" appears for "things that are just called "nano" but aren't really," but these differences are not practical as the starting point for implementing a nanosafety program. Many substances are already regulated because they are dangerous and considered to be toxic or hazardous under law, for example, titanium dioxide. There has been a debate whether these substances are automatically covered by existing law because they are already regulated in many jurisdictions or instead whether special rules regarding the safe use of these products at the nanoscale are also needed. Yet, many industries believe that wholesale extension of existing law without closer scrutiny of data about dangers and risk is unfair and possibly an undue burden on commerce and industry. For materials such as titanium dioxide, there is no legal presumption that these materials will be harmless at the nanoscale despite the absence of any information about their properties and behavior in such small quantities. But, no data exists regarding obvious dangers for use at the nanoscale in order to justify regulation. Human nature resists writing laws with no underlying science, based on fear, but a long history of applying precautionary principles of science because of established toxicity is embedded in many existing national and international laws. Nonetheless, nanoscience establishing clear nanotoxicity is premature. Thus, fear exists, but empirical data to justify use of the law does not.

Since the "fate" of bionano interactions remains unknown for presumably safe materials such as carbon and boron, major health risks could inadvertently be excluded from regulation. Determining whether these results under law are intended, good, or unwanted, and thereby deserving incentives or punishment, however, is a policy decision, derived from political compromise. Policy judgments, not science, will fill this void, without criteria to define nanomaterials or characterize the risks, unless scientists and researchers understand the tools of legislative policy. Conversely, the possibility exists that some substances may be safer at the new level either because of the way in which they are used or because of poorly understood properties that are expressed at the nanoscale.

There is remarkably little consensus regarding the correct method to apply precautionary principles and whether regulation is needed at all. For example, in 2013 the European government's administrative collaboration agency Safe Implementation of Innovative Nanoscience and Nanotechnology (SIINN) compared dozens of definitions from working groups and found an incoherence when searching for common threads of understanding about nanotechnology and its related terms. As discussion slowly evolves toward regulatory language, however, the difficulty of the task of defining nanomaterials and prioritizing risk for workers, consumers, and the environment becomes clear. Applying these terms in workable definitions requires limits: the term "nanoparticle" could, for example, refer only to a substance that has been engineered for a specific purpose to be used in the nanoscale and not the structures of the same size that occur in nature. These categories are fluid: materials may fit both sets of criteria for their definition, and some policymakers might be more concerned about their interaction or their outcome after use than the origins that give rise to a definition. Using scientifically accepted definitions from key reports, such as the government of the United Kingdom, Switzerland, or the multinational Organisation for Economic Co-operation and Development (OECD), might result in legislation that includes broad and vague terms such as "atomic, molecular and macromolecular scales" or "properties differ significantly from those at a larger scale."

Another problem soon becomes clear: such definitions can be meaningless without background explanatory text and sensible criteria. There are no clear criteria for explaining these terms to limit the application of nanotechnology laws, however, because everything is made up of nanoparticles. Therefore even if everyone understood one unified definition of what "nano" is, the law would not have clear limits and could embrace every operation of daily life as new products evolve.

Instead, many different sources have used similar language to define engineered or MNMs, so it becomes deceptively simple to conclude that "everybody" agrees on "what nano is supposed to be." Yet, the scientific concept of "nano" holds ramifications under law and regulation that are quite important. The EU's commission recommendations, published in its official journal, use treaty-based language to justify its pronouncements, but this clear basis for jurisdiction does not make its technical terms clear: "There is no unequivocal scientific basis to suggest a specific value for the size distribution below which materials containing particles in the size range 1 nm–100 nm are not expected to exhibit properties specific to nanomaterials." A definition of the term "nanomaterial" should be based on available scientific knowledge.

The same EU recommendation would allow regulation even if a substance is suspected of becoming harmful without fitting this sharp definition: "A nanomaterial as defined in this recommendation should consist for 50 % or more of particles having a size between 1 nm-100 nm. In accordance with SCENIHR's advice, even a small number of particles in the range between 1 nm–100 nm may in certain cases justify a targeted assessment. But, it would be misleading to categorise such materials as nanomaterials. Nevertheless there may be specific legislative cases where concerns for the environment, health, safety or competitiveness warrant the application of a threshold below 50 %."

Avoiding the hard choices that must be made by legislatures, this EU recommendation throws the question into the realm of hard science without giving the scientists training about their unexpected role as policymakers for continent-wide laws. The statement that definitions will rely on science is therefore both obvious and vague at the same time, because of the failure to provide guidance for implementation. This recommendation does not state whose knowledge, at what moment in time, or how one determines availability of data—whether through privately funded research, academic criteria, scientific consensus, published governmental reports, or merely the data available on the web. This statement of the obvious does not explain why these parameters are important or whether there should be legal significance in the discrepancies in methods or results. The rule does not suggest either whether these limits are based on political judgments or a specific well-known readily accessible written text of scientific consensus. Thus, an illusion of certainty emerges without any real connection between the text of the rule and the use of nanotechnologies in daily life. Consequently, working definitions of nanotechnology seem to have one thing in common: at first, their terms seem easy to understand, the subject of wide consensus. But, upon further reflection, any attempt at applying such terms to a broader context than the one originally contemplated reveals that the consensus is shallow.

Consider, the car called "nano" and the Migros grocery store toy called "nano mania". Few scientists might consider these products to be a "pure" example of "nano," and therefore researchers might reject the idea of labeling those products accordingly. But the mere use of the term "nano" conjures an image of science and regulatory governance in the mind of the general public making a purchasing decision. Defining nano-enabled products in such a manner that would make it clear to consumers that such products, which do contain nanoparticles, are not the nanotechnology that is subject to regulation is problematic.

In practice, things that call themselves "nano" might not meet scientific criteria and yet will be viewed as part of the legal framework by the general public. And the criteria to be applied require further scrutiny when making the legislative line-drawing between "what is nano and what is not nano" for the purposes of law. For example, from the standpoint of risk, it could not be argued that a car is without risk to the consumer population or overall society or that the risk from a car is not potentially lethal regardless of dose or time of exposure. From the standpoint of the percentage of engineered nanoparticles used in the product, it is also unclear whether the combined presence of nanotubes in paint coatings, nanosilver for fabric protection, nanoparticles making brighter headlights and a variety of other parts of the car might constitute a total percentage that conformed to a possible threshold level of nanoparticle engineering for "nano" regulation. This dilemma will become more important as new information reveals important scientific distinctions from the standpoint of risk based on the use of the nanomaterials and the size, shape, and characterization of the materials. Thus, the same nanomaterials may have different health impacts based on context, use, and function or based on dose for exposure. This means that regulating the "pure" nanomaterials that everyone knows are scientifically involved in research might understate risk from nonregulated exposures. Or, definitions may bring under the regulatory harness materials that will not pose important risks to the people who are exposed, thereby wasting valuable administrative resources that are needed to monitor exposures with potential for harm. Thus, when drafting text, accidentally overbroad language can become a legislative nightmare and a regulatory burden even to industries that do not consider themselves to be using nanotechnology, unless the criteria are clear.

This raises an important legislative question from the standpoint of protecting public health: is it appropriate or enforceable to have laboratory-quality CNTs treated the same as a car or a toy under nanotechnology laws, since their presence in these objects changes the very nature of the objects, by design. These terms provide only an illusion of clarity outside the laboratory bench. Few substances behave the same way in a small quantity compared to on a larger scale, regardless of whether involving the nanoscale or not. Imprecise definitions can bring organic substances also into the regulatory regime—whether or not the legislature wants to include them. The legislative drafter who does not focus on the outcome rather than the origins of the nanomaterial may soon find that the purely scientific "real nano" to be regulated is elusive, if it exists at all.

1.1.3 Labels: How Big Is Nano, How Small Is NOT Nano?

Well, everything is made up of nanoparticles. The term is really a mere unit of analysis, not to be confused with a special entity. While it is true that nanoparticles behave differently at the nanoscale compared to the same substance in microns or larger bulk units, it is inaccurate to state that nanoparticles never existed before. That is not why nanotechnology is so special.

Nanoparticles always existed, but humans could not see them and scientists could not understand them until the 21st century

(Fig. 1.2). The discovery of how to alter nanoparticles and use them in a variety of new applications is indeed the magic gift of nanotechnology. The revolutionary understanding about the properties of matter at the small nanoscale and the ability to harness that learning into wonderful new products or stronger old products can reinvent many industries, especially in the realm of medical care. But describing this new aspect of old products and nanomaterials is not easy from the standpoint of consumer labeling, regardless of whether the consumer is an individual, a family, or a major multinational corporation.

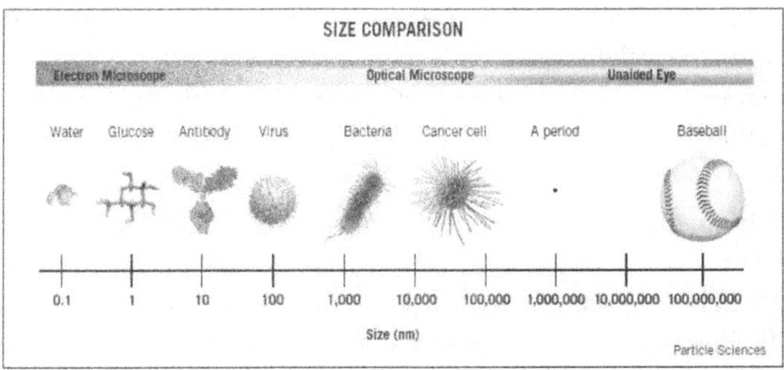

Figure 1.2 Nanoparticles always existed. Photo source: Dr. John Howard, Director, US National Institute of Occupational Safety and Health, PowerPoint presentation at the United Nations ILO, Geneva, Switzerland, November 2008.

The term "nano" may have taken on new cultural or commercial meanings, but regardless of whether those meanings are positive or negative, there is nothing especially informative about a label that reads "Contains Nanoparticles." Size therefore is a deceptively simple criterion for distinguishing between nano-enabled products and other products offered in commerce.

Important health risk information should be disclosed to avoid liability and to allow consumers to make choices about the products they buy. But in an era of unquantified risk, when civil society is just beginning to consciously apply nanotechnology, it is not clear what information should be disclosed as a matter of law. According to one commentator, "A definition is urgently needed, especially for particulate nanomaterials . . . A definition is required for labeling purposes, and would assist industry and regulators in identifying where specific safety assessments might be necessary. We acknowledge that it would need revision in line with fresh scientific evidence." Since everything is made of nanoparticles, the question is really "Which nanoparticles are worth listing on a label?"

This is problematic. The legislative drafting question whether labeling is important has two possible answers:

- Labels are important because consumers will demand them, depending on consumer perception of risk, which may be unrelated to known risk and instead may require labeling to reassure consumers even if a product is safe.
- Labels are designed to actually inform consumers about their known risk and therefore must contain known information that is accurate and cannot unintentionally mislead end users and supply chain consumers, even if contemporary labels must note that the information may change in the event that risks may later emerge.

So, too, the question of how much information to disclose and in what format may be the same for a wide range of consumers, such as individuals and commercial entities. If the desired outcome is to satisfy the inquisitive consumer and avoid liability there may be a large quantity of text to disclose. Some retailers may fear that too much text will scare people away or that a court will determine at a later time that the label did not have accurate information in the text, as disclosed. Science, too, may reveal new problems or solve old ones before the text of the label has been printed, and then the question whether the label should be revised will become very important. Accuracy of the text of the label may therefore change across time. For example, a major breakthrough in the state of the art of measurement or empirically based risk changes our understanding of danger, long after the law has been written. Risks may be less important for some specific uses compared to others; nontoxicity may be less important for drug delivery to a terminal cancer patient than for protecting toddler clothes. Legislatures and regulators may debate about the extent of disclosure, how simple should the wording be about sophisticated scientific questions when very little is known, and how to reasonably streamline this information so that it can fit on the side of a tube of toothpaste or a box for a medical device.

But these questions are not novel for nanotechnologies: what to disclose about known risks that are unquantified or unknown remains a perennial issue in the medical community and for label writers and policymakers. Absorbing new data can be built into a flexible framework under law, but implementation is a matter of political will, beyond the lab and crossing impermeable borders, into the hearts and minds of people throughout civil society. Therefore any meaningful legal criterion requiring disclosure by labeling reflects a societal policy judgment about "how much risk" is acceptable in order to determine which products fall under regulations and what features of those products must be disclosed.

Labels: What Information Should Labels Disclose?

Figure 1.3 Labels according to the GHS. Photo by the United Nations Economic Commission for Europe.

Scientists may view information on the label as simplistic.

Consumers may need reassurance or actual data about handling, storage, and use.

Lawmakers may have different goals, depending on the statutory mission of their agency.

In-house lawyers may wish to prevent accidental harm and prevent liability.

How much information? More than "contains nanoparticles"

Key components whether listed as dangerous or not

Dangers or risks of mixing that could cause combustion, explosion, fire, or corrosion

Depending on the size of the product, possible Safety Data Sheet (SDS) or link to website SDS

Warnings, as appropriate

The European Cosmetic Products Regulation of 2009 and a European Parliament legislative resolution on food information adopted in July 2011 both stated that nanomaterial ingredients should be labeled without stating the parameters of the information to be disclosed nor the limits on the percentage of nano-enabled components in a product that qualifies under law. Legislators and voters will make policy judgments about how to treat products that are self-proclaimed as "nano" but not perceived as genuinely "nano" in the opinion of scientists, and their decisions may express political will based on popular perceptions. For example, it cannot be argued that cars are without risk to consumers or society or that cars are not potentially lethal. Measuring the percentage of nanoparticles used in a product does not give a bright line for determining regulatory jurisdiction either. The policy dilemma of what to do if manufacturers or retailers might call a product "nano" to make it appealing to consumers is also important. If a product is called "nano" but remains unregulated, any purchaser—large corporation or individual consumer—may have a false illusion of also being "safe" once purchasers know that "nano" products in general are subject to a vast apparatus of regulatory oversight. Consumers might be lulled into a false sense of security if labels allow consumers to believe that such "nano" products had passed the rigors of regulatory testing or monitoring, when, in fact, they were not included under the law. Yet, scientists might not consider these products to be a "pure" example of "nano." In a context where the functionality impacts thousands or millions of people, or where people believe that a product applies nanotechnology even when it does not, the difference between a genuine scientific use of "nano" and the popular use of the term can soon become meaningless and therefore should be avoided in favor of a bundle of criteria that describe the harm that society wishes to avoid. Thus this small matter becomes the touchstone of an international debate: what is the responsibility of scientists, government, and business in relation to new discoveries that have no proven threshold for harm but a history of regulated toxicity in larger quantities?

There is no easy answer for this seminal regulatory dilemma. Lawyers advising business may fear that in the event of catastrophic harm, juries and the popular will would base punishment for any wrongdoing on the well-established known toxicity of some substances and punitive damages would follow, as in the case of litigation about asbestos. Although asbestos use is still a thriving industry, the substance acquired a notorious reputation when its use in industry was not accompanied by disclosure of potential harms that scientists considered well known but were unknown to the general public. The disconnect created by this discrepancy nearly crippled the industry, even though far more dangerous substances were successfully used in commerce without litigation or public outcry. The questions about the convergence of perception of risk and popular will for legislation and regulation are discussed in the context of lessons learned from the absence of regulations for asbestos.

The major lesson learned from the sad asbestos experience of the 20th century is that failure to disclose was the source of harm for industry. Without disclosures, proving that use at the nanoscale of known toxins, such as titanium dioxide, was indeed harmless would be difficult, if not an insurmountable evidentiary task, even if the facts proved that the nanomaterials had been used responsibly or had not caused harm. Therefore legislators and regulators should focus on the function when applying nanotechnology and on the monitoring methods that are appropriate for avoiding undesirable outcomes. Of course, the decision of which outcomes are beneficial and which outcomes must be avoided is also a legislative concern, a product of collective political will.

1.1.4 Capturing Discovery: The Limits of Lists and Numbers as Regulatory Criteria

A so-called bright-line principle for legislation is as alluring as eyelashes by Lancome or Estee Lauder but probably not as valuable a commodity for international commerce. What if there is a major breakthrough in the state of the art of measurement or the state of the art regarding understanding of empirically based risk between the time that the law is written and the time that the law will be applied? The hallmark of clear law involves sharp, line-drawing definitions when such distinctions are practical, but exceptions are always possible under law. The recurring phrase that nanoparticles and therefore nanoproducts are "100 nm or smaller" has been used in references and textbooks worldwide. Yet, no one knows whether in applications to industrial processes and exposures in the real world, this arbitrary cut-off is important as a determinant of significant health effects upon humans or the environment. According to UN agencies, "The current state of knowledge about the unique properties of engineered nanomaterials does not permit identifying exact criteria that present 'bright lines' for inclusion, or exclusion, for nanospecific risk evaluation .... use of 100 nm as a cut-off point for particle size does not have a biological basis." Despite the reality that the scientific community references "one to one hundred nanometer" scale as a definition of nanotechnology, it is not practical to put a numerical cut-off into the text of laws surrounding scientific discovery. A major, potentially catastrophic harm due to the use of engineered nanomaterials at 1 to 100 nm or more would fall beyond the scope of regulation and leave people without redress. And safe or reasonable low-risk activities under 100 nm would be subjected to extensive scrutiny, even though they do not pose a threat of harm. Using this well-established numerical cut-off would therefore have an unintended unfair effect: technology that established testing proves is harmless would be subject to regulation nonetheless!

Applying number-based standards invites a numbers game; inevitably, something will be designed to avoid the number in the legislation, whether the regulators write about nanoparticles or taxes. Avoiding a "bright line" definition based on size enables the people writing the law to craft criteria for determining which problems are important enough to merit the bureaucratic resources and attention of the legislation, without inadvertently also regulating minor situations that would distract the implementation of the law. This approach also has the advantage of avoiding the numbers game: it is not possible to engineer away from a threshold or cut-off if is there is none, because the trigger for jurisdiction of regulations is based on function. This avoids the messy questions of fairness that would occur by adopting the approach based on quantity without regard to proven harm. One logical approach is offered by the Royal Commission on Environmental Pollution, instead, would like regulators to focus on "functionality." Their criteria for whether or not a chemical is regulated as nanotechnology would depend not on size as a nanoparticle or form but more importantly on how the nanomaterial is used. The answer to questions about regulation would be examined on a case-by-case basis, and the answer may even be different with each use of the same substance. Since there is probably no single criterion that will define "nano," no term defining nanotechnology and its related sciences may be the wisest use of language. What? Write legislation for nanotechnology without defining the subject?

Yes. This is an old legislative trick that is used when trying to solve a ubiquitous social problem, such as protecting people with disabilities or ending racial discrimination. Legislators frequently follow this path: avoiding definition while using flexible criteria when writing legislation about a subject fraught with uncertainty. Typically, this is accomplished by avoiding defining the key term in the law entirely. Many laws that overcome thorny social problems do not have clear definitions, because the harm is shaped by outcomes (i.e., not obtaining or retaining employment, unequal pay or failure to provide fair opportunity to employment, fair access to voting or services or housing). For example, disability laws do not list a range of diseases or injuries that qualify for the protection against discrimination that are afforded to people with disabilities. Instead, the law protects people who have "significant impairment of one or more major life activities" [4]. One can see the fireworks exploding around each of these words in the magic incantational phrase that gives some people special protection. Litigation has parsed and defined each part of this phrase: "significant impairment" means more than a mere diagnosis regardless of how deadly the disease is [5], "one or more" means that several minor impairments taken together because of genetics, comorbidity, or happenstance can give rise to significant impairment, and "major life activities" are as important as seeing, walking, or reading with comprehension or reproduction of healthy offspring, even though the activity of reproducing offspring does not occur every day [6].

When the law is concerned with eradicating prejudice by providing equal opportunity, for example, it is not practical to list the possible harms, because people who are prejudiced and wish to discriminate can cleverly invent new obstacles to defeat the goals of the law. To achieve the goal, for example, of equal opportunity, therefore the legislature avoids the definition of acts of discrimination by stating the outcome that is desired and not defining the object of the legislation at all. Instead, people who can demonstrate to the court that there is a disparate impact that harms their protected group can claim redress for discrimination.

Legislatures also use the tool of failing to define key terms when they wish to avoid a problem that is so politically controversial that it could destroy the bill before it becomes law. For example, in the United States' Occupational Safety and Health Act of 1970 (OSH Act) [7], "safety" and "health" are not defined. "Health" is defined under the international law of the World Health Organization Constitution [8], but there is no federal right to health under US law. Many legislators who fought for the OSH Act feared that insisting on a right to health by defining the term "health" would result in defeat of the bill and there would never be an occupational health law. "Safety" is used all around the world in a variety of contexts but also has no precise meaning under US law. Instead, the US Congress listed "safety" and "health" among the act's purposes "to preserve our human resources" and avoided the question of how to determine when there is health or safety. Dangers, the US Congress wrote, involve "recognized hazards" as defined by the scientific community. Case law teaches that "recognized hazards" need not be described in the plain text of existing written standards and can be found in the collective wisdom of a vibrant research community [9].

Especially important from the standpoint of social values regarding health and illness that will experience a paradigm shift following widespread use of nanomedicines for early detection of disease, international law that does define "health" does not define "disability." Both the UN Convention on the Prevention of Discrimination against Persons with Disabilities (2006) and the US Americans with Disabilities Act (ADA), which was the model for the international law, do not define "disability." Theoretically, it is not possible to know which people are protected under the law absent a definition of "disability." But the opposite is true.

By defining the rights and the methods for determining whether rights have been protected, these legislative bodies have avoided the messy list of questions and left the door open that any individual can be protected if he or she can prove his or her need for protection against harm. Instead, these laws define a host of contexts in which people are treated as disabled—regardless of what disease or illness they have and, in some cases, regardless of whether they are "healthy" or not. Thus, neither of these laws offers a precise definition dividing people into populations, such as healthy (presumably not protected against discrimination based on disability) or "sick," and both staunchly refuse to label people with "disability." Rather than regulating illness, these disability laws prohibit certain types of activities that are obstacles to equal opportunity for people with disabilities and set forth the steps to be followed to overcome the limits of disability using accommodations and assistive technologies, including nano-enabled technologies.

Legislation avoids the definitional issues, thereby expressing a deliberate purpose: to end the sad history of segregating people on the basis of whether they are "sick" or "healthy." Filling the void with reasonable accommodations, the laws prohibit discrimination based on disability and then provide positive incentives, promoting different activities that achieve the societal goals of the law. Significantly, criteria for determining whether someone is entitled to legal protections is shaped by the context and the individual's ability to function and not the type of disease or the severity of prognosis. And, these goals are achieved under the law without blocking the protection of people against new forms of discrimination or new illness or disease, because of gaps in the law's coverage that might appear if there were too narrow a definition.

These well-respected examples of national and international law offer the hope of flexible regulatory governance for nanotechnology, too. It is tempting to use a "list" approach in order to clarify the regulatory framework's limits, determining which aspects of nanotechnology applications are regulated and which are exempt. But be careful: Lists look easy to use but are a foe to progress! A list of substances, a list of key characteristics or sizes, a list of products below a cut-off size all sound easy to enforce. But a list is also static and cannot be altered once old problems have been cured or new problems come along. Things change, and the list stays the same, unless a committee of experts is convened with the power to change it.

Successful risk management brings together many components of law and science by crafting rules that are workable at the time they are written but also flexible to embrace new developments in science and fold those new developments into the compliance programs. Successful risk management programs are multidisciplinary and, when blended with partnerships among industry, civil society stakeholders, and government at the international level, their acceptance across industries makes them work particularly well to cut costs and save lives. For this reason, some people have argued that the best approach is not to define nanotechnology under law but instead to set forth procedures for risk management with due diligence.

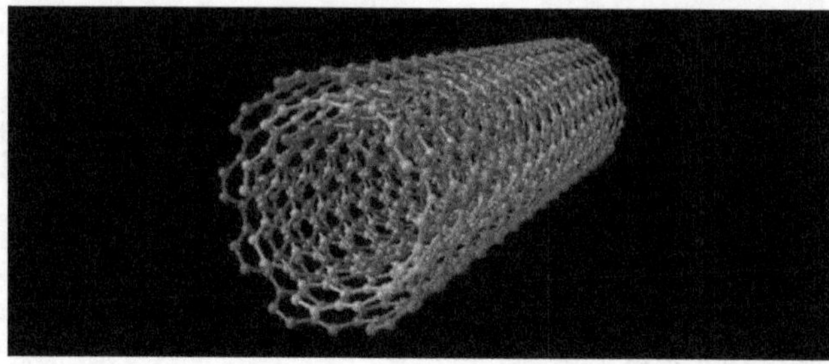

Figure 1.4 Photo by the US National Nanotechnology Initiative (NNI). Source: US government.

The Royal Commission on Environmental Pollution called for functional analysis of nanomaterial applications for this same reason as long ago as 2008. After taking testimony from experts in many walks of life, the commission was baffled about whether nanotechnology is so new and novel that it requires new laws. Among possible criteria for regulation, the commission favored regulation based on function in commerce and society. The same nanoparticle formulation can contain one or more different types of nanoparticles that vary in their structure, function, and chemical composition; examples include liposomes, nanoshells, metal oxides, quantum dots, nanocrystals, and polymer-based nanoparticles. Avoiding a specific definition but providing flexible criteria for determining the subject of regulation enables stakeholders to participate in creating a regulatory framework—the best way to have their buy-in for the final law is to include them in the legislative process. The same product may have a different regulatory outcome on the shelf of a grocery store or a pharmacy compared to the mass quantities of workplace exposure in industry. Therefore, the UK Royal Commission on Environmental Pollution report about nanotechnology in 2008 was wise to emphasize functionality rather than precise terms whose importance or interpretation may change over time, and later morph into meanings that the legislative drafters never intended or that the public will not tolerate even if lawful. Such concerns can be built into a flexible framework under law, but implementation is a matter of political will, beyond the lab and crossing impermeable borders, into the hearts and minds of people throughout civil society.

The goal of applying nano-enabled products to commerce while balancing fair protection from harm can be achieved using risk management systems without actually defining nanotechnology itself. A system of regulation based on key findings about risk in preliminary studies, with heavy penalties for those who fail to undertake those studies or engage in failure to disclose when toxicity is found, may protect the public without worrying about whether a car named "nano" meets the criteria for size at the nanoscale. An example of a flexible legal framework for sorting out and equitably resolving scientific problems that impact daily life can be found in the Globally Harmonized System of Classification and Labeling of Chemicals (GHS), which does not regulate toxic or hazardous substances but requires that information be disclosed to people involved in the handling, use, or transport of those substances. The system does allow outside sources such as international registries of toxic substance to be used as a guide for whether disclosure is required, but the law is equally clear that a toxic or hazardous substance may require disclosure even if it is not found on any list.

In conclusion, there is no law requiring that nanotechnology laws define "nano" terms and no clear scientific data to justify a "bright line." But there may be a set of consequences that the legislature wishes to avoid. Criteria, instead of straight, bright lines, are a better choice for legislative drafting when addressing new technologies that offer the promise of wonderful innovation but also hold unquantifiable risk. By contrast, line-drawing may accidentally include a group that fits the definition but has no place being regulated under the law. In such cases, there is plenty of work for lawyers in litigation. Overbroad legal terminology runs the risk of clogging the regulatory system with a litany of inventions and products that would distract regulators, compliance officers, inspectors, and researchers from attacking the major emerging issues in nanotechnology's applications and can produce unintended economic harm. A definition that is too narrow may exclude important sources of potential harm for no clear or justifiable reason. Therefore, the most important criteria defining the scope of regulations governing nanotechnology may not be scientific at all; it may be political and based on a social policy judgment about the importance of a given function in relation to desired or undesirable social outcomes. In this context, the impact of sound health programs that successfully engage in risk management is much bigger than a few nanoparticles.

1.2 What Is Nanotechnology?

1.2.1 Sample Legal Definitions

The attempt to craft legal definitions for nanotechnologies and related activities is in an embryonic phase. Many experts have met to define the terms pertaining to nanoparticles and nanotechnology, and many meetings have been convened to create legislative text. The state of the art has not yet matured, however, to critique the definitions together in order to make a recommendation for a unified definition.

One government's definition offers that "nanotechnology is the understanding and control of matter at dimensions between approximately 1 and 100 nanometers, where unique phenomena enable novel applications." There are several other definitions floating around the web, but they are awkward and have evoked much debate without clear scientific consensus. One research center defines nanotechnology as "the ability to measure, see, manipulate, and manufacture things usually between 1 and 100 nanometers. A nanometer is one billionth of a meter, a human hair is roughly 100,000 nanometers [in diameter]." Another regulatory model defines or "categorizes chemical substances based on molecular identity, not on physical properties such as particle size, thereby arguing that their jurisdiction already exists to regulate nanomaterials." Another definition views nanotechnology as "encompassing nanoscale science, engineering, and technology, nanotechnology involves imaging, measuring, modeling, and manipulating matter at this length scale."

In 2010, the International Organization for Standardization (ISO) Technical Committee 229 on nanotechnologies (ISO 2010) issued a definition of nanotechnology: "the application of scientific knowledge to manipulate and control matter in the nanoscale range to make use of size- and structure-dependent properties and phenomena distinct from those at smaller or larger scales." The nonprofit nongovernmental ISO uses the definition "material with any external dimension in the nanoscale or having internal or surface structure in the nanoscale." In 2011, the ISO stated, "Nanotechnology is the understanding and control of matter and processes at the nanoscale, typically, but not exclusively, below 100 nanometers." The term "size-dependent phenomena" is hard to understand and therefore would require additional definition in order to shape jurisdiction under law.

One local municipality has attempted to protect its citizens and therefore has joined the regulatory dilemma: According to the survey sent by the Cambridge Mass City Council, "Definition: In this survey 'engineered nanoscale materials' include materials composed of particles or structures with at least one dimension between one and one hundred nanometers (one nanometer ' one billionth of a meter). These materials are referred to as 'engineered' because they are manufactured and used purposefully to make use of size-related properties." Several widely quoted definitions of "nanotechnology" are inconsistent with each other; all are at first blush simple but later prove to be overbroad, and most important for this analysis, few of these supposedly scientific definitions can move easily from one context of nanotechnology applications to another in order to create a fair and predictable regulatory model that can be followed from product development until marketing. One government-based definition claims, "Nanotechnology is the understanding and control of matter at dimensions between approximately 1 and 100 nanometers, where unique phenomena enable novel applications."

According to the strategy set forth by the European Commission, "Originating from the Greek word meaning 'dwarf, in science and technology the prefix nano signifies 10-9 ie. One billionth, One nanometer is one billionth of a meter... The term nanotechnology (for the purposes of the EU strategy) is a collective term, encompassing the various branches of nanosciences and nanotechnologies." Although fascinating, it is not clear that one could apply this poetic definition to a specific place or operation in manufacturing as a matter of law: "Conceptually, nanotechnology refers to science and technology at the nanoscale of atoms and molecules, and to the scientific principles and new properties that can be understood and mastered when operating in this domain. Such properties can then be observed and exploited at the micro- or the macroscale, for example, for the development of materials and devices with novel functions and performance." But, if international regulations were to apply this definition and the types of uses covered, there is hardly any operation of life that one could exclude from the law. EU recommendations 2011 also raise this question about whether a numerical criterion is viable, reserving the right to regulate based on impact, regardless of size. For the purposes of the EU strategy, "'Nanotechnology' is a collective term encompassing the various branches of nanosciences and nanotechnologies."

But it is not clear whether this definition is really workable in daily life: If international regulations were to apply this definition and the types of uses covered, there is hardly any operation of life that one could exclude from the coverage of laws under this definition (Figs. 1.5 and 1.6).

Material that is either a nano-object or is nanostructured	***OECD WPMN: Considerations on a Definition of Nanomaterial for Regulatory Purposes, 2010***

Figure 1.5 Definition of "nanomaterial."

One definition views nanotechnology as "encompassing nanoscale science, engineering, and technology, nanotechnology involves imaging, measuring, modeling, and manipulating matter at this length scale." Another regulatory model defines or "categorizes" chemical substances on the basis of molecular identity, not on physical properties such as particle size, thereby arguing that their jurisdiction already exists to regulate nanomaterials. Although these laws start from conceptually different perspective from a juridical standpoint, both Registration, Evaluation, Authorization and Restriction of Chemical Substances (REACH) in the EU and the Toxic Substance Control Act (TSCA) in the United States have a very similar scope of jurisdiction over chemicals. Many of the well-established toxins used to engineer or manufacture nanoparticles are already on their list of regulated substances, and therefore the question exists whether either of these statutes needs to be modified in order to establish regulation over nanomaterials. These laws provide two examples of regulatory models that categorize chemical substances based on molecular identity, not on physical properties such as particle size. Many nanomaterials are composed of chemical substances subject to the TSCA in the United States or REACH in Europe or are part of international registries of toxic substances.

Nanomaterials based on chemical substances already on the list are not treated as new materials under law, and this has sparked much controversy. First, these materials operate at the nanoscale. Thus the old adage that a small quantity "won't hurt you" is now challenged. Amounts that were previously regarded as safe simply because they caused small or minuscule exposures would represent massive exposure to some nanomaterials, although the same exposure might be considered harmless if it were other materials.

Thus the potential of unquantified risk that scientists believe may exist but cannot prove empirically creates a regulatory dilemma: how to deploy precautionary principles to prevent reasonably foreseeable harm. And because the harm cannot be quantified, is it justified to use the entire regulatory apparatus for every application of nanotechnology that might cause harm?

Definition	Source
Any form of a material composed of discrete functional parts, many of which have one or more dimensions in the nanoscale	***EU SCENIHR The EU Scientific Committee on Emerging and Newly Identified Health Risks (SCENIHR)*** *– 29 November 2007*
Material with any external dimension in the nanoscale or having internal or surface structure in the nanoscale Note: This generic term is inclusive of nano-object and nanostructured material.	***ISO-CEN : ISO TS 80004-1***

Figure 1.6 Definitions of "nanomaterial."

Some of the definitions in respected references, such as the Royal Commission on Environmental Pollution, easily expand into additional industrial activities if applied in a workplace or to consumer products. For example, "nanoscience" is the study of phenomena and manipulation of materials at atomic, molecular, and macromolecular scales where properties differ significantly from those at a larger scale. It is hard to posit a type of substance that does not differ significantly depending on its size, regardless of whether it is inanimate or organic. Applying this definition in a model framework therefore would include regulation of a wide variety of substances and end products that may have nothing to do with the expected hazards posed by nanotechnology, thus forcing many producers to comply with a law whose implementation makes no sense in their business.

1.2.1.1 Example 1: EU NanoReg2

EU NanoReg2 offers an excellent example of a criteria-based definition that does not fall prey to the traps of number-based definitions or a list approach. One of the greatest challenges facing regulators in the ever-changing landscape of novel nanomaterials is how to design and implement a regulatory process that is robust enough to deal with a rapidly diversifying system of MNMs over time. Not only does the complexity of new MNMs present a problem for regulators, but the validity of data also decreases with time as more is understood and as new products are developed. Therefore it is challenging to design a regulatory system flexible enough to embrace new targets and requirements in the future, while solving old problems, too.

To this end, the EU has endorsed "Safe by Design (SbD)" principles. The credibility of such a regulatory system is essential for industry. Accepting the need for regulation, commercial stakeholders demand that regulation be written in a cost-effective manner and hopefully rapidly so that the regulations can be reflected in their business plans. On the basis of an understanding that the regulatory imperative may increase in the future, the EU NanoReg2 program is tasked to chart the path for registering, licensing, and certifying the nanosafety of a wide variety of substances and end products. Built around the challenge of coupling SbD to the regulatory process, NanoReg2 will be a key tool for establishing new principles and ideas based on data from implementation studies. The goal is to establish SbD as a fundamental pillar in the validation of any novel MNM. The EU hopes that grouping concepts developed by NanoReg2 can be regarded as a major innovation; therefore, guidance documents on nanomaterial grouping will not only support industries or regulatory agencies but also strongly support the commercial launch of a new nanomaterial, because so much will be known about the material before it has been applied for commercial use.

'NanoReg2 will establish safe by design as a fundamental pillar in the validation of a novel manufactured nanomaterial'

NanoReg2 is a project funded by the EU's Horizon 2020 research and innovation program.

WP1: Regulatory orientated activities establishing a framework of grouping approaches

Work Package (WP) 1 will:

- Identify, define, and harmonize current and prospective regulatory requirements and needs for MNM risk assessment in order to implement grouping approaches and SbD. Information requirements in different regulatory frameworks such as REACH, biocides and pesticides regulations, cosmetics directives and the Scientific Committee on Consumer Safety (SCCS), and feed and food regulations will be reviewed.
- Develop and apply grouping concepts for MNM and ultimately provide tools applicable for regulatory purposes. First, grouping concepts from different sources will be compiled and reviewed. Through cooperation and partner contributions, information on data management, testing protocols, and MNM characterization will be shared and grouping criteria and concepts will be defined, taking also into account regulatory needs, including analyses of alternatives and socioeconomic analysis. A minimum number of tests will be performed only if important data gaps need to be addressed.

  Grouping criteria will be verified using various datasets, including those collected and made available to the project by industry.
- Combine grouping concepts with other nontesting concepts such as read-across and quantitative structure–activity relationships (QSARs) or quantitative structure–permeability relationships (QSPRs) under development in several ongoing projects or European initiatives. Then, to associate these nontesting concepts with already existing testing concepts like high-throughputs screening, in vitro models, and

in vivo short-term assays, all these testing and nontesting concepts will be associated in a general principle named as Intelligent non-Testing and testing Strategy (ITS).

- Build on data management on well-established expertise developed in NanoReg and the ISA-TAB-Nano template approach.

WP2: Nanomaterials for industrial markets and their corresponding value chains

WP2 will:

- Identify and select existing and new materials as candidates for value chain demonstrators in collaboration with WP1, WP3, and WP4. For each of the selected existing and new nanomaterials, develop life cycle maps, identifying existing and potential exposure scenarios.
- Develop, where possible, a structured approach allowing grouping approaches of similar release and exposure scenarios, taking into account the nanomaterial, the product, and the process and environmental conditions.
- Evaluate any relative change in environmental and human health risk following the SbD (for production process safety and product safety) and carry out risk profiles of selected materials and/or products before and after application of SbD approaches in the case studies, aiming to evaluate the impact of the SbD applied on the product and/or production process in terms of (potential) health and environmental risks and then to estimate any residual risk following SbD and recommend any additional risk mitigation/management measures.

WP3: Safe by Design

The general aim of this WP is to define a system of tools, guidance, and checklists, which in its coherence is referred to as the safe innovation approach, to be used by various actors along innovation chains, supporting improved dealing with new safety issues of innovative nanomaterials and products, on the one hand, but also supporting improved regulatory preparedness on the other hand. The safe innovation approach envisages to be a future-proof approach, able to deal with upcoming generations of nanomaterials. Ultimately, this safe innovation approach should also benefit the public by improved warranting of safety of new products. More specific objectives of WP3 are:

- To develop a coherent set of supportive tools and structures for identifying and integrating efficient hazard and functionality testing during the innovation process (SbD approaches) that aims to minimize uncertainty about health risks to workers, consumers, and the environment, to analyze present approaches of SbD over the whole line from basic research to market, and to illustrate the added value of experimentally obtained safety data during the innovation process and create coherency among present and developed SbD approaches.

- To expand currently applied stage-gate innovation models stimulating interaction between innovation oriented and regulatory oriented stakeholders, from an early stage on, thereby supporting the most efficient way to prove safety of new and relevant nanomaterials as well as the nanomaterials defined in NanoReg1 and their inspired products. The adapted stage-gate innovation model will be used as a core for the safe innovation approach in which the various tools (guidance, structures, and checklists) are described in a coherent way and in view of the roles and incentives of the various stakeholders, like academia, industry, risk assessors, and regulators.

Analyses conducted in this WP will apply grouping concepts developed in WP1 as well as materials identified as relevant for industrial markets in WP2. The defined properties and parameters will be fed into WP4 in order to define barriers of the concept of design of safe innovations.

Moreover, the development of the safe innovation approach will gain views on barriers other than technology-driven barriers to warrant the safety of innovative products like nanomaterials.

See Fig. 1.7.

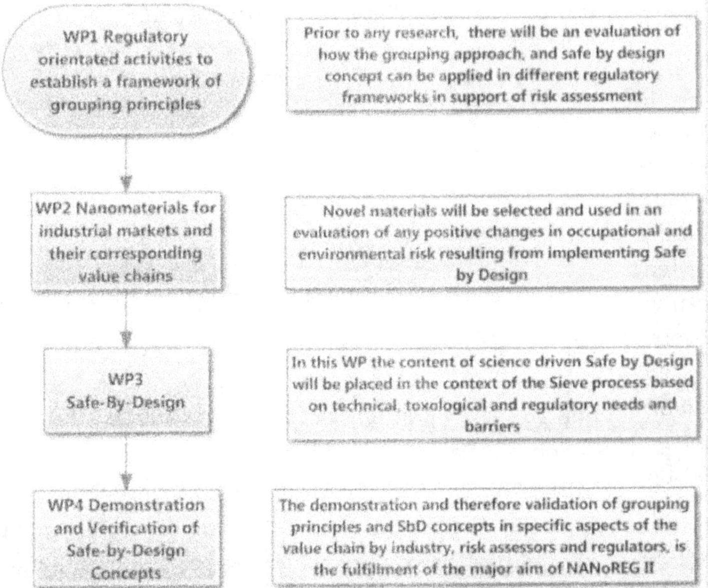

Figure 1.7 Flowchart of NanoReg2. Funded by the European Union's Horizon 2020 research and innovation program.

1.2.1.2 Example 2: EU Commission Recommendation

1.2.1.2.1 *Definition of a nanomaterial*

A material should be considered as falling under the definition in point 1, 2011/696/EU, where the specific surface area by volume of the material is greater than 60 m²/cm³. However, a material that, based on its number size distribution, is a nanomaterial should be considered as complying with the definition in point 1, 2011/696/EU, even if the material has a specific surface area lower than 60 m²/cm³.	***JRC Reference Report 2012, EUR 25404 EN*** ***Requirements on measurements for the implementation of the European Commission definition of the term "nanomaterial"***
A material with one or more external dimensions, or an internal structure, on the nanoscale, which could exhibit novel characteristics compared to the same material without nanoscale features NOTE: Novel characteristics might include increased strength, chemical reactivity, or conductivity.	***British Standards Institution, PAS (Publicly Available Specification) 71: Vocabulary Nanoparticles*** ***EU SCCP: EU Scientific Committee on Consumer Products (SCCP),*** *18 December 2007, Safety of nanomaterials in cosmetic products*

Figure 1.8 Definition of "nanomaterial" from the European Commission.

The EU adopted a definition of a nanomaterial in 2011 (recommendation on the definition of a nanomaterial, 2011/696/EU) (Fig. 1.8). Its provisions include a requirement for review "in the light of experience and of scientific and technological developments. The review should particularly focus on whether the number size distribution threshold of 50% should be increased or decreased."

The commission was expected to conclude the review in 2016, following the consultation of its draft findings with the stakeholders toward the end of 2015 and a public consultation planned to be published in summer 2016 or shortly after.

According to the recommendation, "nanomaterial" means:

> A natural, incidental or manufactured material containing particles, in an unbound state or as an aggregate or as an agglomerate and where, for 50 % or more of the particles in the number size distribution, one or more external dimensions is in the size range 1 nm - 100 nm.
> In specific cases and where warranted by concerns for the environment, health, safety or competitiveness the number size distribution threshold of 50 % may be replaced by a threshold between 1 and 50 %.
> By derogation from the above, fullerenes, graphene flakes and single wall CNTs with one or more external dimensions below 1 nm should be considered as nanomaterials.

The definition will be used primarily to identify materials for which special provisions might apply (e.g., for risk assessment or ingredient labeling). These special provisions are not part of the definition but of specific legislation in which the definition will be used.

Nanomaterials are not intrinsically hazardous per se, but there may be a need to take into account specific considerations in their risk assessment. Therefore one purpose of the definition is to provide clear and unambiguous criteria to identify materials for which such considerations apply. It is only the results of the risk assessment that will determine whether the nanomaterial is hazardous and whether or not further action is justified. Today there are several pieces of EU legislation, and technical guidance supporting implementation of legislation, with specific references to nanomaterials. To ensure conformity across legislative areas, where often the same materials are used in different contexts, the purpose of the recommendation is to enable a coherent cross-cutting reference. Therefore another basic purpose is to ensure that a material which is a nanomaterial in one sector will also be treated as such when it is used in another sector.

1.2.1.2.2 *Nanomaterials in REACH and CLP*

Classification and labeling identify hazardous chemicals and inform users about their hazards through standard symbols and phrases. They need to be harmonized to ensure good worldwide understanding and to facilitate the free flow of goods.

In the EU, the classification and labeling of hazardous chemicals is governed by Regulation (EC) No. 1272/2008 on classification, labeling, and packaging (CLP) of substances and mixtures (the CLP Regulation). REACH is the overarching legislation applicable to manufacturing, placing on the market, and using substances on their own, in preparations or in articles. Nanomaterials are covered by the definition of a "substance" in REACH, even though there is no explicit reference to nanomaterials. The general obligations in REACH, such as registration of substances manufactured at 1 ton or more and providing information in the supply chain apply as for any other substance.

Information on the implementation of REACH for nanomaterials, including guidance and the application of the REACH evaluation processes, can be found on the European Chemicals Agency (ECHA) website.

1.2.1.2.3 *Revision of REACH technical annexes*

The Commission Communication on the Second Regulatory Review on Nanomaterials (October 2012) as well as the REACH Review (February 2013) concluded that REACH and CLP offered the best-possible framework for the risk management of nanomaterials when they occur as substances or in mixtures. However, within this framework more specific requirements for nanomaterials have proven necessary.

Therefore the commission is considering modifying some of the technical provisions in the REACH annexes. The commission is presently finalizing the impact assessment. Its conclusions and the corresponding proposal for modification are being discussed in the CASG Nano, a subsidiary working group to the Competent Authorities for REACH and CLP (CARACAL).

1.2.1.2.4 *CLP*

Nanomaterials that fulfill the criteria for classification as hazardous under Regulation 1272/2008 on the CLP of substances and mixtures must be classified and labeled. This applies to nanomaterials as substances in their own right or nanomaterials as special forms of the substance. Many of the related provisions, including Safety Data Sheets (SDS) and classification and labeling, apply already today, independently of the tonnage in which the substances are manufactured or imported. Substances, including nanomaterials, meeting the classification criteria as hazardous should have been notified to ECHA by January 3, 2011. Any further update to the classification must also be notified without undue delay. On the basis of the information received under REACH registrations and CLP notifications, ECHA published a classification and labeling inventory.

Applicability of provisions of UNECE Globally Harmonized System of Classification and Labeling of Chemicals (GHS) for nanomaterials was being assessed in the biennium 2015–2016 by the subcommittee of experts (see Program of Work). As appropriate, the outcomes will be reflected under CLP.

1.2.1.1.5 *CARACAL Subgroup on Nanomaterials*

In close cooperation with the CARACAL subgroup on nanomaterials ("CASG Nano," composed of member states and stakeholder experts), the commission prepares advice on how to manage nanomaterials in accordance with REACH and the CLP regulation. The first paper, "Nanomaterials in REACH," provides an overview of how the provisions of REACH apply to nanomaterials. The second paper, "Classification, Labelling and Packaging of Nanomaterials in REACH and CLP," focuses on the classification of nanomaterials in accordance with REACH and particularly the CLP regulation. CASG Nano also recommended RIPoN reports as a preliminary advice and basis for ECHA in its further work on relevant guidance documents.

Last updated: 08/08/2016

1.2.1.3 Example 3: US FDA Definitions for the Purpose of "Guidance for Industry"

Offering definitions as a prelude to future regulation, the US Food and Drug Administration (FDA) published guidelines about nanotechnology in June 2011 that unequivocally use a number-based bright-line approach to determine the scope of regulatory jurisdiction but, at the same time, refuse to commit the language used to permanent rulemaking status. The FDA cautiously stated, "These terms are discussed out of context, and without regard to the role of FDA proposed guidance as law or as mere suggestions to be followed . . . FDA's guidance documents, including this guidance, do not establish legally enforceable responsibilities. Instead, the guidances describe the Agency's current thinking on a topic and should be viewed only as recommendations, unless specific regulatory or statutory requirements are cited." The proposed FDA definitions are needlessly complex. The FDA uses several ambiguous terms when creating pseudo criteria for defining nanotechnology. These terms include "Engineered material or end product 'at least one dimension in the nanoscale range' . . . Exhibits properties or phenomena . . . that are attributable to its dimensions . . . Size range of up to one micrometer." This definition also might exclude dangerous nanoscale materials that do not "differ significantly from those at a larger scale," despite their impact upon the environment or public health.

The FDA's notice states, "The use of the word should in Agency guidances means that something is suggested or recommended, but not required. II. SCOPE. This guidance document does not establish any regulatory definitions. Rather, it is intended to help industry and others identify when they should consider potential implications for regulatory status, safety, effectiveness, or public health impact that may arise with the application of nanotechnology in FDA-regulated products... Nor does this guidance document address the regulatory status of products that contain nanomaterials or otherwise involve the application of nanotechnology, which are currently addressed on a case-by-case basis using FDA's existing review processes." This draft guidance, when finalized, will represent the FDA's current thinking on this topic. It does not create or confer any rights for or on any person and does not operate to bind the FDA or the public. You can use an alternative approach if the approach satisfies the requirements of the applicable statutes and regulations. If you want to discuss an alternative approach, contact the FDA staff responsible for implementing this guidance. If you cannot identify the appropriate FDA staff, call the telephone number listed on the title page of this guidance."

1.2.1.4 Example 4: US TSCA of 2016

Summary of the Toxic Substances Control Act, New Chemical Law Updates

Figure 1.9 (Left) In 1970, President Richard Nixon signed the Clean Air Act and the Occupational Safety and Health Act (OSH Act). (Right) On June 22, 2016, President Obama signed the Frank R. Lautenberg Chemical Safety for the 21st Century Act, which updates the Toxic Substances Control Act (TSCA). Photo source: US government.

15 U.S.C. §2601 et seq. (1976)

The TSCA of 1976 provides the Environmental Protection Agency (EPA) with authority to require reporting, recordkeeping, and testing requirements and restrictions relating to chemical substances and/or mixtures. Certain substances are generally excluded from the TSCA, including, among others, food, drugs, cosmetics, and pesticides.

Quick links:

- The official text of all of the TSCA is available in hard copy from the U.S. Government Printing Office. Online versions of the text can be found from the Library of Congress website and from the office of any elected member of the US Congress.

The TSCA provides authority to the EPA:

- Require, under Section 5, premanufacture notification for "new chemical substances" before manufacture (https://www.epa.gov/reviewing-new-chemicals-under-toxic-substances-control-act-tsca/regulatory-actions-under-tsca#section%205).
- Require, under Section 4, testing of chemicals by manufacturers, importers, and processors where risks or exposures of concern are found (https://www.epa.gov/assessing-and-managing-chemicals-under-tsca/industry-testing-requirements-under-tsca-section-4).
- Issue Significant New Use Rules (SNURs), under Section 5, when it identifies a "significant new use" that could result in exposures to, or releases of, a substance of concern (https://www.epa.gov/reviewing-

new-chemicals-under-toxic-substances-control-act-tsca/regulatory-actions-under-tsca#SNUR.

- Maintain the TSCA Inventory, under Section 8, which contains more than 83,000 chemicals. As new chemicals are commercially manufactured or imported, they are placed on the list (https://www.epa.gov/tsca-inventory/about-tsca-chemical-substance-inventory).
- Require those importing or exporting chemicals, under Sections 12(b) and 13, to comply with certification reporting and/or other requirements (https://www.epa.gov/tsca-import-export-requirements).
- Require, under Section 8, reporting and recordkeeping by persons who manufacture, import, process, and/or distribute chemical substances in commerce.
- Require, under Section 8(e), that any person who manufactures (including imports), processes, or distributes in commerce a chemical substance or mixture and who obtains information that reasonably supports the conclusion that such substance or mixture presents a substantial risk of injury to health or the environment to immediately inform the EPA, except where the EPA has been adequately informed of such information.

Source: https://www.epa.gov/assessing-and-managing-chemicals-under-tsca/frank-r-lautenberg-chemical-safety-21st-century-act

Figure 1.10 Farmland in 2007. Photo by Dr. Ilise L. Feitshans.

1.3 Invitation to Discovery: The Way Forward with Nanotechnology

According to the Royal Commission on Environmental Pollution, Paragraph 1. 43, "The governance of emerging technologies under two conditions that pose serious constraints on any regulator. First is the condition of ignorance about the possible environmental impacts in the absence of any kind of track record for the technology. Second is the condition of ubiquity – the fact that new technologies no longer develop in a context of local experimentation but emerge as globally pervasive systems – which challenges both trial-and-error learning and attempts at national regulation." Consistent with this view, "Physically confining materials at the nanoscale alters the behaviour of electrons within them, which in turn can change the way they conduct electricity and heat, and interact with electromagnetic radiation. Moreover, materials engineered at the nanoscale can enter into places that are inaccessible to larger materials, and can therefore be used in new ways. These behaviours also have potential consequences on the abilities of synthetic nanomaterials to cause harm in novel ways." Wise people therefore try to foresee inevitable but presently unknown nanotechnology risks and will try to address risks with best practices, codes of conduct, and scientific principles to prevent harm, using the rule of law.

Law and Health Policy Impacts of the Bio–Nano Interface

To recognize the need to provide a scientific understanding of the vital policy question, Should nanoparticles be regulated as biological entities?, the question should be the subject of a special law and health policy analysis, including assessments of economic impact on medical care, international trade, and consumer products.

"The solution to this dilemma is not simply to impose a moratorium that stops development, but to be vigilant with regard to inflexible technologies that are harder to abandon or modify than more flexible ones . . ." These limits on the present state of the art for quantifying risk sharpen the edge of the dilemma that regulators, industrial stakeholders, and all civil society must courageously examine on the cutting edge of science. Insufficient data exists about actual risk in order to make key policy judgments; history is replete with instances where such assumptions were shown to be flawed too late to avoid serious consequences, even though "science can never definitively prove that something is safe."

Whether one believes that the global commerce using the new knowledge from nanotechnology research will be used for good or ill, the impact of nanotechnology on commerce and the products used by consumers holds importance for the individual health of everyone from the diplomat to the foreign migrant, from the Nobel laureate scientist to the lab technician at the bench using CNTs to develop new medicine. Caring about the impact of nanotechnology on human health—in the workplace and in the greater environment—therefore concerns preserving the work, health, and survival of all civilization. Solving these issues involves hard choices and expensive decisions about health policy that touch the lives of everybody. Harmonizing the law of nanotechnology will therefore require stakeholder participation. With careful forethought, law and science have often achieved similar goals with new materials created for end use in nuclear weapons, for genetic testing and screening, and for the aerospace industry. A transnational flexible framework for regulation, embracing many types of nanotechnology applications across industries and applying across many substances, is therefore essential so that workable laws can address risks without blocking new industries and commerce.

References

1. Peter Hayes, *Nanotechnology Tort Litigation: A Potential Sleeping Giant*, Bloomberg BNA, Dec. 7, 2016.
2. National Science and Technology Council Committee on Technology Subcommittee on Nanoscale Science, Engineering and Technology, *National Nanotechnology Initiative: The Initiative and Its Implementation Plan*, July 2000 Washington, D.C., report to the President of the United States.
3. World Health Organization, *Women and Health: Today's Evidence Tomorrow's Agenda*, WHO report, November 2009.
4. Americans with Disabilities Act (ADA) of 1990, 42 U.S.C. 5000.
5. *Toyota v. Williams.* 534 U.S. 184 (2002) was a case in which the US Supreme Court interpreted the meaning of the phrase "substantially impairs," as used in the Americans with Disabilities Act of 1990. See also Ilise Feitshans, *More Than Just a Diagnosis*, OEM Press Newsletter, OEM Publications, 2002.
6. *Bragdon v. Abbott.* 524 U.S. 624 (1998) was a case in which the US Supreme Court held that reproduction does qualify as a major life activity according to the Americans with Disabilities Act of 1990.

7. Ilise Feitshans, *Designing an Effective OSHA Compliance Program*, Thomson Reuters (available on Westlaw.com).
8. World Health Organization Constitution: "Health is the state of physical, mental and social well-being and not merely the absence of disease or infirmity."
9. *American Smelting & Refining Co. v. Occupational Safety and Health Review* Com'n, 501 F.2d 504, 2 O.S.H. Cas. (BNA) 1041 (8th Cir. 1974), discussed in detail in Ilise Feitshans, *Designing an Effective OSHA Compliance Program*, Thomson Reuters (available on Westlaw.com); *ASARCO v. OSHRC* is a case of first impression that addressed the applicability of the general duty clause to a nonobvious hazard in order to enforce a future modification of OSHA's existing lead regulation. In that case, the term "recognized hazard" was construed to include hazards one can "taste, hear, see, or smell," as well as hazards less easily recognized by conventional testing or monitoring. Subsequent cases define the obligation of employers. According to *Teal v. DuPont*, every employer owes a duty to protect employees from exposure to serious hazards, regardless of whether the employer controls the workplace, is responsible for the hazard, or has the best opportunity to abate the hazard. Even if an employer determines that the specified means of compliance is infeasible, it must affirmatively investigate alternative measures of preventing the hazard and implement prevention to the extent feasible" (quote used with permission from the author).

Chapter 2

Lessons Learned from the Legend of Asbestos

2.1 The Legend of Asbestos That Haunts Nanotechnology Law

As in the case of any new research, the unknown is fraught with peril. Part of the excitement of any new discovery is the encounter with risk as well as the joy of finding something unexplored. In the 20th century, many innocent scientific discoveries caused unexpected bad consequences, such as unintended cancers, dangerous side effects of pharmaceuticals that had a negative impact on health status or longevity, and destruction of property. So, too, asbestos was considered a miracle natural fiber (Fig. 2.1), used for insulation and fireproofing that saved lives and billions of dollars of property throughout the 20th century, until large, unprotected populations became ill with mesothelioma—lethal forms of lung cancer that resulted in massive tort litigation around the world, including the United States. The unresolved policy dilemma surrounding the use, need, and social costs of asbestos brought about a "radical change in the legal landscape" due to a "proliferation of tort claims associated with asbestos exposure" [1]. In 2009 , the World Bank Group issued a "good practice note" to (i) increase awareness of asbestos hazards, (ii) present information on available alternative construction

Global Health Impacts of Nanotechnology Law
Ilise L. Feitshans

ISBN 978-981-4774-84-0 (Hardcover), 978-1-351-13447-7 (eBook)
www.panstanford.com

materials, and (iii) advise adherence to international guidelines when in-place asbestos materials are disturbed [2]. Throughout the 20th century, "Widespread use of asbestos, followed by slowly unfolding tort damage forged and allegiance between science and law that never existed before," but that was predicted in 1995 as "likely to endure into the next century." According to David Rosenberg, by 1994, "Approximately 20,000 damage actions by asbestos disease victims, mostly workers, are pending in state and federal courts across the country, and the number is increasing by several hundred new claims each month… these claims frequently raise exceedingly difficult questions of causation and damage apportionment". Companies spend millions of dollars to defend or settle these claims, even if the companies win most cases" [3]. Rosenberg noted, "The number and expense of these cases is dwarfed, however, by the litigation between asbestos manufacturers, downstream users including retail conduits, asbestos removal projects and toxic waste management companies, who litigate against their insurers and re-insurers" [3]. Internationally, asbestos became the subject of global health standards [4].

Perhaps, engineered nanoparticles such as nanofibers and carbon nanotubes (CNTs) may impact human lungs in much the same manner as asbestos [5], although evidence also exists that engineered nanoparticles pose a risk to the liver and spleen [6]. The World Health Organization (WHO) noted, "Key issues; namely, a likely increase in public and environmental exposure, a documented public concern stemming from hearing scientists acknowledge data gaps and learning about the availability of an increasing number of products, a perceived lack of transparency" make nanotechnology a sensitive area of liability concern [7]. For these reasons, experts quoted in popular articles on the web and in serious law articles have asked whether "nanotechnology is the next asbestos" [8]. Some claim "nanotechnology torts" [9] are a sleeping giant. One engineer posited the dilemma in *Newsweek*, "Is [nanotechnology] the next best thing since sliced bread—or the next asbestos?" [10].

Lessons learned from the era without regulation in the asbestos-using industry may therefore be important for creating precautionary regulations when applying nanotechnology. But, sometimes the perils that are projected or feared are not nearly as big as they have been imagined to be. Unfortunately, it is deceptively simple to glob together two or three complex disciplines, such as law and regulatory governance research, bench science, and litigation strategy, in order to create a discourse about a problem that does not exist yet. Forethought beats afterthought. Mixing bits of complex information into a large cauldron of ideas can be productive only if people first think carefully about the issues by researching carefully those components of information in the cauldron that are poorly understood . For example, there is the danger of accidents with any weapon system, but none of the existing systems—nuclear bombs, missiles, or drones—blew up the world. The existence of nanosafety regulatory mechanisms for testing materials and informing the public and the global harmonization of chemical safety (Globally Harmonized System of Classification and Labeling of Chemicals [GHS]) that makes a level playing field for disclosures throughout the world thus provided a stark contrast to the situation of asbestos litigation when the court found that "no manufacturer ever tested the effect of their products on the workers using them or attempted to discover whether the exposure of insulation workers to asbestos dust exceeded the suggested threshold limits .... The failure to give adequate warnings in these circumstances renders the product unreasonably dangerous" [11].

Several analytical problems are raised by the recent vogue of assuming that the absence of regulation for asbestos prior to its introduction in commerce is a valid paradigm that foreshadows litigation about nanotechnology. This confusion between the state of the art for nanosafety studies and the lessons learned from failure to regulate or provide oversight for asbestos regarding disclosure of dangers is exacerbated by the fact that disciplines rarely interact, and therefore it is possible to distract thoughtful researchers with poor data that they cannot critique easily because it is outside the realm of their immediate expertise. For example, few lawyers understand the occupational and environmental epidemiology of mesothelioma or the natural history of disease. Nor do lawyers understand the key differences that distinguish the substance asbestos from the various processes that apply nanotechnology. Conversely, most scientists are unaware of the regulatory history of asbestos in the United States and abroad that provides the rationale for many mandatory disclosures and product warning labels under national and international laws. Public policy may therefore fall prey to this confusion, with complicated bad results that can only be resolved through litigation.

To avoid confusion between asbestos litigation and nanotechnology risk management, researchers, lawmakers, and policymakers involved in decisions about nanotechnology should be informed concerning asbestos litigation, nanotechnology processes, nanosafety regulations, and the global governance of chemical safety. The basic tools for avoiding this confusion are set forth in this chapter:

- Explaining the heritage of asbestos litigation whose linchpin is the failure to warn of risk
- Extracting the lessons learned from that time before the litigation, when there was no regulation, which is the genesis of the global system for use of toxic and hazardous substance in commerce
- Operationalizing those lessons learned in order to create a new safe and healthful paradigm for risk communication and risk mitigation for nanotechnology

Since 1915, if not before, the common law viewed the fundamental importance of disclosure, as follows: "First the defendant is bound to show that the plaintiff knew and apprehended the danger in question. If it is not clearly apparent, notice of the danger must be given and this notice must be brought home to the plaintiff. Whether the plaintiff has been thus notified . . . is a question of fact and usually for the jury" [12]. This vital criterion is remarkably unchanged in modern case law. Failure to disclose known dangers explains the proliferation of massive asbestos litigation, once the general public realized that key dangers had never been disclosed, thus depriving millions of people of their opportunity to engage in risk mitigation to prevent harm. As a result thousands of claims were filed, resulting in changes in court procedures to handle asbestos cases, new regulations were crafted regarding use of asbestos in industry [13], and a billion dollars in claims were paid out between 1973 and 1982 [14].

Figure 2.1 Asbestos fibers. Source: US government.

In 1984, the Rand Institute for Civil Justice studied the costs of asbestos litigation. Its statistical analysis of the costs for asbestos litigation identified the price of attorney fees for defense as well as plaintiffs as a key cost of thousands of claims against asbestos-using companies. The Rand report used methods that were state of the art at the time to examine the costs of litigation itself. The report examined the dollar value of defense attorney fees as well as plaintiff attorney fees, compensatory damages for claims, and punitive damages when cases went to trial. Their cost analysis showed that plaintiff attorneys worked on contingency fees, earning their money based on a percentage of the payout for successful claims. Defense attorneys who worked on these cases, however, included staff of the defendant corporations and outside counsel in large law firms who were hired expressly to defend against claims. The Rand report found that although the cost per case was less for the defense than the plaintiffs on a per attorney basis, many more defense attorneys were involved in the defense of a claim because plaintiffs sued on average 15 defendants (from different links in the chain of commerce), complaining about harm that caused their claim. Significantly, the Rand report also estimated total defense expenses "which we call organization-versus organization litigation expenses . . . to determine who should pay those claims" [14]. The Rand report therefore shows that defense against asbestos liability claims provided well-paying work that was important as a source of income among asbestos defense lawyers for decades.

It is no surprise therefore that defense lawyers and researchers have begun questioning whether innovations applying the new knowledge from nanotechnology will follow the toxic tort path of asbestos litigation because risks are unknown [15]. Several articles and presentations have addressed this issue in the past decade [16]. Some have examined the size and shape of CNTs and nanofibers

[17] or attempted to develop a preliminary approach for assessing whether problems of deposits of material in the lung, spleen, or other organs will have an asbestos-like pathogenicity [18]. Dr. Diana Boraschi, Institute of Biomedical Technologies, ITB-CNR, Pisa, Italy, stated, "The fact that animal experimentation does not necessarily predict effects on human health. They may be however taken as indication that caution should be applied in handling the materials" [19], possibly requiring new precautionary approaches [20].

As recently as in December 2017, WHO Guidelines on Protecting Workers from Potential Risks of Manufactured Nanomaterials stated from the beginning that the guidelines project aims to avoid the catastrophic effects of occupational exposure without underlying regulations because "recourse to precaution should be used to reduce or prevent exposure as far as possible. This was seen as an important underlying approach in the interest of protecting workers' health, especially given previous experience with asbestos."

Previously, in December 2016, the question of whether nanotechnology is a new asbestos made headlines again [9]. The article suggested that nanotechnology now, like asbestos when it was first being developed as insulation, is "a situation where innovation has outpaced investigation into the potential health and safety hazards of the technologies." But, the development of asbestos insulation in the late 19th century is not a proper comparison with nanotechnologies, because the modern tools for risk analysis and risk assessment did not exist over a century ago. Modern epidemiology and public health monitoring methods also did not exist when insulation methods were developed, even though some efforts to study the impact of asbestos exposure on human health began in parallel with widespread use.

For people who are unfamiliar with nanotechnology, a comparison with asbestos is deceptively simple. "I think we are in an early pre-litigation stage," another lawyer commented. "Because of the potential for nanotechnology to touch upon every aspect of our lives—consumer products such as sunscreens, medical treatments such as drug delivery devices inside human bodies, food packaging, etc.—the foreseeable scope of litigation is vast" [9].

Although both asbestos use and nanotechnology applications may be considered ubiquitous in commerce, the parallels may end there following successful efforts by European Union (EU) Safety by Design (SbD), NanoReg2, and Sustainable Nanotechnologies (SUN) nanosafety regulations, as well as new laws emerging in the United States. Ignoring key disclosure principles of asbestos regulations [4] and their antecedents in case law that was scarred by the failure of disclosure about risks, however, provides an unprecedented opportunity for people who are looking for work. Such professionals may accidently jump to well-intended conclusions with little data to support their conclusions or deliberately confuse bits of facts tied together in order to claim nanotechnology toxic torts are soon on the horizon. For this reason, the Bloomberg BNA article announcing that nanotechnology is a "sleeping giant" for toxic torts misleads its readers; its working assumption that liability will surely arise from the gradual deterioration of the commonly used material results in "the potential for a broader population to be exposed," and, after its further degradation, the potential of the "substance to contaminate the environment was high," understates the systemic failure of mechanisms for disclosure about risk before harm occurred in the case of asbestos litigation [9] and ignores international collaboration toward nanosafety regulations.

According to information accepted by courts in real-life case law [21] from the 1970s, however, "Asbestosis has been recognized as a disease for well over fifty years . . ." In 1924, Cooke in England discovered a case of asbestosis in a person who had spent 20 years weaving asbestos textile products [22]. An investigation of the problem among textile factory workers was undertaken in Great Britain in 1928 and 1929. In the United States, the first official claim for compensation associated with asbestos was in 1927 (https://law.justia.com/cases/federal/appellate-courts/F2/493/1076/4552/#fn6). By the mid-1930s, the hazard of asbestosis as a pneumoconiotic dust was universally accepted (https://law.justia.com/cases/federal/appellate-courts/F2/493/1076/4552/#fn7).

Cases of asbestosis in insulation workers were reported in this country as early as 1934 (https://law.justia.com/cases/federal/appellate-courts/F2/493/1076/4552/#fn8). The US Public Health Service fully documented the significant risk involved in asbestos textile factories in a report by Dreessen et al. (1938). «A controversial but important survey of asbestos insulation workers by Fleischer-Drinker et al., in 1945 examined insulation workers in eastern Navy shipyards who had worked with asbestos ten years or less, and found only three cases of asbestosis. According to the court of appeals that reviewed this report, 'Since asbestosis is usually not diagnosable until ten to twenty years after initial exposure,' the authors' conclusion has been criticized as misleading» [21]. Perhaps recognizing this possibility, the authors cautioned that the study did not «give a composite picture of the asbestos dust that a worker may breathe over a period of years,» and that if the study had been given enough time, "the incidence of asbestosis among these workers would have been considerable greater," according to the court of appeals. Also cited and examined in the federal court were guidelines written in 1947, suggesting threshold limit values for exposure to asbestos dust—that there should be no more than five million parts per cubic foot of air. The report was updated in 1968; the threshold limit value was reduced, but 20 years of unchecked risk represented a working lifetime of exposure.

Dr. Irving Selikoff dedicated his career to understanding the natural history of asbestos diseases and treating patients who suffered from exposure. In 1965, Selikoff and his colleagues published a study entitled "The Occurrence of Asbestosis among Insulation Workers in the United States" [23] after examining 1522 insulation workers in the New York–New Jersey area. Evidence of pulmonary asbestosis was found in almost half the men examined. Over 90% of the people studied who had more than 40 years' experience working with asbestos also had abnormalities. But that was the tip of the iceberg.

In his testimony before the US Congress, Dr. Selikoff stated, "As I sit here now, I am unhappy to say that unfortunately . . . that conditions and dangers so well recognized and so well described 40 years ago are still with us. It is an unhappy reflection on all of us—government, public health authorities, and my own medical profession—that at this time in the United States in the 1960s 7 percent of all deaths among insulation workers in this country are due to a completely preventable cause, pulmonary asbestosis" [24]. Selikoff further predicted 20,000 insulation workers are expected to die of pulmonary cancer and mesothelioma. Additionally, "3.5 million workers are exposed to some extent to asbestos fibers, as are many more in the general population" [25].

In 1966 Selikoff founded and became the director of the Environmental and Occupational Health Division of Mount Sinai

Hospital in New York, and he but remained active in research even after his retirement. After his death, it was renamed the «Irving J. Selikoff Center for Occupational and Environmental Medicine» in his honor and subsequently named «Selikoff Centers for Occupational Health» [26]. In 1982, the asbestos litigation caused by the deaths and illness unfortunately (but accurately) predicted by Selikoff reached a crisis level: three major asbestos-using companies filed for bankruptcy because of over 20,000 pending claims, with thousands of additional claims expected in the near future. The crisis prompted Congress to hold hearings about the underlying health hazards that caused the litigation, and therefore testimony concerning the long-standing evidence of potential harm from asbestos exposure was offered to the general public for the first time.

Evidence disclosed to Congress in public documents but previously not widely known shocked the general public that had been relying on asbestos insulation for heating, cooling, and roofing in their own homes. Thousands of claims for potential harm from asbestos exposure were filed in the wake of unmasking the legendary risks of asbestos, costing a billion dollars for litigation in an era when household incomes were about $20,000 or $30,000 per family. According to the Rand Institute for Civil Justice report in 1984, "Variation in Asbestos Litigation Compensation and Expenses" (which states it was funded by insurance companies), after showing exposure to the harm, plaintiffs "must be prepared to show that the defendant knew or should have known that the exposure to asbestos subjected them to a health risk . . . that the defendant failed to warn them of the risk or that the warning was inadequate, and that the failure to provide an adequate warning was responsible for their harms" [14].

This is consistent with standard principles of torts and product liability in the common law [27]. Since all of these criteria must be met, the final link between the failure to warn and the actual proven harm is the most important criterion. Failure to warn therefore was the crucial link that made possible the phenomenon of massive asbestos litigation—a phenomenon that gave rise to new legislation, new regulations requiring disclosure to operationalize the "right to know," followed by global notification of potential hazards and liability for toxic torts across a wide range of workplace and consumer exposures [28]. Failure to warn was the linchpin of liability.

Without disclosure of potential harm, the plaintiff was robbed of the opportunity to limit exposure and therefore the plaintiff cannot be held responsible for the consequences from engaging in dangerous activities. Under this construct, although the material is dangerous, it can be used with little or no liability in the event that proper warnings have been provided and the best practices were applied to prevent harm. Thus, disclosure followed by training in the best practices to prevent harm and the methods of risk mitigation is the best antidote to potential liability, even in the event of harm. Fastening together the link between the failure to warn and the consequences of that failure is therefore the most important criterion before obtaining any compensation for harm, and this link was secured in asbestos litigation.

In their defense, the asbestos-using industry often claimed that workers and the general public assumed the risk by using or working with these materials [21] and thereby provided informed consent. This, too, is an area where the ubiquitous character of asbestos in commerce and nanotechnology provides a parallel that is deceptively simple. Surely, consent is a problematic tool for asbestos litigation analysis because asbestos is ubiquitous in society; there are so many possible opportunities for exposure that no one could refuse them all. Courts consistently found that exposed people did not provide informed consent to the dangerous exposure. Therefore it was impossible to capture the data about exposures already experienced by workers, consumers, and their families, such as housewives who washed asbestos workers' clothes, in order to institute risk mitigation methods. No one could guestimate whether there had been short-duration exposure that was acute or a small, long-term exposure; therefore there was no way to determine whether the people who were exposed had crossed a threshold of exposure leading to harm, and if they did, in what context. As passersby to a construction site in the city on their way to the subway? As a home improvement hobbyist trying to insulate the house to give a better life to the family? As a worker in an office where asbestos had been used but the pipes and flooring had not been properly maintained? The list goes on, and therefore so does the litigation.

Directly opposite from the extremely sad history of pain suffering and near bankruptcy that were part of the asbestos litigation before the regulations that govern asbestos use in commerce today, nanotechnology's revolutionary approach to manufacturing and production opens a refreshing window of opportunity for a new paradigm of risk governance [29] that takes into account unknown but possible dangers and requires disclosure followed by steps toward reducing risk. Although the state of the art of the understanding regarding the emerging risks of nanotechnology in general and the potential adverse health effects on the skin, lungs, and reproductive health of workers from occupational exposure to CNTs and nanofibers in particular require closer scrutiny, leaping to the conclusion that nanotechnology and asbestos hold the same risk can lead to bad policy. Researchers have observed nanoparticles in the lung [30], and scientists are studying whether translocation of nanoparticles in the alveoli can impact other organs, but this does not mean that nanotechnology must echo the asbestos paradigm for risk mitigation and effective measures to prevent harm.

In 1984, the Rand report acknowledged the importance of the asbestos litigation as a potential precedent for product liability and toxic tort cases, stating, "The asbestos calamity may be unique in scale and intensity, or it may be the forerunner of a wave of environmentally based injury claims" [14].

The notion that history repeats itself can be useful if it forces policymakers and businesses to grapple with hard questions for risk management, protecting public health under international law. The real key is that failure to disclose the risks and dangers of exposure meant that there was no risk mitigation process in place compared to the new regulations that require best practices for risk mitigation in the 21st century. New methods have tried to create a more integrated approach to theory and reality surrounding important but difficult environmental and public safety and health questions. To apply these principles with forethought, there should be positive incentives to create a flexible framework for nanotechnology laws that will be attractive to stakeholders so that people will want to comply with the law—just like they stop at red lights without first reading the statute. There is a kernel of truth in the claim that like asbestos, nanotechnology is ubiquitous and dangerous. Unfortunately, it is dangerous to make this comparison without forethought:

Asbestos is a substance. Nanotechnology is a science. Both have risks.

The driver for asbestos litigation was not the ubiquitous nature of the product nor the danger itself. Indeed the common law has long had a tradition of protecting enterprises who engage in ultrahzardous activities, if it can be shown that such activities are essential to society. Explosives are an easy example; whether fireworks or demolition of an old building, special law governs use of explosives [27]. Failure to disclose the hazards of using asbestos made juries angry. As a result people who were exposed in the workplace or as consumers could not take precautions to mitigate risk. There are credible arguments that the massive asbestos litigation would not have occurred had regulatory protections been in place before use by industry.

The legendary costs of asbestos teach civil society that regulations are essential to protecting industries and employers as well as the public health and that regulations for risk mitigation and risk management must have a flexible framework that includes follow-up throughout the life cycle of the product, from discovery and bench science to commercialization and distribution to end users to disposal and recovery of usable materials that may be present even in hazardous waste. Mixing together several strands of complex disciplines without clear thinking can generate plenty of confusion and thereby generate income regarding dangers and a cause of action that presently does not exist. But, ignorance of these key principles under law cannot justify a random approach to nanotechnology governance.

The risk posed by ignorance of risk management and the danger of applying policies without forethought is a far greater danger to public health and democracy than the dangers presented by nanotechnology sciences. A candid discussion of known risks was missing in the early history of asbestos use, leading to litigation. Disclosing that problems are unquantified is not the same as hiding known dangers; disclosure of the known but unquantifiable risks can avoid potential liability so long as the disclosure is accurate. Candid disclosure, including accurate statements that risks are unknown, followed by documented efforts toward risk mitigation, is the criterion that will distinguish any effort to shackle nanotechnology in toxic torts compared to the ghosts of asbestos litigation that haunt nanotechnology researchers, commercialization, and end users.

In sum, the absence of any history of deliberately hiding documentation about the risks and dangers surrounding nanotechnology is a key difference between nanotechnology applications and the asbestos precedent. Although it remains true that anyone can sue anybody anytime, it is inappropriate for policy to be dictated by ignorance of the processes involving nanotechnology or a misunderstanding of the reasons behind existing regulations about asbestos and other toxic or hazardous substances or an odd conglomeration of both of these policy elements. It must be noted that not only plaintiff lawyers but also defense lawyers can profit from this opportunity, too, hiring experts and generating the documents and reports required when effectively defending against claims. Toxins that are far more lethal than asbestos did not face such rigorous liability claims and successful litigation because one unique feature of asbestos litigation is that courts consistently found the deception was willful. Information that could have protected people and saved lives without economic disruption was deliberately hidden, according to courts in thousands of cases. It must therefore be underscored that in the asbestos cases, the defendant's act of hiding was punished, not necessarily the magnitude of risk involved in the exposure that caused harm. Far more dangerous risks than asbestos exposure have passed through commerce without the notorious litigation of asbestos fame.

2.1.1 Policy Dilemma: No Substitute Works as Good or Harms as Bad as Asbestos

Mesothelioma is a death sentence. If nanotechnology generates even the same fears, this technology may be crushed before even the simple applications can get off the ground It sounds like they need a nanodrug now, one to fight nanothelioma [16b].

Asbestos is a naturally occurring substance that continues to save lives and protect property, but the dilemma it poses for policymakers is a modern paradox: Durable goods and reliable products created using asbestos throughout the 20th century remain in use and working properly over half a century later, but asbestos misuse has created "one of the longest running and deadliest toxic torts in U.S. history" [9].

The legendary property of being fireproof is a co-equal part of the asbestos litigation history. This property keeps asbestos use essential across a wide variety of industries that require insulation for heat, energy, and fireproofing for buildings and storage facilities, creating jobs, protecting property, and touching the daily life of billions of people. To date, no nation has effectively banned asbestos. Although several have legislated a ban on its import or export, these limits have not been effective, neither in implementing procedures within the administrative infrastructure (to operationalize the political will of the legislatures), nor in practice [31]. Instead, the fiber that is infamous for killing thousands of citizens also saves millions of human lives, protects public health by preventing fires, and protects property and has not been replaced successfully. Thus, asbestos use remains highly prevalent in civil society. For this reason, too, asbestos remains important in brake linings for cars, trucks, trains, and airplanes. Therefore, despite its bad reputation, asbestos in the 21st century continues to sustain commerce, and its use represents a multibillion-dollar industry. Therefore, the legendary conundrum surrounding the ubiquitous use and social costs of asbestos that brought a "radical change in the legal landscape" [1] haunts nanotechnology today [9].

Legendary for fireproofing and excellent insulation, asbestos is equally notorious as being synonymous with danger because of the massive "avalanche" [15] of litigation arising from exposure-related deaths from lung cancers and mesothelioma.

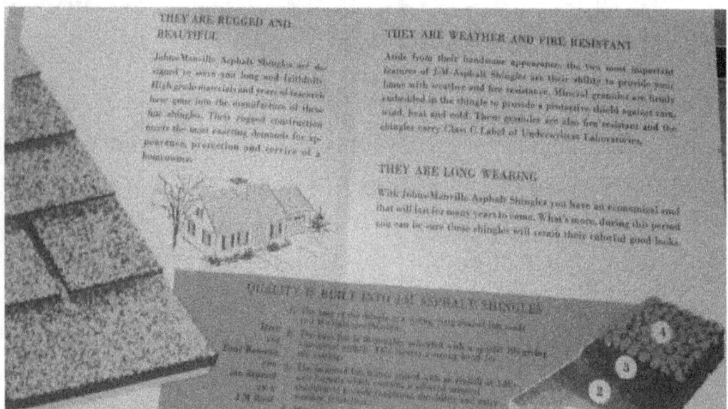

Figure 2.2 Advertisement for asbestos for roofing and insulation in houses in the United States, circa 1950s.

Asbestos has been known to humanity since ancient times. As a generic term, it applies to several inorganic, fibrous, silicate minerals that possess a crystalline structure. Asbestos is incombustible in air and separable into filaments. It was used as an insulator against heat as early as 1866; asbestos cement was introduced in about 1870 [21].Asbestos insulation material has been commercially produced since at least 1874 [21], and some of those structures created so long ago using asbestos cement remain standing and useful. According to the *New York Times*, in 1982, "The health hazards of asbestos result from the tendency of its tiny fibers, when brought into contact with air, to dislodge and float freely. Once inhaled, these fibers can cause cancer or severe respiratory diseases, and asbestos has been blamed for tens of thousands of deaths over the last 40 years" [32]. Yet, when examining the possibility of using safer substitutes, the same article stated, "Asbestos, a relatively inexpensive substance because it is plentiful in nature, has provided simple solutions to a host of problems. Asbestos consists of microscopic fibers that can be processed into other materials to make them stronger, more flexible and more resistant to heat and chemical reactions." In 1978, "619,000 metric tons were sold, but the tonnage used last year (1981) still had a wholesale value of about $110 million" [32]. The article further offered an expert opinion that "there are substitutes in practically every application, but they are inferior, more costly and contain unknown health hazards of their own," according to Robert Clifton, asbestos commodities specialist for the National Bureau of Mines. And, the article concluded quoting an asbestos sales manager: "We haven't seen anything that works as well or makes economic sense."

Asbestos in the 20th century therefore enjoyed an undisputed place as a benefit championing safety in the building plans and engineering of key structures used for offices, homes, hospitals, museums, storage, and every possible construction project designed with public safety as its primary goal. Asbestos is equally important as a miracle of modern technology that made possible advances in food processing (heating and storage of canned goods without burning the cooking facilities), clothing that enabled firefighters and first responders to enter lethally dangerous places with significantly less risk of harm, shipbuilding to prevent loss of life and damage of goods from fires at sea during World War II, and affordable insulation to control the temperature in millions of workplaces and homes. The countless uses of asbestos across a wide variety of applications that bring exposure to the general public means that people are exposed to it whether or not asbestos is part of the work process where they are employed, the housing they choose, or the products they buy.

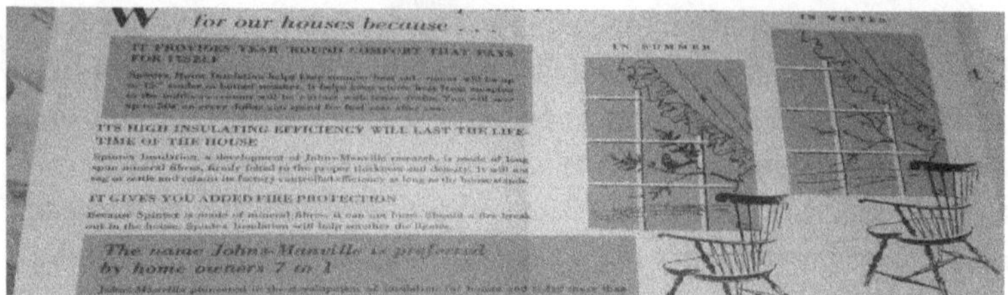

Figure 2.3 Advertisement for asbestos for roofing and insulation in houses in the United States, circa 1950s.

Absence of an effective substitute for asbestos exacerbated this dilemma. Asbestos exposure is associated with pleural plaques, pleural calcification, pleural effusion [33], asbestosis [22a, 34], lung cancer [35], pleural mesothelioma [36], peritoneal mesothelioma [37], and cancers of the larynx [38]. Good surveillance programs seek to diagnose these conditions before they manifest clinically, taking into account the long latency period for disease. Many programs also screen for existing disease and monitor the progress of disease, cognizant of the impact of cumulative effects from exposure across many years.

Case Law: Evidence of Known Dangers about Asbestos Exposure That Were Hidden from the Plaintiff who Died of Asbestosis

In 1973 the federal court of appeals in Texas [39] stated:

"The disease of asbestosis, was widely recognized at least as early as the 1930s. An expert witness, Dr. Hans Weill, testified that prior to 1935 there were literally 'dozens and dozens' of articles on asbestos and its effect on man. Dr. Clark Cooper, an expert witness for the defendants, stated that it was known in the 1930s that inhaling asbestos dust caused asbestosis and that the danger could be controlled by maintaining a modest level of exposure. Dr. Cooper testified as follows:

Q. The state of knowledge in the 1930s, let's say, in your opinion was asbestosis as a disease known about and recognized as a danger caused by inhaling asbestos dust?

A. Yes.

Q. And would you say that would have been rather common knowledge known in the 1930s?

A. Yes, I would say that. The answer to [that] would be yes.

Asbestos litigation that strikes fear in the hearts of so many insurers and risk managers has spanned several decades, starting in the 1970s and remaining unfinished in the 21st century [14]. For many claimants who were workers who had been exposed to asbestos in their work, the first avenue for redress was the system for workers' compensation. But asbestos-related diseases were not recognized by those systems at the time, and therefore, plaintiffs who were sick and dying needed another theory to sustain a claim that would address the harm. One plaintiff who had succeeded in the workers' compensation system nonetheless succeeded in suing 10 defendants who were held accountable for their contribution to his fatal asbestos exposure.

The breakthrough case that opened the floodgates of litigation was based on product liability. In *Borel v. Fibreboard Paper Products Corporation et al.* [21], the federal court of appeals stated clearly for the first time that there existed product liability for a manufacturer's failure to fulfill the duty to warn industrial insulation workers of dangers associated with the use of asbestos. Clarence Borel, an industrial insulation worker who won his workers' compensation claim, successfully proved that he had contracted the diseases of asbestosis and mesothelioma as a result of his exposure to the defendants' products over a 33-year period beginning in 1936 and ending in 1969. A unique provision of Texas workers' compensation law allowed the worker to litigate a claim against a third party if the worker could prove to a court that the injury that had been compensated was not caused by the employer. Thus, Borel sued the manufacturer of asbestos products on a theory of product liability that ultimately would reimburse the Texas workers' compensation system for claims paid. The jury in the trial court returned a verdict with a large award in favor of Borel on the basis of strict liability for failure to warn of the potential harm from the product used in the workplace and in commerce. The court found that failure to warn resulted in depriving the worker of the ability to refuse hazardous work or to engage in risk mitigation. And the court of appeals in Texas then upheld that award in 1973.

Half a century later, the text of the testimony remains so compelling that it can't be summarized and is reprinted here from the court's opinion: "Clarence Borel began working as an industrial insulation worker in 1936. During his career, he was employed at numerous places, usually in Texas, until disabled by the disease of asbestosis in 1969. Borel's employment necessarily exposed him to heavy concentrations of asbestos dust generated by insulation materials. In his pre-trial deposition, Borel testified that at the end of a day working with insulation material containing asbestos his clothes were usually so dusty he could 'just barely pick them up without shaking them.' Borel stated: 'You just move them just a little and there is going to be dust, and I blowed this dust out of my nostrils by handfuls at the end of the day, trying to use water too, I even used Mentholatum in my nostrils to keep some of the dust from going down in my throat, but it is impossible to get rid of all of it. Even your clothes just stay dusty continually unless you blow it off with an air hose.'"

Borel testified that he had known for years that inhaling asbestos dust 'was bad for me' and that it was vexatious and bothersome, but that he never realized that it could cause any serious or terminal illness: "Yes, I knew the dust was bad but we used to talk (about) it among the insulators, (about) how bad was this dust, could it give you TB, could it give you this, and everyone was saying no, that dust don't hurt you, it dissolves as it hits your lungs."

And then, Borel made the magic statement that represented the situation of future plaintiffs: "**You just never know how dangerous it was. I never did know really. If I had known I would have gotten out of it**" [21]

This statement cannot be considered empty litigation posturing, because the judge and the jury believed him. By proving that the workers were not given respirators, the evidence presented underscored the point that there was no reason to believe that any protection was needed. The court found, "Although respirators were later made available on some jobs, insulation workers usually were not required to wear them and had to make a special request if they wanted one" [21]. Sadly, Borel died before the case reached trial. The court allowed his widow to continue the case. The court found [40], "We cannot say that, as a matter of law, the danger was sufficiently obvious to asbestos installation workers to relieve the defendants of the duty to warn." The court further noted, "This necessarily included a finding that, had adequate warnings been provided, Borel would have chosen to avoid the danger" [41].

The court weighed this evidence and found, "The utility of an insulation product containing asbestos may outweigh the known or foreseeable risk to the insulation workers and thus justify its marketing. The product could still be unreasonably dangerous, however, if unaccompanied by adequate warnings. An insulation worker, no less than any other product user, has a right to decide whether to expose himself to the risk." Significantly, "The manufacturer is held to the knowledge and skill of an expert. This is relevant in determining (1) whether the manufacturer knew or should have known the danger, and (2) whether the manufacturer was negligent in failing to communicate this superior knowledge to the user or consumer of its product" [42]. Thus, it was not merely the failure to warn standing alone but the fact that it was proven before the court that the manufacturer had information that it deliberately failed to disclose that deprived Borel of the ability to refuse the work or engage in risk mitigation. The court then provided detailed descriptions of the studies of workers who used asbestos and became ill, starting in 1924, and continuing its discussion of the scientific literature throughout the 20th century until the time of trial and concluded, "There is strong evidence in the record that Borel never actually knew or appreciated the extent of the danger involved."

Defense counsel against Borel's claim argued that the worker had assumed the risk by accepting the terms of employment. Under this construct, accepting the job and performing the work constituted a deliberate contribution to the harm that led to his own death. The court of appeals disagreed, stating, "We have held, contributory negligence or assumption of risk is a defense to a strict liability action only if the plaintiff's conduct is both voluntary and unreasonable. Messick v. General Motors Corp., supra; Restatement (Second) of Torts, 402A, comment (n); Prosser, Law of Torts, 102. The trial court's charge was overly favorable to the defendants. Despite this error, the jury still found that Borel had not assumed the risk even under the harsh volenti doctrine." Regarding assumption of the risk, the court held, "Nor can we say that the danger was so obvious that Borel should be charged with knowledge as a matter of law" [43]. And, the court refused to accept any theories that held the plaintiff responsible because "we are not confronted with a failure to follow adequate instructions or warnings. Indeed, the evidence tended to establish that the defendants gave no instructions or warnings at all."

The court's rationale was supported by the common law. "In reaching our decision in the case at bar, we recognize that the question of the applicability of Section 402A of the Restatement to cases involving 'occupational diseases' is one of first impression. But though the application is novel, the underlying principle is ancient. Under the law of torts, a person has long been liable for the foreseeable harm caused by his own negligence. This principle applies to the manufacture of products as it does to almost every other area of human endeavor. It implies a duty to warn of foreseeable dangers associated with those products. This duty to warn extends to all users and consumers, including the common worker in the shop or in the field. Where the law has imposed a duty, courts stand ready in proper cases to enforce the rights so created. Here, there was a duty to speak, but the defendants remained silent" [39].

The absence of any opportunity for applying risk information was key for the court in the Borel decision: "A product must not be made available to the public without disclosure of those dangers that the application of reasonable foresight would reveal. Nor may a manufacturer rely unquestioningly on others to sound the hue and cry concerning a danger in its product. Rather, each manufacturer must bear the burden of showing that its own conduct was proportionate to the scope of its duty."

This principle that misconduct could not fulfill the duty to disclose echoed throughout trial courts for decades after 1973, and the principle provides the rationale for due diligence practices that have become internationally accepted as standard operating procedures across many industries since the 1990s. In 1974, a year after the Borel decision, Paul Bordeur, an established writer for *New Yorker* magazine, published his book [44], about a then-unexplained epidemic of mesothelioma in Tyler, Texas. Brodeur documented disruption in the life of the community as a consequence of disease and asserted that such illness could only be attributed to the activities of asbestos mining and the factory owned by Johns Manville there, which used the mined asbestos in the manufacture of housing insulation, roofing, floor tiles, and a wide variety of applications that were purchased by downstream suppliers to make safer and stronger buildings.

Brodeur described the public health impact of unchecked asbestos exposure in consumer use, workplace exposure, and secondary exposure to families. The state of the art was too immature to describe the epidemiology of global health impacts such as disease burden among end users and caretakers, although such topics are studied by health economists in the 21st century. But Brodeur's work was influential because he accurately described the extent to which greater population outside of the factory in Tyler, Texas, was impacted by asbestos exposure, beyond factory workers, engulfing administrators from the asbestos-manufacturing sector and embracing housewives who washed their husbands' clothes after exposure in the workplace. Epidemiological methods were unsophisticated at that time, relying more on observation and less on statistical analysis of data, because neither the computer systems for collecting big data nor the methods for analyzing and interpreting big data existed in that time. By the 1990s, treating asbestos-related illness and resolving surrounding litigation had become a local industry, too [45], providing work for major environmental and public health research at the University of Texas, Tyler, that continued into the 21st century.

In 1982, the *New York Times* reported, "The nation's largest producer of asbestos products, the Manville Corporation, filed for reorganization under Chapter 11 of the Federal bankruptcy code. Manville said it had no alternative because of what it maintained was a devastating volume of lawsuits facing it from the victims of asbestos-related injuries, some of them fatal, over the last 40 years. Manville's move prompted broad debate on product safety standards, the proper use of bankruptcy laws and the limits of a corporation's responsibility for backing the safety of its products" [32]. According to the Rand report about the costs of litigation concerning asbestos, which was written in the wake of the filing for bankruptcy by three major multinational corporations involved in the manufacture and distribution of asbestos-based products, over a billion dollars had been paid in claims about asbestos by 1982 [14], less than 10 years after the court of appeals decision in 1973. The expense of paying the judgments from juries who were outraged by the secrecy that caused the harm in those claims, plus the uncertain costs of pending litigation, provided the rationale for the bankruptcy filings, which became the subject of US congressional hearings [46]. The Rand report accurately predicted that additional claims would be filed and that expensive asbestos litigation was "likely to endure into the next century" [3].

Despite the cost of litigation and the pain and suffering from the disease burden upon human health, a countervailing interest in using asbestos was compelling at the time of emerging litigation and explains why asbestos use remains important for industry in the 21st century: Viewed by many as a miracle fiber that prevented shipboard fires in World War II, asbestos provides excellent insulation for roofing, temperature control, and construction materials in millions of offices and homes and crucial brake linings for cars, airplanes, trains, and other transit. *The New York Times* reported that asbestos "use is widespread in making cement pipes for sewage systems, the linings for brakes and clutches for motor vehicles, roofing and flooring materials, and the insulation devices for electrical circuits" [32]. Questions about quantifying risk and estimating the impact of possible exposures, whether past or future, represent a tangent to the real issue at the bottom of the litigation that distinguishes nanotechnology applications as a new art and science from the old but hidden hazards that were associated with asbestos exposures. *Risk communication did not exist.*

By contrast, there is little dispute that the risks associated with discoveries using nanotechnology are unquantified; the societal debate before litigation is which are the correct methods for control and who shall bear the burden of responsibility for the implementation of those protections: industry and commerce, individuals whether corporate multinationals or end user humans, or the government? Applying nanotechnology before risk data exists may be necessary, but it means that it is particularly difficult to limit risks with a reasoned and rational plan. This justifies risk management programs. There is no such thing as «zero risk;» but risk management programs alert staff throughout every level of the enterprise to the many compelling reasons to cooperate in programs that increase awareness of risks, publicize the training that will prevent injuries and illnesses, and thereby demonstrate proactive concern for protecting safety and health in the face of unquantified significant risk. No such plans were available to the workers whose decades of exposure in the 20th century resulted in the asbestos litigation that followed. Practices that destroyed the individual ability to reduce risk outraged judges, juries, and eventually legislatures much more than any of the associated risks from exposure.

Researchers, industry leaders, governments, and the general public are all stakeholders with shared interest in the use of the most beneficial applications of nanotechnology with minimum public health and environmental risk. The emerging technology represents a major departure from the asbestos precedent because nanosafety constituents have the opportunity to craft regulations in advance of exposure [47], and manufacturers and distribution chain users have the opportunity to craft in-house compliance programs that use due diligence tools that will fulfill their duty to warn. Potential sources for additional risk are not easy to measure or control, but they have an impact on global health. The combined effect of all the exposures has an impact on health, well-being, and the costs for caring for disease in each nation and around the world. Plaintiffs who cannot pinpoint which source caused the harm that they have experienced, therefore, sue several defendants at the same time, hoping that the courts and liability attorneys for defense will sort out their respective responsibility. Therefore asbestos litigation is expensive, bringing into court as defendants people who should not be litigants, and sometimes the wrong results are obtained because of these difficult evidence issues.

2.1.2 The Case That Never Ends: Key Asbestos Litigation Including "Cancerphobia"

Risk or danger is not actually the key criterion that determined the outcome of asbestos litigation. Instead, litigation that resulted in large judgments against asbestos using enterprises following asbestos exposure to workers and consumers in the general public is more likely an outgrowth of failure to warn or disclose well-known dangers of potential harm.

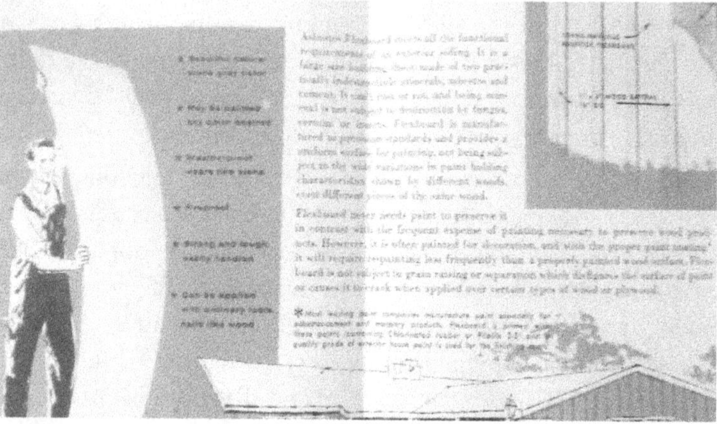

Figure 2.4 Advertisement for asbestos for roofing and insulation in houses in the United States, circa 1950s.

Even prospective harm that could not be clearly proven, such as the emotional stress from "cancerphobia," became compensable in order to fill the void in asbestos regulations. In *Norfolk v. Ayers* [48], several asbestosis patients claimed that their former employer, a railroad, had negligently exposed them to asbestos, which caused them to contract the occupational disease asbestosis, one of few diseases that was specifically discussed by members of Congress in the legislative history of the Occupational Safety and Health Act (OSH Act). Next, they feared they would develop cancer. They sought recovery for mental anguish based on their fear of developing cancer as part of their occupational disease damages, under the Federal Employers' Liability Act (FELA) [49]. In a novel view of the law of significant risk, the US Supreme Court upheld the rights to compensation for these employees on the basis of the present fear of future illness.

The Supreme Court opined that mental anguish damages resulting from the fear of developing cancer may be recovered under law. Thus an injured worker who has asbestosis but not cancer can recover damages for fear of cancer under FELA without proof of physical manifestations of the claimed emotional distress after proving that the fear is genuine and serious. There are several implications of this decision for future occupational health nanotechnology standards that will face a challenge from potential liability after harm. First, this decision underscores that the popular misunderstanding about the natural disease process can be more important than contracting the disease itself. Although cancerphobia is not a recognized "occupational disease," the Supreme Court accepted it as a compensable harm. Second, this case proves that an employer who acts without due diligence for protecting health will be held accountable even if no law is broken and no illegal activity has occurred, simply because the employer has overlooked protections, thereby creating fear of harm.

These cases explain why the key facet of asbestos litigation that fascinates the defense bar and that makes their clients shudder at the sound of the word "asbestos" involves the wide range of exposures to the general public, hurting "maintenance workers and residents" [31]. In Japan, it has been suggested that unquantified risk from asbestos will recur as a major public health issue, because encapsulated asbestos was set free when buildings were destroyed by the 2011 tsunami and earthquake. The antidote for the hemorrhaging litigation that threatened to bankrupt the asbestos industry was clear and enforceable regulation. Not until the general public was satisfied that companies had acted with due diligence, as demonstrated by documented adherence to established asbestos regulations, was it possible to limit liability in order to protect commerce. Confusion between exposures in one worksite for one employer combined with exposures in another workplace or in daily life is a major barrier to finding the true cause of illness. This so-called synergy is complex in the cases of workers who also smoke and therefore cannot prove to a judge and jury which exposure caused the damage to their lungs. "Most workers who sue have been exposed in more than one employment context and to a variety of asbestos products" [3a], as well as environmental toxins.

This uncontrollable variable for exposure fractured the notion that employers are responsible for providing a safe and healthful workplace, because if an employer behaved responsibly it would be difficult to disentangle the employer's role from the bad actions of other employers who had caused harm. From the standpoint of public health–planning strategies for screening and treatment and financing delivery of health care, the confluence of several sources of exposure is problematic. The cumulative impact of different workplace exposures both increases the size of the population at risk and at the same time makes it more difficult to blame a specific exposure as the cause of harm. Different employment exposures are complicated further by environmental exposures, consumer exposures, and lifestyle questions, such as smoking. Such evidentiary questions are complex and not easily resolved. Uncontrollable sources of exposures in daily life are both a sword for the plaintiffs' attorneys, who can credibly complain that exposures caused harm, and a sword in the defense of manufactures and downstream producers, who claim the harmful exposure came from other products. Plaintiffs' attorneys can get attention from a court if they can prove there was exposure to asbestos, but defense attorneys can cut down the credibility of liability claims by showing that even if asbestos caused the harm it is not their product that was involved. Answering such claims is very expensive and keeps the bar association active for millions of highly paid billable hours that translates into entire careers!

2.1.3 Law of Significant Risk Limiting Dangerous Exposures

No federal law in the United States required risk mitigation practices during Borel's lifetime. It was the Wild West in Texas way back in the days when the asbestos-mining companies began applying their miracle material for fireproofing, insulation, roofing and flooring, and shipbuilding to win the World War and the Cold War and build housing offices and public works throughout the midtwentieth century. In the 1930s, when there was the Great Depression, anyone who worked was lucky to have a job. In the 1940s , leaving a job as an insulation worker would have been unpatriotic. In the 1950s, construction boomed through the next decade with a strong postwar economy, but no law regulated the dangerous exposures.

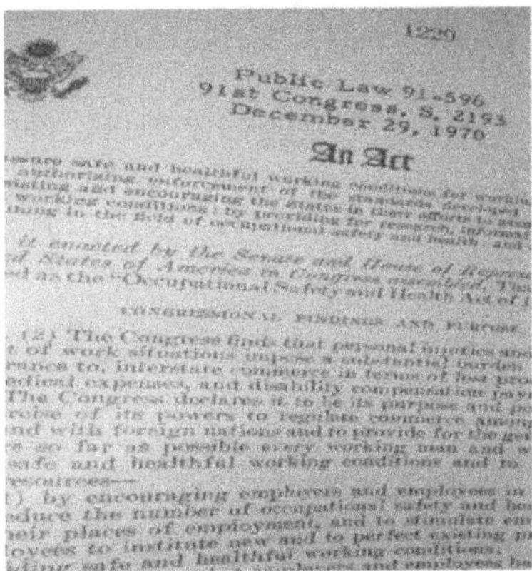

1220

Public Law 91-596
91st Congress, S. 2193
December 29, 1970

An Act

Figure 2.5 Text from *Legislative History of the Occupational Safety and Health Act of 1970*, prepared by the Subcommittee on Labor of the Committee on Labor and Public Welfare United States Senate. Source: US government.

Limits upon exposures, disclosure of information, and the right to know about toxic and hazardous materials in the workplace available with a click on the internet were not available. Without law to govern any limits on these activities or mandate disclosure, prudence was cast aside; instead, workers and the general public were faced with risks that courts later determined were avoidable. More importantly, those people who were unwittingly exposed were denied the opportunity to mitigate risks using basic tools of risk management. By contrast, for nanotechnologies, precautionary principles law and science have been woven together to create a safety net for information in the 21st century. Instead, in the late 1960s, several major events, including a very large mine explosion, brought workplace safety and accident prevention to the attention of the entire United States. There was so much public outrage about accidents and occupational diseases that even a conservative president, Richard Nixon, eagerly signed a workplace safety and health bill into law. President Nixon, who was elected because he claimed to represent the "silent majority" of Americans who favored the war in Vietnam, nonetheless understood that his white middle-class power base of blue-collar workers was eroding rapidly because of the high accident and death rates among people in key industries. The National Safety Council, a voluntary organization of industry leaders, "found that there were more disabling injuries and deaths in the work situation than were due to motor vehicles in the first six months of 1967" [50] (see Figs. 2.5 and 2.6).

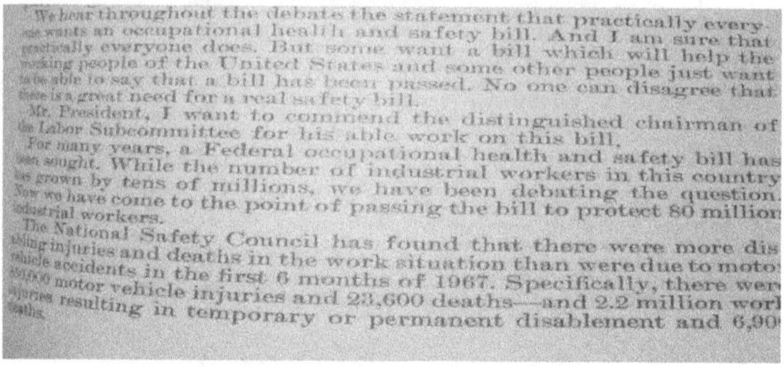

We hear throughout the debate the statement that practically every-one wants an occupational health and safety bill. And I am sure that practically everyone does. But some want a bill which will help the working people of the United States and some other people just want to be able to say that a bill has been passed. No one can disagree that there is a great need for a real safety bill.

Mr. President, I want to commend the distinguished chairman of the Labor Subcommittee for his able work on this bill.

For many years, a Federal occupational health and safety bill has been sought. While the number of industrial workers in this country has grown by tens of millions, we have been debating the question. Now we have come to the point of passing the bill to protect 80 million industrial workers.

The National Safety Council has found that there were more disabling injuries and deaths in the work situation than were due to motor vehicle accidents in the first 6 months of 1967. Specifically, there were [illegible] motor vehicle injuries and 23,600 deaths—and 2.2 million work injuries resulting in temporary or permanent disablement and 6,90[illegible] deaths.

Figure 2.6 Text from *Legislative History of the Occupational Safety and Health Act of 1970*, prepared by the Subcommittee on Labor of the Committee on Labor and Public Welfare United States Senate. Source: US government.

Even more politically shocking in the context of a nation whose federal election issues in 1968 pivoted on whether the United States should be fighting the war in Vietnam, "During the first six months of 1967 in Vietnam, there were 4,800 deaths and 31,913 military personnel wounded . . . there were roughly 30 times as many people injured at work in the United states as were injured fighting in Vietnam during the same six months of 1967" [51]. In the background memo for the OSH Act, in the legislative history this statistic was also highlighted [52], "As former Secretary of Labor Shultz pointed out during the hearings on this bill . . . during the past four years, more Americans have been killed where they work than in the Vietnam war . . . the annual loss to the Gross National Product is estimated to be over $8 billion" (Fig. 2.7).

Furthermore, a study by the Surgeon General of the United States examined 142,000 workers and concluded that 65% of the people were exposed to harmful physical agents, noise, or toxic substances. Asbestos was singled out as a "material which continues to destroy the lives of workers" [25]. Quoting Dr. Selikoff that "it is depressing to report in 1970 that the disease that we know well 40 years ago is still with us just as if nothing was ever done," the legislative history states that 20,000 insulation workers are expected to die of pulmonary cancer and mesothelioma. Additionally, "3.5 million workers are exposed to some extent to asbestos fibers, as are many more in the general population" [25].

In 1967, work accidents and illnesses cost the American economy over $8 billion. Ten times more man-days were lost to injury than were lost because of strikes in 1966.

We read the headlines concerning the strikes that are claimed to damage the economy. We should stop and consider the fact that we lose 10 times as many man-days of work in America every year due to industrial accidents on the job than we lose in strikes, lockouts, and walkouts all combined.

During the first 6 months of 1967 in Vietnam, there were 4,899 deaths and 31,913 military personnel wounded or a total of 36,812 injuries and deaths; that is, there were roughly 30 times as many people injured at work in the United States as were injured fighting in Vietnam during the same 6 months of 1967.

The legislation is long overdue. The devastating statistics and the bad record we have in industrial accidents in this country show that this country should have had such legislation years ago.

Yet in 1968, when Congressman O'Hara and I introduced in to the House and Senate the first comprehensive occupational health and safety bill, we drew the critical fire of several organizations who claim that legislation of this type is too expensive.

One may well ask too expensive for whom? Is it too expensive for the company who for lack of proper safety equipment loses the service of its skilled employees? Is it too expensive for the employee who loses his hand or leg or eyesight? Is it too expensive for the widow trying to raise her children on meager allowance under workmen's compensation and social security? And what about the man—a good hardworking man—tied to a wheel chair or hospital bed for the rest of his life? That is what we are dealing with when we talk about industrial safety.

We should have uniform standards so that no one industry would gain an advantage over any other industry. It would be far cheaper for our economy. It would save money in workmen's compensation premiums.

We are talking about people's lives, not the indifference of some cost accountants. We are talking about assuring the men and women who work in our plants and factories that they will go home after a day's work with their bodies intact. We are talking about assuring our American workers who wok with deadly chemicals that when they have accumulated a few years seniority they will not have accumulated lung congestion and poison in their bodies, or something that will strike them down before they reach retirement age.

We know that our worker's lives can be protected. It's good business to protect workers. And yet in 1966, there were 14,500 industrial deaths—2.2 million disabled—and a total of 7 million who sustained some injury in industrial accidents.

That is why I introduced my occupati[illegible] why I was happy to def[illegible] tor Williams [illegible]

Figure 2.7 Text from *Legislative History of the Occupational Safety and Health Act of 1970*, prepared by the Subcommittee on Labor of the Committee on Labor and Public Welfare United States Senate. Source: US government.

So there it was: the same workers in heavy industry who supported the war and voted for President Nixon were dying in exponentially greater numbers than the people fighting the war (Fig. 2.8).

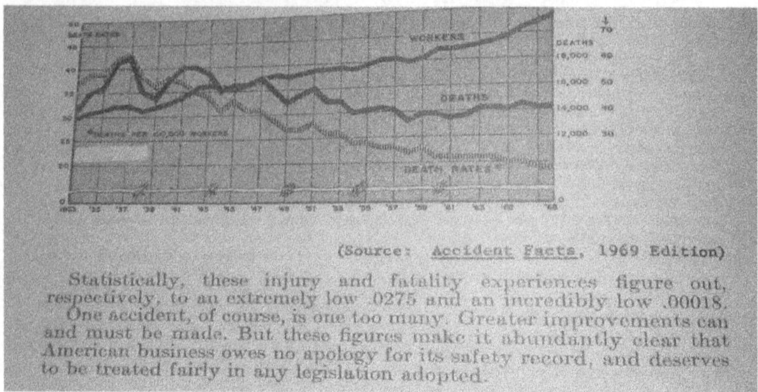

(Source: Accident Facts, 1969 Edition)

Statistically, these injury and fatality experiences figure out, respectively, to an extremely low .0275 and an incredibly low .00018.

One accident, of course, is one too many. Greater improvements can and must be made. But these figures make it abundantly clear that American business owes no apology for its safety record, and deserves to be treated fairly in any legislation adopted.

Figure 2.8 Text from *Legislative History of the Occupational Safety and Health Act of 1970*, prepared by the Subcommittee on Labor of the Committee on Labor and Public Welfare United States Senate. Source: US government.

Something had to be done.

The legislative history of the United States' OSH Act is unique because it reflected a popular will to end the era without any regulatory governance and put precautionary principles into place for enforcement of public health protections [53]. The OSH Act later became the model for the International Labour Organization (ILO) Convention C155 regarding occupational health programs at the national level, signed by many nations. And, as nations adopted ILO C155 as the template to build an infrastructure protecting safety and health, the OSH Act's response to the crisis in asbestos injury and illness indirectly became a model for the world, whose traces can be found in the international right to know programs, such as the global programs that require training in chemical safety when using specific toxic substances. Passage of the OSH Act therefore was a game changer because it created requirements for safety and health protection and disclosure of recognized hazards where no law had gone before.

The OSH Act requires that the Secretary of Labor Law create and enforce rules by inspection to protect every man and woman in the United States. The OSH Act is small enough to be memorized.

It fits onto 22 pages of text; it is succinct and clear compared to environmental legislation that uses hundreds of pages of text or the Affordable Health Care Act that uses over 800 pages. Although the OSH Act has been frequently criticized by labor organizations and employer trade associations, and offers for reform appear in Congress every year, despite all that brouhaha, no Congress has changed or modified the OSH Act text. The text of its legislative history is as clear and direct as the purpose of the law itself.

How Does Anyone Know What the Legislature Wanted to Do When It Wrote the Law?

The answer to the classic question about legislatures, "What were they thinking?," can often be found in the "legislative history." Although the term can apply generically to a wide variety of texts, testimony, and draft bills before the bills have been signed into law, most legislatures have kept a tradition of providing a summary of the new law and parts of the legislative debates and testimony in one book, for example, *Legislative History of the Occupational Safety and Health Act of 1970* (Fig. 2.9). These documents are typically preserved for posterity along with various draft versions of the bill and a final version that has been signed into law. Although the use of legislative history to glean the meaning and intention of new law can be controversial and is always subject to many interpretations, the legislative history nonetheless does provide insights into the key issues that motivated the legislature to act and often also provides a window into popular views of major problems at the time of the writing of the law.

Section 5(a)(1), the General Duties clause, is the heart of the statute (Fig. 2.10). It is well established under US law that the Secretary of Labor is allowed to promulgate and enforce standards that provide "employment and places of employment that are free from recognized hazards," under the auspices of the Occupational Safety and Health Administration (OSHA). Once OSHA determines that a significant risk exists and that such risk can be reduced or eliminated by a proposed standard, section 6(b)(5) requires it to issue the standard on the basis of the best-available evidence that "most adequately assures" employee protection, subject only to feasibility considerations. As the Supreme Court has explained, in passing section 6(b)(5), "Congress . . . place[d] worker health above all other considerations save those making attainment of this benefit unachievable." The Supreme Court has recognized such protective measures may be imposed in workplaces where chemical exposure levels are below that for which OSHA has found a significant risk. Furthermore, the OSH Act allows the Secretary of Labour to "modify" and "revoke" existing occupational safety or health standards: Section 6(b), 29 U.S.C. 655(b). This statutory language has withstood litigation even as high as the Supreme Court [13b].

Figure 2.9 Cover of *Legislative History of the Occupational Safety and Health Act of 1970*, prepared by the Subcommittee on Labor of the Committee on Labor and Public Welfare United States Senate. Source: US government.

The general duty clause clearly requires that information must be disclosed about "recognized hazards" as part of analyzing the risk, and failure to do so can incur fines and statutory penalties up to and including shutting down operations or closing a business. There has been little case law defining the scope of the General Duties clause, probably because it is written so clearly that its requirements are easily understood. Although employers who are subject to enforcement fines and penalties often try to claim the General Duties clause does not apply to the situation before the court, few cases have challenged whether the clause creates a standard that can be enforced under the OSH Act. Confronted with an epidemic of deaths and lost productivity in an era when the United States still cared about heavy industry and needed strong physical labor to perform heavy industry tasks, Congress made an unusual decision in order to get the administrative wheels turning in order to start up the new laws.

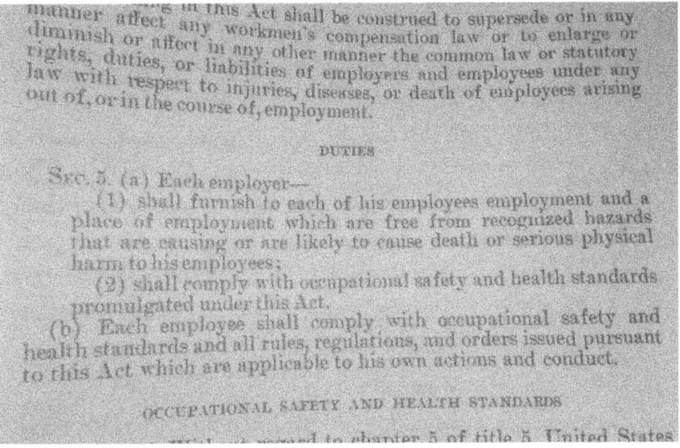

...in this Act shall be construed to supersede or in any manner affect any workmen's compensation law or to enlarge or diminish or affect in any other manner the common law or statutory rights, duties, or liabilities of employers and employees under any law with respect to injuries, diseases, or death of employees arising out of, or in the course of, employment.

DUTIES

SEC. 5. (a) Each employer—

(1) shall furnish to each of his employees employment and a place of employment which are free from recognized hazards that are causing or are likely to cause death or serious physical harm to his employees;

(2) shall comply with occupational safety and health standards promulgated under this Act.

(b) Each employee shall comply with occupational safety and health standards and all rules, regulations, and orders issued pursuant to this Act which are applicable to his own actions and conduct.

OCCUPATIONAL SAFETY AND HEALTH STANDARDS

Figure 2.10 Text from *Legislative History of the Occupational Safety and Health Act of 1970*, prepared by the Subcommittee on Labor of the Committee on Labor and Public Welfare United States Senate. Source: US government.

Specifically, the OSH Act provided for an array of start-up standards, drawn from the controversial matrix of voluntary compliance—rules that had been written by industry for use by industry leaders themselves. Such informal guidelines became rules that were anointed by the OSH Act as "national consensus standards" to take effect immediately upon passage of the law until the Secretary of Labor created new rules to preplace the standards. Because they were written without the backing or legitimacy that can be invested by a government authority, no one pretended that they were rooted in sound science or met the administrative standards for substantial evidence to prove risk of harm. Instead, the rules were written by people in industry, well intended, who saw the high percentage of death and disability and felt bad about these results. They wrote rules that were rooted in activities that were politically expedient in their own company but not necessarily rules that were defensible scientifically or in a court.

Although the successful plaintiff litigation around asbestos had not yet begun, much harm had been done and people were already dying from exposures that could not otherwise be explained. Congress understood that these primary wage-earning voters, taxpayers, and their families were immediately impacted by the emerging asbestos crisis, and Congress also understood the importance of addressing asbestos within the rubric of the OSH Act. The alternative would have required a special set of white-lung legislation for disability, parallel to the legislative mandate for black lung, the disease that cripples and kills coal miners. Indeed the Mine Safety and Health Act (MSHA) [54] was passed specifically to provide special compensation to black-lung victims just one year before the passage of the OSH Act. Rather than addressing the economy piece by piece, the law offered an opportunity to protect workers in the largest private industrial sectors by requiring that employers provide "employment and places of employment that are free of recognized hazards" [53].

The few cases that have opposed the application of this phrase have not been successful because scientific findings and the literature in the scientific community can be included within the scope of evidence to demonstrate that a problem is a "recognized hazard." Substances that were addressed by national consensus standards, including asbestos, were embraced by the General Duty clause from the very start. Given this clear mandate to create standards that embedded precautionary principles into daily work, one of the first activities when OSHA opened for business, therefore, was to address the rapidly emerging issues of occupational disease from workplace exposure to asbestos [55].

In 1974, in *IUD v. Hodgson* [56], the court of appeals examined whether OSHA had authority to require recordkeeping and medical examinations regarding occupational exposure to asbestos. From the standpoint of clinical care and medical diagnosis, asbestosis is not cancer, but it is true that many people who have asbestosis eventually develop cancer for reasons that are unknown to modern medicine. The precise mechanism of the natural history of these two diseases are unknown, but the popular perception of them is so close that people often confuse the two, and this has caused courts to consider the two illnesses as if they are the same. The court reluctantly accepted OSHA's exercise of authority to establish such controls and to set a deadline for implementation. Eventually, these controls proved important in saving the life not only of workers but also of the asbestos industry itself, once juries saw corporate employers using asbestos had acted with due diligence using state-of-the-art protections required by federal law.

Yet, when the same court reviewed OSHA's asbestos regulations 15 years later in *Building Trades of ALF-CIO v. Brock*, new complexity and sophistication in scientific data transformed the judicial review [57]. Rather than adopting *Hodgson's* simplistic view that "[t]echnological progress in industry appears not to have been accompanied uniformly by corresponding reductions in the health hazards," the court instead made judgments about epidemiology throughout its opinion. In the newer case, the court itself discussed such nonlegal concepts as odds ratios, standard mortality, and proportionate mortality. It also included in its opinion a chart showing relative risk for smokers versus nonsmokers who were occupationally exposed to asbestos over a working lifetime. The court stated, "Even under the assumption of a 20 year working life, [OSHA] found the risk at the existing PEL to be 44 extra cancer deaths per 1000" [58]. It accepted the agency's threshold determination that even at reduced levels there exists a significant risk, which would be measurably reduced by implementation of the modified standard, a narrow view of "significant risk." The court deferred to the agency's findings except when the terminology used by the agency was unduly vague: the court rejected a ban on the spraying of asbestos-containing products, and it required for a clearer definition of the phrase "any kind of construction work" when determining the scope of exemptions from the standard. The court agreed with OSHA, however, that despite the synergistic relationship between asbestos and smoking, which increased the workers' risk, there was no justification to require that employers ban smoking by everyone with occupational exposure to asbestos.

The true test of the US congressional intent to provide a wide gamut of protections against diseases from workplace exposures involved an unrelated substance, benzene [59]. In *Industrial Union Department v. American Petroleum Institute ("The Benzene Case")*, the Supreme Court vacated the OSHA standard for benzene because it was not supported by evidence of significant risk. The Supreme Court viewed the OSH Act as requiring the agency to do its job correctly by setting the exposure limit "at the lowest technologically feasible level that will not impair the viability of the industries regulated." But, the Supreme Court was not convinced that OSHA had done so. The plurality of the Supreme Court wrote that the authorizing statute did indeed require OSHA to demonstrate a significant risk of harm (although not with mathematical certainty) in order to justify setting a particular exposure level. The Supreme Court wrote that the need for a quantifiable risk assessment was not a "mathematical straightjacket," but nonetheless required substantial evidence of risk of potential harm to meet the procedural requirements for the standard's justification.

In the benzene case the administrative agency, OSHA, acting under statutory requirements to create standards under the OSH Act, had applied a national consensus standard but later determined the standard was too weak. OSHA then held hearings to revise the benzene standard, but the hearings were merely a circus of political pressure. Agency conclusions must be "supported by substantial evidence in the record considered as a whole": OSH Act Section 6(f), 29 U.S.C. 655(f). OSHA must use the "best-available evidence," which includes "the latest scientific data in the field"; "research, demonstrations, experiments, and such other information as may be appropriate"; and "experience gained under this and other health and safety laws": OSH Act Section 6(b)(5), 29 U.S.C. 655(b)(5). The agency ultimately reduced the exposure limit significantly but had not reduced it as low as labor organizations wanted. Trade associations for the petroleum industry were unhappy with OSHA's decision, too, because they feared they could not comply with the new stronger standard that which reduced benzene exposure without justification of the costs in terms of lives saved and injuries avoided. Consequently, OSHA was sued by labor and management.

The Supreme Court was baffled by the evidence presented, because the agency could not provide evidence to justify its actions to satisfy either side. As a result, the Supreme Court crafted the principles for the law of significant risk, a concept that did not exist under law at the time, in order to set very preliminary limits on agency actions, while endorsing the principle that the federal government has a proper role in controlling toxic and hazardous substances in the workplace. The Supreme Court held that OSHA is not required to support its finding of significant risk "with anything approaching scientific certainty" and that the determination of whether a particular risk is "'significant' will be based largely on policy considerations." Nonetheless the mishmash of data and medical information presented by the agency to the Supreme Court did not support the agency's conclusions. The Supreme Court therefore vacated the benzene standard, requiring the agency to rewrite the standard after better evidence had been gathered by new hearings.

The Supreme Court's strict attention to science policy and the evidentiary steps required to prove risk set a new administrative standard for safety and health rules, one that is immune to attacks about undue expense. The Supreme Court's staunch refusal to engage in cost–benefit analysis because there is no statutory language requiring the agency to do so has withstood scrutiny decades later, even in the wake of omnibus budget limits and subsequent regulatory requirements to demonstrate cost-effectiveness under the Paperwork Reduction Act. According to the ruling of the Supreme Court, in addition to the "attainment of the highest degree of health and safety protection for the employee," other considerations shall be the latest available scientific data in the field, the feasibility of standards, and experience gained under this and other health and safety laws. Whenever practicable, the standard promulgated shall be expressed in terms of objective criteria and of the performance desired [13b, 60]. In sum, the Supreme Court set forth the elements of due diligence found within the contours of in-house compliance programs, a blueprint that can be applied to nanotechnology regulations and international compliance efforts to ensure nanosafety.

According to this legal standard for preventing significant risk of harm from exposure to toxic substances, once OSHA of the federal US Department of Labor determines that a significant risk exists and that such risk can be reduced or eliminated by a proposed standard,

OSH Act section 6(b)(5) requires it to issue a standard that is based on the best-available evidence, that "most adequately assures" employee protection, subject only to feasibility considerations. As the Supreme Court has explained, in passing section 6(b)(5), "Congress . . . place[d] worker health above all other considerations save those making attainment of this benefit unachievable" [13b, 60b, 61]. Therefore any standard promulgated under this subsection need not be constrained by a cost. Under this ruling too, OSHA may require warning signs, as needed, to ensure that employees are apprised of all hazards to which they are exposed, relevant symptoms and appropriate medical treatment, and proper conditions and precautions of safe use or exposure. Labeling and employee warning requirements provide basic protections for employees in the absence of specific permissible exposure limits, particularly by providing employers and employees with information necessary to design work processes that protect employees against exposure to hazardous chemicals in the first instance. The Supreme Court has recognized such protective measures may be imposed in workplaces where chemical exposure levels are below the level where OSHA has found a significant risk [62].

All of these questions were part of a bigger package of rights and obligations that was created across the decade from 1970 to the mid-1980s, following the birth of the OSH Act. Even though the OSH Act clearly states that the law did not create new rights of action and that the law could not have an impact upon state law or be used as evidence of harm in state courts, the need for capacity building to have clear steps for preventing recognized hazards resulted in the evolution of new regulations about toxic and hazardous substances. Therefore the OSH Act did not create the right of action for asbestos litigation. The OSH Act is silent about the right to litigation and, instead, mandates that employers provide workers safety and health information, a right to information. The absence of a private right of action parallel to the citizen suits in environmental law is perhaps the most tragic flaw of all OSH Act limitations, because it means that only living workers who are fully employed by the creator of hazardous conditions can complain or bring a cause of action to enforce standards, but there is no redress for harm to the worker, the community, or the family that is injured by the increased global burden of injury or disease.

The greatest flaw in the OSH Act statutory scheme is that does not grant workers or their families a private right of action to prevent harmful activities before an accident or long-term exposure that holds risk of occupational disease. Workers or their unions may complain about hazards to their employer or to OSHA, but they cannot take their complaint to court to start a process of hazard abatement to mitigate risk or reduce foreseeable harm. This limitation not only bars claims in workers compensation but also prohibits workers and their families from bringing a claim before harm occurs when the OSH Act has been violated. Therefore the burden falls entirely on the governmental administration to prioritize complaints and take action to prevent hazards from becoming injuries or workplace fatalities. Instead of allowing the private right of action that would have made the law self-enforcing, the Congress chose to make complaining to OSHA the exclusive remedy, and the agency then, in turn, determines whether to investigate the allegation of danger at the worksite.

Although OSHA is required by law to investigate fatalities, this language has been held to apply only to accidents and catastrophic events but does not apply to long-term chronic exposures that cause occupational disease. This limit on the law, restricting the scope of claims for harm or exposure to known and recognized hazards, is unlike other areas of tort law. The absence of a private right of action for harmed workers and their families or for citizens who fear the negative health impact of exposure to workplace dangers is the greatest flaw in the occupational safety and health statute. This aspect of the OSH Act was written to deliberately prevent litigation of the magnitude that occurred in the asbestos cases, where courts understood that the exposures were so reckless that the hazards went beyond the restrictions for workers compensation laws. But this political compromise made a trade- off that created a huge administrative burden for the enforcement and inspection infrastructure. The impact of this burden is so great that it ultimately influences the distribution of agency resource and impacts policy. Thus, the agency's decision to inspect is a confluence of prioritized complaints and random policy agenda for enforcement. There is good reason to expect that this same procedural limit would apply in the case of nanotechnology. This limitation may prove unfortunate in cases involving nanotechnologies because the time required to address complaints may take so long that problems can fester or resources may be wasted in inspection, if the complaint requesting inspection is based on popular fear and not rooted in risk that the employer has failed to disclose.

2.2 Lessons Learned from Asbestos

Worldwide, regulatory actions since the late 20th century have consistently required disclosure, training in handling transport and storage of materials, medical surveillance for occupational exposure to asbestos, and a commitment to moving forward in the capacity building for national government efforts to regulate and monitor asbestos use [63]. Redefining risk communication, implementing risk management to reflect the slowly emerging information about nanotoxicity, and preventing liability that occurred in the wake of harm require recognizing the inextricable link between work, health, and the survival of civil society. The sad history of asbestos litigation in the absence of sound, scientifically based regulation underscores the precarious balance that must be maintained between work and health, because even once-healthy corporations can face bankruptcy when they fail to pay attention to this inextricable link.

The sad legend of asbestos teaches the lesson that regulation saves money and avoids injury and that the failure to regulate has large costs in terms of public health burdens and the economic instability from potential bankruptcy of industry. Therefore, if nanotechnology might be the next asbestos, as some have complained, then lessons learned from asbestos case law can be invaluable to design sound occupational health policy and implementation of environmental health protections.

An important outgrowth of regulation in response to massive unprecedented harm that hurt the companies as well as people is maturation of the concept of regulatory governance called the "right to know." This concept actually embraces a bundle of rights that, taken together, codify the methods for performing the duty to warn in a manner that is meaningful for risk mitigation and preventing avoidable injury. A contemporary example of international law that addresses the right to know is the Globally Harmonized System of Classification and Labeling of Chemicals (GHS) partnership between multinational corporations and governments. Due diligence practices that are performed by recognizing the right to know, disclosing information about safe handling, storage, and transport of dangerous materials, and in-house reporting of these steps using documented disclosure of known risks are potent for protecting against potential liability. International laws, US laws, and the GHS agreements are examples modern law that enables use of toxic and hazardous substances so long as safeguards are in place and the methods of avoiding known dangers that have been disclosed are followed. So doing may threaten the viability of an emerging regulatory framework for nanosafety as well as emerging commerce and industry.

National and international laws [64] have expanded gradually but exponentially in the time after asbestos litigation and before the birth of nanotechnologies in industry. The infrastructure in which nanosafety is scrutinized probably would have prevented many of the excessive exposures to asbestos if the same issues were presented today. And, the presence of a strong regulatory infrastructure is responsible for the ongoing vibrant use of asbestos in industry and commerce in the 21st century. In stark contrast to the careful monitoring of new processes created by applying nanotechnology, the hallmark of the asbestos litigation is that many juries were shocked by evidence accepted in courts, which revealed that enterprises using asbestos in manufacturing, mining, and milling were aware of the well-known risk of harm leading to asbestosis, lung cancer, and mesothelioma, which had been discussed in the scientific literature since the late 19th century but deliberately failed to make that information available to exposed people in the workforce, among consumers, and in the general public. Judges and juries therefore concluded that people who were exposed to asbestos were deprived of their opportunity to take precautions to mitigate risk. Subsequently, juries also found that those enterprises deliberately hid such information from end users and the general public and agreed that the remedy for this secrecy was to award punitive damages, often amounting to very large sums of money. By contrast, there is comparatively little litigation about radium or radioactivity or plutonium or a wide variety of known toxic substances that are found on registries worldwide. Few are as notorious as asbestos.

Some dangers from asbestos exposure were documented in the medical literature decades before asbestos became widely used, but companies that used the substance in commerce did not disclose risks to the workers, their families, and the general public. Many risks remained unquantified, not subject to precautionary principles codified in regulation. Once an epidemic of mesothelioma and asbestos-related disease overtook the working population, the absence of accessible toxicology data meant that injured workers and their families could not get medical care for a disease that was unrecognized, and treatments were therefore unknown. By contrast, a proactive effort by nanomaterial manufacturers, suppliers, distributors, the government, and other stakeholders offers the prospect of revealing potential harm so that end users can exercise a modicum of control to prevent harm. The Sustainable Nanotechnologies Project (SUN) funded by the European Union (EU) under the Seventh Framework Programme exemplifies the importance of translating emerging information into the decision support system for risk management. SUN operates on the following hypothesis:

"Industries can avoid potential liabilities if they adopt an integrated risk assessment and management approach addressing the entire life cycle of nanotechnology products" [65].

This approach is in stark contrast to the asbestos experience and perhaps based on a desire to avoid repeating the errors that the Rand report so artfully analyzed in 1984. The pain and suffering of tens of thousands of successful plaintiffs, however, resulted in lessons learned about the necessity for underlying regulation before a product is allowed to enter into commerce or when it is already on the market. The notion that one line of toxic tort cases could threaten to bankrupt an entire industry was far-fetched until the Johns Manville filing in 1982, when the company sought relief as a plaintiff on the grounds that toxic tort damage claims threatened the viability of the major multinational company. Even if the move was avoidable it was not frivolous, and the filing was accepted by the courts. The new reality of this potential economic catastrophe of an unchecked global health impact sent shock waves throughout the business community across the world. As a result, for the first time, corporate decision makers took the lead in lobbying government to create regulations.

"The real lesson of the Reagan years is that some regulation is absolutely unavoidable and necessary," according to Sigler and Murphy [66]. Regulatory procedures, combined with monitoring to reduce exposure, have therefore altered the course of occupational epidemiology within the asbestos industry so that workplace risks do not impede flourishing commerce. This lesson learned can be applied to nanotechnologies, too. The regulations clearly spelled out the obligation to disclose in exchange for a limit on disclosure that protected proprietary information such as trade secrets. Their efforts also meant that their competition would be compelled to disclose risks and dangers, thereby creating a "level playing field." The Manville bankruptcy filing proved that any employer or manufacturer was at risk of the harms caused without a regulatory safety net; on balance, even corporate stakeholders in civil society need the protection that a regulatory framework provides.

2.2.1 The Road Not Taken for Asbestos Regulation

One can only wonder whether the notorious asbestos exposures would ever have happened had there been a GHS for the classification and labeling of chemicals [67], requiring disclosure and training in the safe use of dangerous materials a century ago. The GHS is an indirect product of the asbestos experience. A multinational treaty whose implementation was crafted in partnership with industries, the GHS provides unified symbols for coding materials as flammable corrosive or poison in a manner that can be understood across cultures, languages, customs, or borders. Following an international mandate, a coordinating group comprising countries, stakeholder representatives, and international organizations was established to manage the work. This group, the Inter-Organization Programme for the Sound Management of Chemicals Coordinating Group for the Harmonization of Chemical Classification Systems, established overall policy for the work and assigned tasks to other organizations to complete. Their work was divided into three main parts: classification criteria for physical hazards, classification criteria for health and environmental hazards (including criteria for mixtures), and hazard communication elements, including requirements for labels and Safety Data Sheets (SDS). The criteria for physical hazards were developed by a United Nations Subcommittee of Experts on the Transport of Dangerous Goods/International Labour Organization working group and were based on the already harmonized criteria for the transport sector with input from the Organisation for Economic Co-operation and Development (OECD) [68]. The GHS remains a valid risk governance framework under law that can be either used as a model or used to include risk governance handling, transporting, or storing nanomaterials.

Oversight to ensure compliance is made possible by better-quality statistical data, long-term epidemiological information, sound industrial hygiene practices codified into the laws, and a level playing field that requires everyone to provide information in order to engage in commerce. Transmitting SDSs to users along the supply chain is required, thereby creating a fail-safe mechanism for achieving disclosure without local government inspections. Companies that have focused on sustainable characteristics like emphasizing long-term business strategies, not just short-term goals, strong corporate governance of workforce health and productivity, and sound risk management practices for occupational and environmental health and safety outperform their peers following the 2008 financial crisis [69]. Internationally harmonized hazard communication therefore is an important lesson learned for averting an asbestos litigation catastrophic scenario, because it offers a fail-safe method for disclosure. Nanotechnology and applications of nano-enabled products for consumer use are likely to fit neatly into this pre-existing conceptual matrix crafted for the GHS. Regulations thus hold the added benefit for manufacturers and downstream users that failure to disclose hazardous conditions and known dangers cannot give a competitive advantage to anyone who hides dangers and offers a false pretense of safety. And the clear mandate to disclose is very handy when appearing before a jury.

2.2.2 Components of International Regulation of Asbestos Exposure

Internationally and in most nations, the law is clear about requirements to disclose dangers, their possible consequences, and the methods for reducing risk and preventing avoidable harm in a manner that simply did not exist when asbestos litigation began, half a century ago or nearly a century ago, when the dangerous exposures occurred. Regulatory procedures, combining disclosure during training with monitoring to reduce exposure, have altered the course of occupational epidemiology within the asbestos industry so that notoriously dangerous workplace risks do not impede flourishing commerce. Balancing this policy dilemma was addressed in international standards regarding occupational exposure to asbestos in 1986, the International Labour Organization (ILO) Asbestos Convention (C162) [4]. This international standard exemplifies the international scientific consensus that asbestos exposure in the workplace is dangerous, requiring a regulatory apparatus designed to minimize risk, while enabling its use at work. The format of the convention adopts a flexible approach to asbestos regulation. Instead of setting a specific exposure limit, the standard enables national and local jurisdictions to adjust their requirements under law in order to meet the special needs unique to their situation. Although this approach has been criticized as offering "easy" standards for compliance in some nations and more rigorous standards in others, ILO C162 offers a positive incentive for compliance among employers from a wide gamut of economic situations, thus saving lives and reducing hazardous exposures worldwide. Consistent with ILO C155, the ILO Convention on Workplace Safety and Health, the requirements for disclosure about health and safety hazards, and training for the proper use, handling, and transport of asbestos, as determined under the auspices of national law, embodies a clear protection against the failure to warn of potential harms and codifies the right to know.

This lesson learned can be applied to nanotechnologies, too. According to the ILO convention, "The term asbestos means the fibrous form of mineral silicates belonging to rock-forming minerals of the serpentine group, i.e. chrysotile (white asbestos), and of the amphibole group, i.e. actinolite, amosite (brown asbestos, cummingtonite-grunerite), anthophyllite, crocidolite (blue asbestos), tremolite, or any mixture containing one or more of these" [4]. The key feature of a flexible framework is that although many important details of the program are not expressly stated, the key elements of a good program are included within its purview. In so doing, the regulatory structure that was created in partnership between corporate decision makers and government saved the life of the industry. Transitioning away from the Wild West approach of voluntary disclosure not required by regulation that moved to demonstrating full compliance with new laws was not easy. ILO C162 Article 3 allows great leeway to regulators to create doable programs: "National laws or regulations shall prescribe the measures to be taken for the prevention and control of, and protection of workers against, health hazards due to occupational exposure to asbestos.... National laws and regulations drawn up in pursuance of paragraph 1 of this Article shall be periodically reviewed in the light of technical progress and advances in scientific knowledge" [4]. This provision anticipates the cyclical review of laws and their implementation, consistent with precautionary principles of risk management that can apply equally to bulk materials and substances applying nanotechnologies.

Older ILO standards ratified before ILO C162 concerning asbestos exposure reflected a political agreement to limit toxic exposures at a certain level specified in a given convention. By contrast, ILO C162 has no specific exposure indices. For the first time in the ILO regulatory history, the national legislature's act of ratifying the asbestos convention was not a commitment to a specific level of exposure, based on scientific findings that could be outdated by new research regarding risks or new treatments for exposure or new methods for prevention. The new approach for performance rather than quantifiable exposures was a breakthrough in legislative techniques requiring risk analysis, because the flexible approach enabled each member state to set its own limits, consistent with its economic context, reflecting flexible criteria that change with state-of-the-art science. Member states may exclude particular branches of economic activity or particular companies from application of provisions of convention after taking into account the frequency, duration, and level of exposure, type of work, and conditions at any workplace. This provision was written with the intention to create a flexible framework, adjustable for different workplaces, across nations of differing levels of development, so long as exemptions are justified. Theoretically, this exemption runs the risk of becoming an exception that renders the rule meaningless by granting variances to employers or large sectors of the economy, while maintaining a facade of compliance with international standards. Nonetheless, the ILO also has an extensive oversight apparatus, and thus flexibility is an innovative approach that has the advantage of not becoming outdated. New methods can be folded into the legal framework for regulations and risk governance if they provide more effective indicators for measurement and medical surveillance that are applicable across any exposure, including the nanoscale.

2.2.2.1 Medical surveillance

Medical surveillance has become a vital tool for risk mitigation, and this will remain constant for occupational exposure to applications of nanotechnology. "Medical surveillance" of asbestos workers under ILO C162 refers to monitoring health status. Under ILO C162, Part IV, Article 21, "Surveillance of the Working Environment and Workers' Health," applies to all activities involving exposure of workers to asbestos. ILO C162 Article 21 provides that medical surveillance shall be comprise five components: (i) medical examinations, (ii) monitoring without cost to workers, (iii) information and "individual advice" to workers regarding results of medical examinations (the so-called right to know), (iv) alternative jobs for workers for whom asbestos exposure is "medically inadvisable" [4], and (v) a national infrastructure for notification of disease. These five components are important for all types of occupational health risk management. ILO C162 enables competent authorities in member states to authorize and verify that programs have these key components. Medical surveillance provisions must be viewed as but one small component of an overarching, cohesive administrative scheme for inspection, engineering controls, and enforcement of a host of occupational safety and health laws. Recognizing the long-term health impacts of exposure, ILO C162 wisely requires medical surveillance after the end of employment [4]. Regulatory agencies can therefore use medical surveillance as a systemic indicator of the effectiveness of engineering controls to prevent work-related disease or to detect longitudinal changes in the prevalence and incidence of occupational disease. Therefore ILO C162 offers a good blueprint for occupational medical surveillance programs [63] that may be useful for monitoring exposures to manufactured nanomaterials (MNMs) in industry.

2.2.2.2 Due diligence

Risk management systems were invented in the late 20th century to control risks that cannot be quantified but are nonetheless real. Embedding compliance values into the corporate culture and its infrastructure may be viewed as an essential step toward constructing and implementing a system of mandatory safety and health procedures integrated into operations, designed to ensure that proper compliance with the law is actually achieved to the maximum extent practical. If a risk management system works perfectly, embedded values will also function as a safety valve, preventing the flow of inappropriate concepts from gaining currency throughout the system. Significantly, if the correct values have been embedded into the infrastructure, illegal or irresponsible activities will be prevented by the system itself. At the same time, programming can consciously focus on achieving specific goals, such as gender equity, racial parity, or prioritization of specific substances for targeted compliance activities.

Due diligence requires developing committee structures and then the regular use of a smooth-flowing, ongoing infrastructure for improved communications for training and for reporting compliance problems. Each in-house compliance program must decide for itself the area of emphasis that will best foster compliance specific to the workplaces and the demands of the tasks at hand. A clear organizational structure allocates in-house compliance tasks to various groups that specialize in their areas of expertise and communicate their findings to the larger committee. This requires both buy-in among staff and rules that make in-house staff accountable for the achievement of compliance program goals and objectives. These goals and objectives should be stated in written policies, followed by action. The in-house compliance program for safety and health cannot be treated as if it were a trade secret; it must be open to everyone, and their complaints as well as praises must be heard and taken seriously without reprisal. The concept of embedding compliance programming into the employer's infrastructure values for disclosure that fulfill the duty to warn, as exemplified by the GHS. When operating properly, in-house systems improve working conditions and at the same time offer valid documentation of good-faith efforts to do the right thing.

Due diligence recognizes that there is no such thing as "zero risk" or a "risk free" workplace. Effective prevention using reasonable methods exist; implementing those protective strategies through compliance programs can alert staff throughout every level of the enterprise to the many compelling reasons to cooperate in programs that increase awareness of risks and prevent injuries and illnesses. With visible support and full commitment to safety and health from executive-level management, best practices may range from the use of internal occupational safety and health statistics in comparison to national or international statistics for the same economic sector, job description, or industry in quarterly reports and reporting on workplace safety issues to the board of directors on a regular basis. There should also be a well-documented infrastructural decision to include safety engineers and industrial hygienists among the vice presidents and executive staff who have hands-on decision making and also participate in risk communication and training about handling hazardous substances by directly meeting with staff, including people who handle the dangerous materials.

One key tool for maintaining due diligence is the internal audit, such as the steps required in the Environmental Health and Safety Audit Privilege Act. Learning from the tragic failure to disclose in asbestos cases, the Texas law requires regularly scheduled environmental or health and safety audit [70] documents to be generated, following a "systematic voluntary evaluation, review, or assessment of compliance with environmental or health and safety laws or any permit issued under those laws conducted by an owner or operator, an employee of the owner or operator, or an independent contractor of: a regulated facility or operation; or an activity at a regulated facility or operation." In this sense the term "voluntary" is slightly misleading: Employers and facility owners are required to conduct some internal review under law, but the term "voluntary" is used because they are free to choose the method for review on their own. Although the audit is required by law, the system is considered voluntary because the employer is free to choose any methods for maintaining a consistent system that its staff deems appropriate, noting that the system may be subjected to further scrutiny by inspectors from the state. Also the law requires in-house audits but does not provide immunity from liability or administrative penalties in the event of system failure. The facility cannot claim as a defense in court that merely having a review is adequate to meet the requirements of compliance with the law; the system must also be effective. Effective programs under this law use the "best practices" and develop internal incentives for compliance, including [71]:

- Setting procedures that are capable of reducing criminal acts
- Assigning high-level personnel to compliance
- Preventing the delegation of authority to people who are likely to engage in illegal activity
- Effectively communicating standards and procedures to all employees
- Implementing reasonable compliance measures such as monitoring, auditing, and reporting systems
- Using discipline to enforce standards
- Taking appropriate steps to detect and correct offenses, including changing the program, as necessary

Repeated compliance activity requires diligent monitoring to prevent harm.

Thus, effective compliance programs seek to motivate participants on the basis of philosophies and incentives that come from within the corporate structure itself, even though everyone knows there are penalties for failure to create any program at all. Following these principles under the Texas law, employers are able to embrace best practices without seeking administrative approval from the inspectors and can implement new procedures rapidly, with no excuse for failure to provide key components of effective programs.

2.2.2.3 Cost of medical surveillance

As a benchmark for international laws, ILO C162 requires that medical examinations be "free of charge" to workers, and this is consistent with practice in most countries. ILO C162 states that medical examinations should take place during working hours, without loss of pay. Access to care and to exam results is part of the right to know that embodies the fundamental precautionary principles for reducing risk. Medical exams remain a sticking point for all occupational health services delivery and research, even in nations with outstanding health care insurance programs and reliable medical surveillance. ILO C162 provides that member states shall require medical examinations determined by the competent national authority, within a larger program [72]. Member states are free to determine the frequency (e.g., annual or biennial), place (e.g., at the worksite or in a governmental health facility), and extent of such examinations. Many nations require preplacement screening and periodic examinations, although the length of time between examinations may vary. Nations are allowed to determine the content of medical examinations, which may include occupational and smoking history, a physical examination, a pulmonary function test, and a chest radiograph. Quality control in pulmonary function tests remains difficult because of variation in instruments and measuring techniques [73]. Ideally, results for medical surveillance should be discussed by two independent specialists (a board-eligible/board-certified radiologist or an experienced physician with expertise in pneumoconiosis). If the readings differ, a third reading should be obtained and there should be a discussion of the reasons for the different findings.

2.2.2.4 Information provided as the right to know, including the GHS

Sound occupational safety and health programs that implement disclosures to realize the right to know are the grease in the wheels of powerful economic machinery, not the fat to be trimmed from employer programs in lean times [74]. No society survives without health and without healthy workers to produce necessary goods and services! Asbestos, beryllium, cadmium, decompression, ethylene, florescent lighting, granite, heat, inorganic mercury, jagged edges on desks, ketone, lead, methane, noise, vapors, oxygenated hydrocarbons, quartz, radium, silicosis, tiredness, uranium, wood dust, x-rays, yttrium, and zinc oxide . . . an alphabet full of hazardous agents, all have one thing in common: they threaten the life and well-being of workers throughout the world.

Regarding the methods to be used for workplace exposure measurement, and controls, an internal management system is important for logging the incidence and duration of injuries and illness, performing internal audits, reporting improvements and violations to in-house staff, and informing all staff of the hazards and their right to information. Such disclosures, including a description of the right to know, are widely required under local and national laws but are also the backbone of GHS compliance. Traditionally accepted models for occupational safety and health corporate compliance infrastructure are especially useful for documenting and managing the data for compliance with standard features of regulatory requirements: (i) targeting problem areas before or after formal complaints, (ii) troubleshooting before potential violations of law, and (iii) analyzing job hazards. ILO C162 states that workers are entitled to "individual advice" regarding the results of their medical exams. Occupational health management systems can save the life of marginal employers by preventing costly accidents and obvious hazards that can disrupt, if not bankrupt, any employer, regardless whether the employer is large or small. ILO C162 does not expressly define the scope of information to be provided to the physician, but implicit in the requirement that workers have "individual advice" is an underlying concept: that physicians have detailed health information about the worker who is the object of medical surveillance. Worker training embraces best practices for the safe handling of hazardous materials, a description of possible long-term and acute health effects, and a clear statement outlining the right to information in an accessible format according to ILO C162. Taken together with ILO C155, there is a wide range of information that must be provided to workers and therefore cannot be hidden from consumers or the general public.

International laws and conventions requiring disclosure of hazards are neatly packaged in the GHS [75]. The GHS mandates disclosure of hazards in a manner that was not available to workers and consumers in the early days of asbestos use when there was no legal framework for exposure monitoring and risk mitigation. The substances need not be produced by the workplace; they may be components of upstream products in a pass-through. Another area of cooperative global work under the GHS involves developing practical tools for controlling exposures to chemicals, particularly in small and medium-size businesses. Control banding is one of the tools currently under development by the World Health Organization (WHO) and the ILO to use the agreed hazard classifications of chemicals identified through implementation of the GHS, together with information about exposure potential to identify broad, simple, and effective control approaches [76]. Therefore regulatory agencies in the United Nations system, individual governments, regional agencies such as the European OSHA, and nongovernmental organizations can speak to stakeholders with one voice under the GHS. If followed, GHS requirements may save the lives of many employers and marginal enterprises, in addition to protecting public health by preventing expensive liability.

2.3 Nanotechnology Revolutionizing Risk Communication

Nanotechnology's revolution offers another marvelous opportunity to humanity: the chance to operationalize an exciting and useful public health paradigm for risk communication. The traditional model for occupational health in general and occupational medicine in particular is exemplified by one image painted by Norman Rockwell in the late 20th century: an old, white gentleman, holding a stethoscope as the symbol of his referent power, with a halo about him, bestows information upon the nuclear family: father, mother, and child. The viewer does not know what is said, but the images convey a sense that important information is being exchanged with great confidence, because all three of them, even the unknowing minor incapable of consent here, is riveted to his every word. The viewer can see a clear power relationship that holds the physician in command: viewers may presume from this image that the doctor knows something that no one else does. The most striking thing about this image, however, is how well it captured the essence of classic risk communication models in medicine throughout the 20th century, if not before.

This famous invented image of doctors ignored the possible risks the physician faces: contracting of contagious disease, fatigue from overwork, and stress from possible mistakes that can lead to aggravating, if not devastating, claims of medical malpractice. The model ignores the risk to the communicator, the information giver, treating the act of communication as something from above, something higher and more important than toil, and certainly not as work. If society is to survive, however, it will be necessary to turn this picture around in order to listen to information provided by patients and then focus on the totality of risk for all social partners, both above and below the informational divide. Scientific questions about emerging technologies and the regulatory framework that will reply to those new inventions therefore must discard the burdens of treating occupational health as an "us and them" situation. New regulatory policies can address the raised by nanotechnology in the workplace and for public health due to consumer and environmental exposure. This requires examining the risks for professionals, also embracing the needs of vulnerable populations by using sound epidemiology to study the health impact on children, older workers, and caretakers in the family. New risk communication systems must then assist manufactures, supply chain consumers, and end users to discuss health risk questions in a manner that is accessible and understandable for stakeholder communities.

2.3.1 Due Diligence Is Your Best Friend

Due diligence is the fundamental concept for crafting and implementing effective in-house management systems for compliance with health and safety laws. When executed properly, due diligence catalogues information, documents accurate and timely reporting of internal audit information, and ultimately avoids liability. In many legal systems, the law requires that every effort be made in advance to protect the public health and manage risk, regardless of whether that risk can be quantified. The ability to prove that such efforts exist on a systematic basis throughout the employer's company, the so-called paper trail, is dependent entirely upon proving this concept of due diligence. In those cases that followed the regulatory path, potential liability claims could be answered by stating that there was full compliance with safety and health laws and defendant companies faced reduced liability or no liability at all. A rule of reason relying upon documented due diligence can therefore support a defense against liability claims because compliance with the law is a fair standard for determining whether there has been deliberate or avoidable harm.

As in the case of asbestos litigation, so, too, in the event of possible toxic tort claims about nanotechnology, the existence (or absence!) of such programs will prove to be a linchpin in any litigation, regardless of whether the plaintiff's claims are exaggerated or valid claims of proven major harm. Thus, creating and implementing such programs are as important as studying the risks and dangers of nanomaterials in some circumstances. There is a long list of compliance tools that can be used, as appropriate, to help achieve good compliance. The cornerstone for any effective compliance system is real, ongoing, and overt commitment. Without a clear and readily discernible focus on these three elementary components, success is extremely unlikely. Beyond simply issuing a written compliance philosophy and saying that everyone must adhere to it, commitment is evinced by actions. A written policy a team including safety engineers and industrial hygienists among the vice presidents of major employers should choose those aspects of all the available tools that will be the most suitable to a given set of operations for a given business. Risk managers, industrial hygienists, and many additional health professionals use well-honed prevention strategies to stop problems from becoming catastrophically large. Such in-house programs and the supporting infrastructure of regulatory agencies work together to do much more for the economy than merely reduce the costs of accidents and the overall burden of disease in society.

2.3.2 Candor, Not Secrecy

Communication is the blood that circulates to turn management commitment into active involvement in compliance programs that provide evidence of due diligence. Avoiding downtime, replacement costs, and displacement of families who lose a major or primary wage earner with the attendant social problems for family survival is among key goals of compliance programs that ultimately benefit all of civil society. Effective programs are custom-tailored to the specific needs of an enterprise and are designed to be flexible [77]. The question of whether the valid reasoning supports a claim based in fear may not be decided until months or years later, after trial or after expensive legal negotiations if the person making the claim can persuade the court that the fear is reasonable in the context of the facts presented to the court, as in the case of the US court of appeals delay for pre-registration of commercial rodenticide for fabrics containing nanosilver [78].

Carbon nanotubes (CNTs), for example, can be formed as single-walled carbon nanotubes (SWCNTs) or multiwalled carbon nanotubes (MWCNTs) that are remarkably strong. This makes them highly desirable in commerce but potentially lethal if they were to remain lodged in the lungs or migrate to other key organs. CNTs are elongated tubular structures, typically 1–2 nm in diameter as SWCNTs. They can be produced with very large aspect ratios and can be more than 1 mm in length. CNTs have very high tensile strength and are considered 100 times stronger than steel, while being only one-sixth of its weight, making them potentially the strongest fiber known. CNTs also exhibit high conductivity, high surface area, distinct electronic properties, and potentially high molecular absorption capacity. CNTs and MWCNTs hold an unquantifiable potential risk regarding possible harm, even though these products are the building blocks for so much of daily life.

Compliance programs that focus on the unquantified risks of CNTs and MWCNTs can offer manufacturers and suppliers concrete proof of due diligence and thus allow an employer to enjoy a presumption of compliance in areas of the law where the limits are unknown by showing that the steps taken to protect public health were reasonable in their context, given the state of the art of nanotoxicity and the requirements under law. To do so, however, requires candid, ongoing, detailed scrutiny of internal problems within the worksite, as well as documentation of every reasonable effort for troubleshooting and creating reasonable solutions to problems. These features of in-house programs provide the best justification for any actions in light of scientific uncertainty. Therefore, well-documented, in-house occupational safety and health programs are essential, especially given the global acceptance for the GHS, which requires disclosures and training across a wide variety of substances as a matter of national and international law.

If communication in-house is the lifeblood of vibrant corporate compliance program, then due diligence is its protective skin. A management system is a very effective tool for ensuring compliance with the law [60]. Effective in-house risk management programs require the best-possible lines of communication in order to be successful. To work, lines of communication must be established long in advance of any emergency. Such efforts can also facilitate compliance with new changes or modifications in the governing law because of their inherent flexibility. The strongest systems are cyclical. The cycle makes reporting normal as well as required, with built-in follow-up as well that ensures long-term accountability. With the help of corporate anthropologists and outside consultants, staff can be briefed about key issues, when designing right-to-know training, tailored to corporate culture of a specific enterprise, thereby creating a specialized system with effective results [60]. Written policies may also be required to withstand scrutiny in a court of law in the event of a violation of national or local health codes, tort litigation, or environmental litigation, and therefore their text should reflect the presence of embedded compliance values. Simply giving staff lectures without backup resources or problem-solving workshops in anticipation of hazards or without considering the new occupational health effects that the sector of the economy may face is shortsighted. Normal operating procedures double-check against system failures. And documentation offering proof of such safeguards against system failure is the key to a solid defense during litigation. But all this is meaningless if, from a practical standpoint, staff members do not get the message or do not hear the information transmitted during training. Cyclical reporting therefore also requires creating a conduit for candid and unpunished feedback about problems as reported to management by staff, regardless of whether the comments are protected by law.

2.3.3 The Question Is Knowledge and Consent, Not the Magnitude of Risk

No lawyer can ever offer a guarantee that following recommended steps for demonstrating due diligence will prevent litigation or unfounded claims. Fear surrounding unknown risks about nanotechnology can make it easy to persuade a judge that there is a valid reason for the court to hear a claim, in an abundance of caution, rather than dismissing a case that scientifically makes no sense. Even though there is no specific equation that will predict how much money is saved by investing in occupational safety and health risk management programs, or how many unfounded claims are avoided, risk management enables employers to provide a reasonable answer to claims and an insurance policy for responding to reasonable sounding claims that are unfounded. For this reason, due diligence dictates that in-house compliance programs be established to avoid liability; reduce lost productivity and down time; avoid administrative fines, penalties, and compensation costs; and prevent injury.

For commercialization of nanotechnology applications in commerce, policy issues regarding the need for robust methods to tease apart causation about asbestos-related illness will arise again if many people fear that important health risks have been deliberately hidden from public view. Uncertainty in science is a wild card that sharpens the edge of the dilemma faced by all potential regulators or legislative drafters but raises the stakes for those who use dangerous substances without the safety net of an in-house compliance program. Beyond simply issuing a written compliance philosophy and saying that everyone must adhere to it, commitment is evinced by actions. In that limited sense, theorists who fear that nanotechnology is the new asbestos are correct! The so-called fate of nanoparticles and their interaction within biological processes in humans are poorly understood. Therefore, downstream users in commerce are likely targets for litigation about harm from nanoparticles, regardless of evidence concerning dose and effect. In the labs where nanoparticles are used, there may be additional mundane issues of noise and vibration emissions from work equipment, which must be controlled whether or not they interact with nanoparticles to impact human health. Thus, in-house compliance with existing regulations and proactive support for applying nanosafety law is an important component of programs designed to prevent or limit potential liability by actually protecting public health through access to robust accurate information about risks and dangers associated with nanotechnology.

In conclusion, the key lesson learned from asbestos litigation is that it is more expensive to delay disclosure of dangers than to be candid about known risks and to disclose those risks promptly. When risk is unclear, state that uncertainty exists without alarmist exaggeration and without sugarcoating to provide a false sense of security. Instead, discuss the risks in a manner that takes into account the needs across many populations and listens to feedback from users and then applies that feedback wisely to work processes and risk communication in training, in social media, and for the public.

References

1. Ilise Feitshans, "Basic Concepts and Public Health Dilemma Surrounding Asbestos Exposure," memorandum in support of draft legislation "High-Risk Occupational Disease Notification Act of 1995," Columbia University School of Law, Legislative Drafting Research Fund, prepared for the NY State Assembly, Richard Gottfried Assembly Member.
2. http://siteresources.worldbank.org/EXTPOPS/Resources/AsbestosGuidanceNoteFinal.pdf.
3. (a) David Rosenberg, The dusting of America: a story of asbestos; carnage, cover-up, and litigation, *Harvard Law Review*, **99**:1693; (b) Paul Brodeur, *Outrageous Misconduct: The Asbestos Industry On Trial*, New York, Pantheon Books, 1985.
4. ILO Asbestos Convention C162 (1986), ratified by 35 nations as of June 2017.
5. Email message from Andrew Cutz, dated Friday, August 21, 2009, on lung damage in Chinese factory workers sparks health fears (nanoparticles).
6. NanoImpactNet3 papers. Dr. Michaela Kendall, an expert in nanoparticle exposure and nanotoxicology from the European Centre of Environment and Human Health (University of Exeter, U.K.), recommends the following approach: long-term, possibly low-volume gaseous collection method that deposits carbon nanotubes (CNTs)/carbon nanofibers (CNFs) onto a substrate that may be followed by a microscopic counting procedure (preferably transmission electron microscopy [TEM] or atomic force microscopy [AFM]), with parallels to the asbestos fiber identification method. If such a method cannot be identified or the scientific community does not reach a consensus on an accepted method, a desk-based risk and hazard assessment of each CNT/CNF should be conducted that, in particular, focuses on the length of the CNT/CNF and propensity of the particular CNT/CNF of interest to occur as single fibers or small agglomerates that are capable of lung penetration.
7. UN Food and Agriculture Organization (FAO) and the World Health Organization (WHO) expert meeting on the application of

nanotechnologies in the food and agriculture sectors: potential food safety implications. Meeting report, p. 26, Rome, 2010.

8. FoodNavigator-USA.com, *Nanotechnology: The New Asbestos?*, www.foodnavigatorusa.com/Suppliers2/Nanotechnology-The-new-asbestos. The safety risks of nanotechnology use by the food industry could make it "the new asbestos."
9. Peter Hayes, *Nanotechnology Tort Litigation: A Potential Sleeping Giant*, Bloomberg BNA, Dec. 7, 2016.
10. Nanotechnology is the next asbestos, www.newsweek.com/next-asbestos-139271. The next best thing since sliced bread or the next asbestos?
11. (a) *Borel v. Fibreboard Paper Products Corporation et al.*, National Surety Corporation, intervenor-appellee, 493 F.2d 1076 (5th Cir. 1973) U.S. Court of Appeals for the Fifth Circuit, Sept. 10, 1973, rehearing and rehearing En Banc denied May 13, 1973, citing *Alman Bros. Farm & Feed Mill, Inc. v. Diamond Lab. Inc.*, 5 Cir. 1971, 437 F.2d 1295; (b) *Davis v. Wyeth Laboratories, Inc.*, 9 Cir. 1968, 399 F.2d 121; (c) *Basko v. Sterling Drug, Inc.*, 2 Cir. 1969,416 F.2d 417; (d) *Sterling Drug, Inc. v. Yarrow*, 8 Cir. 1969, 408 F.2d 978; (e) *Sterling Drug v. Cornish*, 8 Cir. 1966, 370 F.2d 82. Further the court noted, "The rationale for this rule is that the user or consumer is entitled to make his [sic] own choice as to whether the product's utility or benefits justify exposing himself to the risk of harm. Thus, a true choice situation arises, and a duty to warn attaches, whenever a reasonable man would want to be informed of the risk in order to decide whether to expose himself [sic] to it."
12. Thomas Burdick, *Burdick's Law of Torts*, New York, Columbia University, pp. 94–95, Section 86, 1915.
13. (a) ILO Convention and OSHA regulations, 29 CFR 1910; (b) Ilise Feitshans, *Designing an Effective OSHA Compliance Program*, Thomson Reuters (available on Westlaw.com).
14. James Kakalik, Patricia Ebener, William Feitiner, Gus Haggstrom, and Michael Shanley, *Variation in Asbestos Litigation Compensation and Expenses*, Institute for Civil Justice, Rand Publication Series, Santa Monica, California, 1984.
15. Orrick Consultants, *Nanotechnology and Asbestos: Informing Industry's Approach to Carbon Nanotubes, Nanoscale Titanium Dioxide and Nanosilver*, www.orrich.org, 2015.
16. (a) ABC News, *Nanotechnology Is the Next Asbestos, Union Says*, www.abc.net.au/news/2009-04-14/nanotechnology-is-the-next-asbestos;

(b) Jon C. Ogg, *Is Nanotechnology the Next Asbestos?*, 24/7 Wall St., May 20, 2008, 247wallst.com/healthcare-business/2008/05/20/; (c) *Commentary: Is Nanotechnology the New Asbestos?*, http://www.mystatesman.com/news/opinion/commentary-nanotechnology-the-new-asbestos/4pUK6TURaKUIBadYsN6dKN/; (d) *CLE Webinar – Strafford*, https://www.straffordpub.com/products/nanotechnology-the-next; (e) AzoNano.com, *Are Carbon Nanotubes the Next Asbestos?*, https://www.azonano.com/nanotechnology-video-details.aspx?VidID=283; (f) Law360, *Is Nanotechnology the Next Asbestos?*, https://www.law360.com/articles/97912; (g) www.youtube.com/watch?v=6L7xXgWcbrQ.

17. Ken Donaldson, Fiona Murphy, Rodger Duffin, and Craig Poland, Asbestos, carbon nanotubes and the pleural mesothelium: a review of the hypothesis regarding the role of long fibre retention in the partietal pleura, inflammation and mesothelioma, *Particle and Fibre Toxicology*, open access review, 2010, http::www.particleandfibretoxicology.com/content/7/1/5.

18. Craig Poland, Rodger Duffin, I. Kintoch, Andrew Maynard, W. A. Wallace, A. Seaton et al., Carbon nanotubes introduced into the abdominal cavity of mice show asbestos-like pathogenicity in a pilot study, *Nature Nanotechnology*, **3**:423–428, 2008.

19. Quoted in Ilise Feitshans's International Safety Resources Association (ISRA) testimony before the National Institute for Occupational Safety and Health (NIOSH), United States, in response to NIOSH request for public comments on *NIOSH Current Intelligence Bulletin: Occupational Exposure to Carbon Nanotubes and Nanofibers*, Docket No. NIOSH-161, "Legal Basis and Justification: Recommendations Preventing Risk from Carbon Nanotubes and Nanofibers," Revised Feb. 18, 2011.

20. Ibid., prepared in response to the question presented by NIOSH: Whether the hazard identification, risk estimation, and discussion of health effects for carbon nanotubes and nanofibers are a reasonable reflection of the current understanding of the evidence in the scientific literature. Stakeholder response prepared on behalf of the ISRA, 2010.

21. *Borel v. Fibreboard Paper Products Corporation et al.*, National Surety Corporation, intervenor-appellee, 493 F.2d 1076 (5th Cir. 1973) U.S. court of appeals for the Fifth Circuit, Sept. 10, 1973.

22. (a) Cooke, Fibrosis of the lungs due to the inhalation of asbestos dust, *British Medical Journal*, **2**:147, 1924; (b) Cooke, Pulmonary asbestosis, *British Medical Journal*, **2**:1024, 1927.

23. I. J. Selikoff, R. A. Bader, M. E. Bader, J. Churg, and E. C. Hammond, Asbestosis and neoplasia, *American Journal of Medicine*, **42**:487, 1967; I. J. Selikoff, J. Churg, and E. C. Hammond, The occurrence of asbestosis among insulation workers, *Annals of the New York Academy of Sciences*, **132**:139, 1965.
24. H.R. Rep. No. 14816, 90th Cong., 2d Sess. 349, 355 (1968), testimony of Dr. I. J. Selikoff of the School of Environmental Sciences Laboratory, Mount Sinai School of Medicine, City University of New York.
25. US Congress 92d Congress 1st Session Legislative History of the Occupational Safety and Health Act of 1970 (S.2193, P.L.91-596), prepared by the Subcommittee on Labor of the Committee on Labor and Public Welfare United States Senate, June 1971, p. 143, background memo, summarizing the comments of former Secretary of Labor Shultz.
26. https://en.wikipedia.org/wiki/Irving_Selikoff.
27. Thomas Burdick, *Burdick's Law of Torts*, 3rd Edition. New York, Columbia University, p. 97, Section 88. "The plaintiffs, while lawfully upon the highway as spectators and in the exercise of due care, were injured by the fireworks. There was no evidence of negligence by the defendants." Citing *Scanlon v. Wedger* 156 Mass 462, 31 N E 642, 16 LRA 395 (1892).
28. Ilise Feitshans, Hazardous substances in the workplace: how much does the employee have the "right to know"?, *Detroit College of Law Review*, **1985**(3).
29. Vladimir Murashov and John Howard, Essential features for proactive risk management, *Nature Nanotechnology*, **4**, 2009.
30. C. C. Chou , H. Y. Hsiao, Q. S. Hong, C. H. Chen, Y. W. Peng, H. W. Chen, and P. C. Yang, Single-walled carbon nanotubes can induce pulmonary injury in mouse model, *Nano Letters*, **8**(2):437–445, 2008. doi: 10.1021/nl0723634. Epub 2008 Jan 29.
31. Irina Zen, Rahmalan Ahmad, Krishna Gopal Rampal, and Wahid Omar, Use of asbestos building materials in Malaysia: legislative measures, the management and recommendations for a ban on use, *International Journal of Occupational and Environmental Health*, **19**(3):169–178, 2013.
32. Thomas J. Lueck, Technology: has asbestos a substitute?, *New York Times*, Sept. 2, 1982.
33. A. Churg, Nonneoplastic diseases caused by asbestos, in A. Churg and F. H. Y. Green, eds., *Pathology of Occupational Lung Disease*, Igaku-Shoin Medical, New York, pp. 213–278, 1988.

34. S. McDonald, Histology of pulmonary asbestosis, *British Medical Journal*, **1927**:1025–1026.

35. (a) K. M. Lynch and W. A. Smith, Pulmonary asbestosis III: carcinoma of lung in asbesto-silicosis, *American Journal of Cancer*, **24**:56–64, 1935; (b) R. Doll, Mortality from lung cancer in asbestos workers, *British Journal of Industrial Medicine*, **12**:81–86, 1955.

36. J. C. Wagner, C. A. Sleggs, and P. Marchand, Diffuse pleural mesothelioma and asbestos exposure in the North Western Cape Province, *British Journal of Industrial Medicine*, **17**:260, 1960.

37. A. D. McDonald and J. C. McDonald, Epidemiology of malignant mesothelioma, in K. Antman and J. Aisner, eds., *Asbestos: Related Malignancy*, Grune & Stratton, Orlando, pp. 31–56, 1987.

38. (a) M. L. Newhouse and G. Berry, Asbestos and laryngeal carcinoma, *Lancet*, 615, 1973; (b) P. M. Stell and T. McGill, Asbestos and laryngeal carcinoma, *Lancet*, **2**:416–417, 1973.

39. *Borel v. Fibreboard Paper Products Corporation et al.*, rehearing and rehearing En Banc denied May 13, 1973

40. 399 F.2d at 129-130.

41. *Borel v. Fibreboard Paper Products Corporation et al.*, citing *Davis v. Wyeth Laboratories, Inc.*; *Charles Pfizer & Co. v. Branch*, Tex.Civ. App.1963, 365 S.W.2d 832

42. *Borel v. Fibreboard Paper Products Corporation et al.*, rehearing and rehearing En Banc denied May 13, 1973, citing *Wright v. Carter Products, Inc.*, 2 Cir. 1957, 244 F.2d 53

43. Ibid., citing *Halepeska v. Callihan Interests, Inc.*, supra; *Schiller v. Rice*, Tex.Sup.1952, 151 Tex. 116, 246 S.W.2d 607.

44. Paul Brodeur, *Expendable Americans*, McGraw Hill, New York, 1976.

45. Ilise Feitshans, unpublished notes after tour of training facilities for occupational nurses, Tyler Texas Commencement Conference; see also Ilise Feitshans, *Designing an Effective OSHA Compliance Program under the Texas Medical Records Law*, 2003.

46. Richard B. Von Wald, *Oversight Hearings on the Effect of the Manville and UNR Bankruptcies on Compensation of Asbestos Victims*, Subcommittee on Labor Standards Committee on Education and Labor, U.S. House of Representatives, Washington, D.C., Sept. 9, 1982, p. 64.

47. John Howard, *Nanotechnology: The Newest Slice of Global Economic Daily Life*, invited lecture, International Labour Organization (ILO), Geneva, Switzerland, Nov. 28, 2008.

48. *Norfolk & Western Ry. Co. v. Ayers*, 538 U.S. 135, 123 S. Ct. 1210, 155 L. Ed. 2d 261, 19 I.E.R. Cas. (BNA) 1217, 2003 A.M.C. 609, 33 Envtl. L. Rep. 20155 (2003).
49. Federal Employers' Liability Act (FELA), §§1 et seq., 35 Stat. 65 (codified 45 U.S.C.A. §§51 et seq.).
50. US Congress 92d Congress 1st Session Legislative History of the Occupational Safety and Health Act of 1970 (S.2193, P.L.91-596), prepared by the Subcommittee on Labor of the Committee on Labor and Public Welfare United States Senate, June 1971, p. 509, comments by Mr. Yarborough.
51. Ibid., p 510 comments by Mr. Yarborough.
52. Ibid., p 142.
53. Occupational Safety and Health Act of 1970, 29 USC 651 et seq.
54. Ilise Feitshans, Seminar on workplace injuries, no. 1, *Arlington County Bar Association Journal*, **2**, April 1984.
55. Ilise Feitshans, OSHA considers modification of current asbestos standard, no. 7, *Arlington County Bar Association Journal*, **2**, Jan. 1987.
56. *Industrial Union Dept., AFL-CIO v. Hodgson*, 499 F.2d 467, 1 O.S.H. Cas. (BNA) 1631, 4 Envtl. L. Rep. 20415 (D.C. Cir. 1974).
57. Ilise Feitshans, *Designing an Effective OSHA Compliance Program*, Thomson Reuters (available on Westlaw.com), 1990 and 2013.
58. *Building and Const. Trades Dept., AFL-CIO v. Brock*, 838 F.2d 1258, 13 O.S.H. Cas. (BNA) 1561, 1988 O.S.H. Dec. (CCH) P 28134, 18 Envtl. L. Rep. 20507 (D.C. Cir. 1988).
59. Ilise Feitshans, Law and regulation of benzene, *Environmental Health Perspectives*, **83**, Aug. 1989, National Institute of Environmental Health Sciences, Research Triangle Park, North Carolina, publishers commenting upon *Industrial Union Department v. American Petroleum Institute (The Benzene Case)*, 448 U.S. 607 (1980), United States Supreme Court.
60. (a) 29 U.S.C. 655(b)(5); (b) Ilise Feitshans, *Designing an Effective OSHA Compliance Program*, Thomson Reuters (available on Westlaw.com), 2013.
61. *American Textile Manufacturers Institute, Inc. v. Donovan*, 452 U.S. 490, 509 (1981) (``Cotton Dust").
62. *Industrial Union Department v. American Petroleum Institute (The Benzene Case)*, 448 U.S. At 657-58 & N.66.
63. Krishana G. Rampal and Ilise Feitshans, Legal Requirements for Medical Surveillance of Asbestos Workers in Malaysia, the USA and

Under International Law Comparison of the laws in C155, ILO, USA and Malaysia, ILO and NIOSH International Conference on Asbestos Pittsburgh Pa. 1988. Department of Community Health, Medical Faculty, National University of Malaysia and the Johns Hopkins University School of Hygiene and Public Health, Baltimore, USA. Social Science Research Network: papers.ssrn.com/paper.taf?abstract_id'241093.

64. ILO Asbestos Convention C162 (1986), ratified by 35 nations as of June 2017.
65. European Union Sustainable Nanotechnologies Project: A New Integrated Approach for Risk Assessment and Management of Nanotechnologies, Contract Number 604805.
66. Jay A. Sigler and Joseph E. Murphy, *Interactive Corporate Compliance: An Alternative to Regulatory Compulsion*, Quorum Books, Greenwood Press, CT, 1988.
67. The official text of the GHS, which was adopted on June 27, 2007, is available at http://www.unece.org/trans/danger/publi/ghs/ghs_rev00/00files_e.html.
68. Four major existing systems served as the primary basis for development of the GHS. These systems were the requirements in the United States for the workplace, consumers, and pesticides; the requirements of Canada for the workplace, consumers, and pesticides; the European Union directives for classification and labeling of substances and preparations; and the United Nations Recommendations on the Transport of Dangerous Goods. The requirements of other systems were also examined, as appropriate, and taken into account as the GHS was developed.
69. John Howard, OSHA Consultation Grantees Annual Meeting: "The Times They Are A Changin'," Westin Hotel, San Diego, California, April 28, 2009.
70. Ilise Feitshans, *Designing an Effective OSHA Compliance Program*, Thomson Reuters (available on Westlaw.com), 2013, citing Texas House Bill 2473 (codified at Tex. Rev. Civ. Stat. Ann. Art. 4447c and Tex. Gov't Code Ann. §552.124). Thanks to Terrel Huut, who provided a copy of this bill and commentary.
71. Tex. Rev. Civ. Stat. Ann. Art. 4447c and Tex. Gov't Code Ann. §552.124.
72. B. Boehlecke, Medical monitoring of lung disease in the workplace, in J. B. L. Gee, ed., *Occupational Lung Disease*, Churchill Livingstone, New York, pp. 225–240, 1984.

73. R. M. Gardner, J. L. Clausen, R. O. Crapo, G. R. Epler, J. L. Hankinson, R. L. Johnson, Jr., and A. L. Plummer, Quality assurance in pulmonary function laboratories. Official ATS statement, *American Review of Respiratory Disease*, **134**:625–627, 1986.

74. Ilise Feitshans's speech text: "Health at Work: A Basic Human Right Brought to Daily Life by the ILO Encyclopaedia," March 18, 2009, "Moving to Sky," American Society of Safety Engineers (ASSE), Middle East Chapter (MEC) Conference, discussed in Kuwait times, "KOC to Join United Nations SAFEWORK in COS Centre," May 12, 2009.

75. Over 25 UN agencies and regional groups such as the European Union, governments, and individual trade organizations participate in the Globally Harmonized System for the Classification and Labeling of Chemicals (GHS) to promote consistency among hazard and risk assessment within the global system for classification of hazards, implementing the GHS at the national level. This includes harmonizing methods for risk assessments for specific chemicals and preparing Safety Data Sheets.

76. (a) http://wm.unitar.org/.../GHS_Companion_Guide_final_June2010.pdf; (b) http://www.unitar.org/cwm/; (c) http://www2.unitar.org/cwm/ghs_partnership/CT.htm.

77. Ilise Feitshans and Brian Sharpe, The nuts and bolts of compliance programs, in *Bringing Health to Work*, Emalyn Press, 1997.

78. *NRDC v. EPA*, discussed in Chapter 3 of this book.

Chapter 3

Global Health Impacts: Nanotechnology Fighting Cancer and Disability

3.1 Revolutionary Promises of Nanomedicine

Nanomedicines hold a miraculous promise: killing cancer tumors within the tumor cells; regenerating bones and teeth from an individual's own stem cells; creating neural pathways for people who may never have been able to see, hear, or process normally before; or creating artificial blood cells for a wide array of medical uses. Applying nanotechnology techniques at the nanoscale to medicine, "nanomedicine" is defined as the use of nanotechnology in every branch of health care, ranging from monitoring and detection of disease to new organs that can be implanted in patients, medical devices that can study patients at the cellular level, drug delivery systems, and nanosensors to notice changes following treatments. One branch of nanomedicine promises remarkably sensitive diagnostics that will detect metabolic changes, making possible information about disease at a very early stage before symptoms appear. For example, blood analysis at the nanoscale, combined with molecular imaging using nanotechnologies, can detect and trap diseased cells. Therapies targeting diseased organs and cells can use nano-enabled devices to repair damaged tissue and oncological drugs [1]. Standard medical care or aesthetic surgery using nanotechnologies will be able to meet any popular demand to look

Global Health Impacts of Nanotechnology Law
Ilise L. Feitshans

ISBN 978-981-4774-84-0 (Hardcover), 978-1-351-13447-7 (eBook)
www.panstanford.com

like famous stars once replacement parts for use within humans are commercially available on a mass scale.

Nanomedicine also promises an important ability: to apply particle-sized operations with precision. Therefore, manufactured nanomaterials (MNMs) can potentially be designed to cross previously impermeable natural barriers, such as the blood–brain barrier and the placenta. This enables nanomedicines to access new sites of delivery and to interact with DNA or small proteins in blood or within organs, tissues, or cells. Nanosensors can improve monitoring in intensive care units and offer new avenues for monitoring patient follow-up care, transforming the capabilities of people who are considered disabled. Gold nanoparticles can track cells in cancer treatment so that doctors can watch to see if treatments are actually working [2]. Innovations applying nanotechnology to medical care will also be used to create new baseline testing, thereby expanding the concept of check-ups and testing during necessary health care beyond responding to symptoms that can already be seen. Data generated by applying nanomedicine to disease, through imaging, through the use of new medical devices, and because of nanoparticle analysis for blood and genetic testing, will enable more effective therapies that will reduce costly hospitalization and enable some patients with previously untreatable or incurable illness to return to productive work. And the contribution of nanotechnologies to electronics makes possible analysis and interpretation of big data with insights into global health impacts by following trends across populations by region, by disease, by genetic makeup, or by race or gender.

Conversely, it is possible that unintended voyages of nanoparticles may disrupt the natural process, and therefore research is needed about transplacental transfer of MNMs as well as accumulation of nanomaterials in the lungs, spleen, and blood. Coating particles and changing the function of the materials on their surface may increase the biocompatibility of medical treatments, too, with similar questions about whether this is good or desirable. By offering the possibility of biological offspring to lesbian, gay, bisexual, and transgender (LGBT) or infertile individuals, who would not have dreamed of natural parenting a century before, nanomedicine presents the possibility to bring new constituencies under the banner of parenting one's own children. Nanomedicines with increased biocompatibility will reduce the human body's natural resistance, a stark contrast to expensive pharmaceuticals that prevent rejection of artificial materials and transplanted organs from other humans. Nanomedicines therefore may transform long-term disability or terminal illness into a temporary setback on the road of life.

So, some promises sound more like ancient science fiction rather than scientific fact; these developments may give new daily meaning to personalized medicine and precision medicine. For example, stem cells, selected at the right age in a person's life can be stored and later grown in a laboratory to become any needed body part, thanks to applications of nanotechnology: teeth, liver, heart neurological pathways, and bone. Fixing damaged organs with new replacement parts made from one's own stem cells may prolong life beyond natural cell longevity with less risk of immunoresponse rejection. Like canning fruit at the right moment of ripeness, stem cells from an individual may provide the key to prolonging his or her life decades later. Standard medical care or aesthetic surgery using nanotechnologies will also be able to meet popular demand to look like famous stars once replacement parts for use within humans are commercially available on a mass scale.

Nanomedicine therefore offers promising treatments for cancers, broken bones, and invisible disabilities but raises puzzling new bioethical questions at the very same time. These puzzles are an important piece of nanotechnology's big-picture social impact. Applying nanomedicine means that presymptomatic recognition of illness and therefore the definition of disability will expand, at the same time increasing the possibility of returning to mainstream society after treatment. Societal questions of discrimination for presymptomatic disability and payment for treatment among people who appear to be well will loom in parallel with each new test offered to the public. Nanomedicines made available with forethought therefore will increase the size of the disabled population, while offering the promise of splendid opportunities to fold into the fabric of the typical workforce people from so-called vulnerable populations who previously had been excluded from mainstream civil society. Vulnerable populations (people with disabilities [PWD] and the aged, children, and women of childbearing age) will have new avenues for returning to the mainstream of work and greater civil society. New treatments will raise questions about how society will view people who seem healthy but require time off from work and insurance coverage for treatment [3]. Harvesting and storage raises questions of access to care, whether there may be avoidable economic cost for storing materials that are not needed or unused and whether there will become a unified societal vision for the correct teeth; eyes; bone shape; size, including height; or other external characteristics.

Beyond the exciting potential for a better quality of life that sounds like beautiful fiction, nanomedicine can also be seen as part of a tsunami of sweeping changes in the global perceptions of health and disability under law and in the overarching society. The arrival of nanotechnology, praised and heralded as a welcome revolution reshaping industry, provides the perfect opportunity for rethinking workplace design, social constructs of health and disability, and the configuration of natural biological families. The tantalizing possibility of blending into the weave of the workplace fabric antidiscrimination goals that protect people from discrimination and ensure equal opportunity to work regardless of handicap or impairment also means that predicting illness before it occurs and offering treatment may reshape social conventions regarding informed consent, the right to refuse treatment, the social meaning of disability, and the importance of access to health care.

The relationships between society and individuals who have access to nanomedicines and precision medicine will also redefine the terms "health" and "disabled" under law for insurance, protection against discrimination, and employment. Nanomedicine and the prospect of presymptomatic cures will therefore have an important health impact upon civil society's perception of disability and reshape the cultural response to illness.

Although perfecting the technology of nanoparticle manipulation is recent, these problems posed by the impact of innovations from new technology upon health care are not new. There is extensive overlap between nanotechnologies and genetic sciences because much genetic material is less than 100 nm in size, thus revisiting many bioethical issues that were discussed during the Human Genome Project in the 1990s. Potential health impacts of applied nanotechnologies for genetic material and chromosomes exist, whether by accident or by intentional social design. These same pathways are important for the development of nanomedicines, which make possible the transfer of targeted drug delivery across the placenta and tiny nano-enabled medical devices that will make possible new forms of fetal surgery. For emerging genetic therapies in the 1990s, epidemiologist Abby Lippman noted [4], "Geneticists and their obstetrician colleagues are deciding which fetuses are healthy, what healthy means, and who should be born."

This new ability to mix genetic materials and fashion new genetic scripts using nanotechnologies also requires reflecting with forethought about eugenic concerns expressed by Troy Duster [5] regarding the "acceptability" or socially undesirable nature of specific genetic traits. The ability to "choose" fetal characteristics or to manipulate natural features by applying nanotechnology will inevitably reshape the prevailing societal views of "healthy," "acceptable," or even "desirable" traits in new generations. Despite legislative efforts, there remains an inextricable link between societal notions of appropriate race or genetic configuration for people in society, which impacts social norms about which genetic traits constitute "sickness" or "health." Thus, there may be a shifting of social conventions and accepted medical standards for the appropriate level of functioning for a "normal" or "healthy" human being once nanomedicines become widely commercialized and relatively accessible to the general public. This has implications for public health planning strategies and also for expanding the parameters of protections against discrimination for people who are disabled. For example, one pediatric center for nanomedicine has experience with developing personalized and predictive oncology and the development of engineered protein machines for treating single-gene disorders [6].

As in the case of genetic testing, questions raised by presymptomatic testing using nanomedicine for adults or the fetus can be described with words that echo AIDS testing [7], decades earlier: if fear of discrimination deters people from seeking a diagnosis because the prognosis makes them likely to lose their job or their insurance, then benefits of these discoveries will be undermined [8]. For this reason, the definition of disability, or more precisely which types of illness fall beyond the socially acceptable health and into the realm of disabilities needing public health supports, presents an issue that is inextricably related to applying nanomedicines under law.

3.1.1 Nano World Cancer Day

The Nano World Cancer Day [9] is an event organized as part of the annual World Cancer Day on February 4; its dual focus is to increase public awareness about nanomedicine, while bringing people new hope about fighting cancer [10]. Industry leaders in the fight against cancer predict that nanomedicines will play an important role for all health care. In 15 nations across Europe, dynamic and informative events are designed to inform journalists, students, and the general public about the work and perceptions of local researchers, entrepreneurs, oncologists, clinicians, public authorities, and sometimes patients. In each country these stakeholders deliver short speeches and answer questions about the latest innovations in nanomedicine.

Under the slogan "Nanomedicine: Smart Solutions to Beat Cancer," the Nano World Cancer Day is a collaboration by the European Technology Platform on Nanomedicine (ETPN) and the European Project ENATRANS (Enabling NAnomedicine TRANSlation) (Fig. 3.1).

Nano World Cancer Day Website Defines Nanomedicine

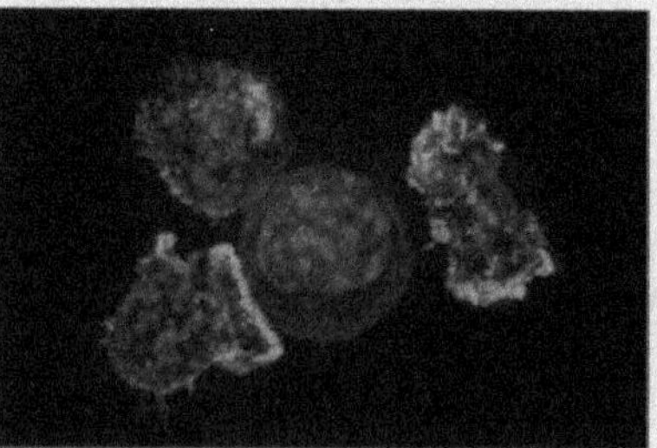

Figure 3.1 Killer T-cells surround a cancer cell. Source: US National Institutes of Health (NIH).

3.1.2 Benefits to Patients: New Bones, New Teeth, and New Organs

The key benefit to patients who apply nanomedicine to a precision medicine strategy will enable prevention strategies to start at a cellular level. Precision medicine, relying upon genetic profiles, epigenetics, and nanomedicines, will be combined to deliver drugs, treatments, and implantable devices to patients with less chance of rejection or wasted drugs by meeting the special needs unique to their bodies. The expanding list of disease categories that will be touched by nanomedicine presently includes cancer, cardiovascular disease, neurodegenerative disease, musculoskeletal disorders, and inflammation and offers patients a better quality of life. Patients are expected to benefit from new applications of nanotechnology to diagnostics and treatment because nanomedicine has the potential to enable early detection at the cellular level. Around 100 nm or less is also the scale of many biological mechanisms in the human body. Nanomedicine holds the prospect of enabling scientists to create artificial blood cells, new types of proteins , and drugs that can interact with DNA and human proteins at the molecular level. Therefore, nanomedicine is expected to play a key role in enabling personalized, targeted, and regenerative medicine. Regulation of nanomedicines offers a new opportunity for patients with their medical teams to rethink existing pathways for rehabilitation, financing of rehabilitation, and the desired outcomes of treatments and to reinforce existing protections against discrimination for PWD.

All of these goals and their implementing tasks can be integrated into the legal framework for oversight and monitoring of patient progress when applying nanomedicines, regardless of whether these issues are examined in existing regulatory governance or under new law. Key factors about the nanoparticles used in detection, treatment, drug delivery, and monitoring, such as characterization of the form, size, and function of manufactured nanoparticles and the differences among categories of nanoparticles from the standpoint of toxicity assessment, can be examined in multistage adaptive rules. Experience in clinical trials, standardized procedures, and the best techniques for assessment can be folded into the regulatory matrix in order to offer the maximum benefits to patients and society.

Nanomedicine therefore makes possible greater lead time to refine diagnosis, discuss options for treatment, and create strategies for monitoring and follow-up across many types of diseases. This new dimension for early disease detection with attendant jobs for health care delivery and management will represent a new sector for the global economy and will also produce the rehabilitated workforce to staff it. But the reliability of such early warnings will also raise again old questions about the scope of insurance coverage, informed consent for procedures, and social stigma for people who wish to exercise their right to refuse treatment. Nanomedicine also provides important new tools to deal with the grand challenge [11] of reducing maternal mortality during pregnancy and childbirth and infant mortality—one intractable medical problem that has plagued humanity since the start of civilization [12]. What a magnificent gift science would give civil society if nanomedicines and applications of nanotechnologies would make enhanced prenatal care, childbirth systems, and high-quality neonatal care available and affordable to all (Fig. 3.2).

Possible Positive Impacts of Applied Nanomedicine

Scaffolds for tissue regeneration

Opportunities to replace organs with nano-enabled organs produced from the patient's own stem cells

Sharper focus of technology upon cancer cells and disease sites in order to Destroy tumors, effectively control disease process, and heal wounds

Reduced toxicity of drugs due to smaller doses and also because of greater control of drug delivery

Potential to reduce side effects compared to larger doses in traditional drug delivery

Social Impacts in the Context of Nano-Applications

Improved access to outpatient care for treatment using nano-enabled approaches

Extended survival curve for tumors that have poor prognosis, especially but not limited to tumors among young patient populations

Reduced economic cost of raw materials for some treatments

Greater availablity of limited health resources when smaller quantities are used

Figure 3.2 Applications and societal impacts of nanomedicine.

3.1.3 New Technologies for HIV-Positive Communities Living with AIDS

Nanotechnology has been applied since 2008 to prevent HIV/AIDS. Medical research has been seeking new ways to enhance the natural autoimmune system for people who have autoimmune deficiencies for decades [13]. For people who suffer from multiple sclerosis, HIV, and a variety of other diseases, nanomedicine's ability to manipulate nanoparticles in the cell provides hope of autoimmune system repair. From the standpoint of HIV prevention, microbicides in vaginal creams and other topical applications of nanotechnology have been used in condoms for over a decade [14]. Prophylaxis for HIV/AIDS remains important because 8 out of 10 pregnant women living with HIV received anitretroviral medicines to prevent mother-to-child transmission of HIV, but there were 170,000 new infections among children in 2014 [15]. No one understands why some babies are born with HIV and others are not, after maternal exposure. The UNAIDS document "We Can Prevent Mothers from Dying and Babies from Becoming Infected with HIV" offers an outcome framework that explains "why this is a priority area" and then sets for the blueprint of "what needs to be done" with a call for "bold results." The framework mandates "integration and linkages" with treatment services, communities, and scaled-up coverage within health systems. Therefore, HIV/AIDS provides an excellent example of interdisciplinary scientific efforts that can benefit from new nanomedicines and from information made available under the umbrella of research using nanotechnologies [15].

3.1.4 The Fair Share in Posterity for All: Assisted Reproductive Technologies Enabling LGBT Families

For centuries, reproductive health has been the subject of cultural, ethical, and legal ambivalence. Although every culture and legal system needs to repopulate in order to continue society, many legal systems do not welcome reproduction for everyone. A possible shift in culture regarding reproductive health among people who are healthy and competent as parents, although legally disabled, may occur because nanomedicine and nano-enabled devices will be able to reduce the global disease burden. By expanding the actual medical basis for "Capability to reproduce," new technologies offer parenting opportunities to populations who in the past could only fantasize about having their own biological offspring and raising them in their own family.

Frozen eggs from humans, stored long in advance of pregnancy, and a variety of new forms of surrogate parenting enable women to continue working in highpowered careers without staring squarely at the face of a biological clock or depending on having a male partner in order to have their own child. Into this mix too, nanomedicines and nano-enabled devices will liberate LGBT individuals from the traditional constraints upon having their own offspring and will activate their "freedom to make informed, free and responsible decisions" about reproductive health involving reproduction" under UN directives and international public law. Access to a range of reproductive health information, goods, facilities and services, as promised in General Comment n. 22 (GC/22), may be achieved using nanomedicines to improve prenatal care, prevent cancer of reproductive organs in men as well as women across all age groups, and protect the fetus from extra-utero harms that impact maternal health, such as loud noise or respirable toxins in the environment. Good science applying nanotechnology may therefore offer solutions to the challenge of documented health disparities that undermine reproductive health for all.

Nanomedicine also makes possible equitably including all groups of people within the same system, such as LGBT and infertile populations using assistive reproductive technologies (ART). The United Nations (UN) International Conference on Population and Development in Cairo (1994) codifies the definition, "Reproductive health is a state of complete physical, mental and social well-being and not merely the absence of disease or infirmity, in all matters relating to the reproductive system and its functions and processes . . . Reproductive health care is defined as the constellation of methods, techniques and services that contribute to reproductive health and well-being by preventing and solving reproductive health problems . . ."

Consistent with this legal mandate, the UN Committee on Economic, Social and Cultural Rights (ECSCR) adopted GC/22 in March 2016, which elaborates upon sexual and reproductive health rights, stating, "[r]eproductive health . . . concerns the capability to reproduce and the freedom to make informed, free and responsible decisions. It also includes access to a range of reproductive health information, goods, facilities and services to enable individuals to make informed, free and responsible decisions about their reproductive behavior" (para. 5 E/C.12/GC/22). The committee in GC/22 further stated that "the right to sexual and reproductive health is also indivisible from and interdependent with other human rights. It is intimately linked to civil and political rights underpinning the physical and mental integrity of individuals and their autonomy, such as the right to life; liberty and security of person." Several key challenges confront the global efforts to implement these beautiful notions about reproductive health and sexual health, which implicitly impact legal, ethical, and cultural questions about who should parent or whether there is a benchmark for the basic standard of living for a "healthy" child.

According to Louis Henkin, in whose memory the Human Rights Clinic at Columbia University School of Law was founded, "Human rights is the idea of our time. It asserts that every human being, in every society, is entitled to basic autonomy and freedoms and respect and basic needs satisfied" [16]. Without defining autonomy, Henkin further noted that human rights "include freedom from mistreatment and undue governmental intrusion." This principle applies to ensuring freedom from constraints regarding reproductive health. Rejecting the notion that there is a trade-off between "privacy" and "public good," Henkin [17] views "privacy" as "freedom from governmental intrusion" and "autonomy" as a zone "of presumptive immunity to governmental regulation" and that such freedom is an essential public good [18]. Henkin's widely accepted approach of including these rights as part of the societal notions of public good reflects leading codifications of international human rights that treat privacy as an important part of society in daily life. Under the United Nation Charter (henceforth UN Charter) [19], the contracting parties state their aspiration to "promote" economic and social advancement and "better standards of life, including the promotion of human rights protections" in Article 13 [20]. The newest UN pronouncements about rights associated with reproductive health offer a new legal argument that reproduction has coequal importance among those well-established rights associated with health, and life, as protected by governmental nonintervention in matters that concern privacy and autonomy. Pretty words. Yet, it remains unclear how the international system will apply the principles described using these words. The task ahead is to create an infrastructure for people who are interested in operationalizing these ideas. Nanotechnology offers humanity the ability, within the rubric of nanomedicines, to give these words potent meaning in daily life [21], thereby opening new avenues of family life for individuals who are LGBT or part of the disability community.

3.2 Global Health Impacts: Improved Quality of Life for Children

Pediatric Nanomedicine Center Links Health care and Engineering, March 2011: First-of-Its-Kind Research Center Includes Physicians and Scientists

Physicians and engineers devoted to pediatric nanomedicine plan to develop targeted, molecular-sized nanoparticles when treating pediatric diseases such as pediatric heart disease and thrombosis, infectious diseases, cancer, sickle cell disease, and cystic fibrosis. According to Nanomedicine Gang Bao, PhD, Robert A. Milton Professor of Biomedical Engineering, Wallace H. Coulter Department of Biomedical Enginnering at Georgia Tech and Emory University, "Nanotechnology can be applied to many diseases, and the application of nanotechnology could have a profound impact on improving children's health" (Fig. 3.3).

Figure 3.3 Children who will benefit from nanomedicine. Photo by Bernd Kulow, Freiburg, Germany.

Fourteen key priority centers have been identified. These are hematology and oncology, immunology and vaccines, transplant immunology and immune therapeutics, pediatric health care technology innovation, cystic fibrosis, developmental lung biology, endothelial biology, cardiovascular biology, drug discovery, autism, neurosciences, nanomedicine, outcomes research and public health, and clinical and translational research.

Emory and Georgia Tech previously had significant successful research partnerships in nanomedicine funded by the National Institutes of Health. These have included a Center of Excellence in Nanotechnology for the detection and treatment of cardiovascular disease, the development of personalized and predictive oncology, and the development of engineered protein machines for treating single-gene disorders.

For example, scientists at the Center for Nucleoprotein Machines are focused on developing a technology to correct single-gene defects that lead to human disease. They hope to use this approach to treat and eventually cure sickle cell disease, first focusing on curing a mouse model of sickle cell. The new technology would then be applied to human sickle cell patients.

"Nanomedicine is expected to dramatically exceed what has occurred in the field thus far, and our belief is that it will revolutionize medicine," says Bao. "We plan to make this new pediatric nanomedicine center a leader in applying these unique discoveries to treating and curing children's diseases."

3.2.1 UN International Convention on the Rights of the Child

Nanomedicine's revolutionary capacity to treat diseases in children also provides an opportunity to uphold the rights of children (Fig. 3.4) to life, education, and the opportunity to become a mature human being under the UN convention that codified those rights as international law. But to do so under international law is not as easy as one might think. Despite its bold attempt to amplify the right to health as articulated by the World Health Organization (WHO) and its ability to describe the need for education, food, and other basic human needs that are intrinsic to sustainable development [22], the UN International Convention on the Rights of the Child does not contains any reference to the natural beginnings of life or the onset of rights as a human being, so there is no starting point to trigger the protections afforded to human children under these documents. Consequently there is only a weak international law to support pediatric interventions that can have a positive impact on infant and child health.

Figure 3.4 (Left) Child in the United States (photo by Ilise L. Feitshans) and (right) portrait of Joseph Issac Feelus, 1917, Charoy Family Archives, used by permission.

Major gaps in the terms of the convention designed to protect the rights of children leaves open questions regarding the legal support for applying nanotechnologies to prenatal care, neonatal care, and pediatric care once commercialized nanomedicine treatments leave the laboratory bench and enter into clinical trials and the marketplace. This is problematic if a child wants a treatment and the parents refuse to allow it or vice versa. The convention is silent about whether children have a capacity for informed consent and, if so, what standards apply to ensure that children have the right to know important information and whether they have understood the information as given. The convention offers no discussion about criteria for determining if children understand medical conditions that they experience, nor does it give them a voice in treatment if they do understand what is happening to them.

Education must be given to all children and therefore implicitly be guaranteed to children who are mentally retarded, but the convention does not discuss whether there is an age limit for the term "child." Efforts to educate mainstream people who under local laws may be children their entire life are qualitiatively different compared to general education, and the convention does not mention whether efforts to educate them stop magically at a certain time such as the national legal age of majority, despite their limited competence. These issues come to the fore when using nanomedicines, because teenage parents may have conflicting rights and obligations, depending on whether the law views them as parent or child. Children who will always be children for life may require supplemental protection under law.

Another important gap in international public law could be filled by the opportunity to provide better health using nanomedicines and nanotechnology. There is a reference to the unborn in the preamble of the document, but there is no clear definition of the term "child" to determine whether a child includes people starting at the time of birth, at conception, at a discernible scientific point of reference (e.g., after the first day of life), or at a magically legislated moment in time (e.g., the time when a valid birth certificate has been issued under local law). Significantly, the UN convention makes no effort to harmonize national laws governing the legal distinction between adult and child, which is a transformative definition from the standpoint of rights and duties under law, but varies according to age in different places. Typically, adulthood begins at either 18 years of age or 21, but some places have a bifurcated system that allows some rights, such as voting or military service, at 18 and other rights, such as drinking alcohol legally, at 21. For this reason, too, teenagers, who need special protection of their rights because their developmental phase is characterized by acquiring but not yet possessing sound judgment, are not clearly protected by the plain language of the convention itself. This increases the burden on the government to protect children from the standpoint of passive exposure to risks and dangerous substances found in a wide variety of consumer products, including nano-enabled products in commerce. Under the international law, therefore, childhood has no beginning and no end. The ability of nanomedicine to legally provide access to care in the spaces created by these gaps in international laws therefore is unclear.

Since it is not clear whether children have the same rights in this convention throughout their early life, protecting their right to health is problematic from a regulatory standpoint. Logically, one could imply duties based on the text of other conventions, but realistically, children should not be held accountable for obligations until they reach adulthood (as defined by law), and therefore require a heightened level of protection and safeguards for their health related decision making compared to other vulnerable populations, which served as the justification for creating this convention. The convention's failure to designate the level of protection afforded to children at different levels of development is a major flaw in the text of the international convention because it nonetheless asserts a firm mandate that each child is entitled to respect, freedom from slavery or forced labor, health protections, and a right to education. But the convention does not state in any tangible form when an individual attains such rights (at conception, birth, birth registry, or a moment chosen by the legislature of a signatory nation) or when he or she has aged out of this special class of protections. Equally vague is whether the state or individuals have any responsibility for children who cannot, for medical reasons, be educated due to incompetence and, if legally determined to be incompetent, who may maintain the status of a child throughout their lifetime, even into old age [23].

The ancient notion of incompetence preventing an individual from assuming the full responsibility in society and thereby legally remaining a child for his or her entire life (protected by legal guardians who engage in the individual's medical decision making for him or her) is commonly found in most legal systems, but it is not discussed in the convention. The practical notion that some people are too handicapped to help themselves remains constant, even if new treatments and nanomedicines shift the line drawn by the law between such individuals and the rest of society. The common law, for example, traditionally recognized a "rule of sevens" that distinguishes between levels of development. The principle of the rule of sevens was very useful in determining state obligation and responsibility for a child and the child's obligations as a citizen. The rule is that under age 7 a child is presumed under law not to understand any laws or obligations. From age 7 to 14, there is a presumption in favor of not holding a child accountable or responsible for willful acts, but the presumption can be disproven. From age 14 until majority (often but not always age 21), there is a presumption that the child does understand the consequences of acts and undertakes responsibilities, but in contrast to full adults, that presumption can be disproven. This last presumption is never put into place in the case of individuals who are legally determined to be incompetent, and therefore most legal systems treat them as children, with surrogate guardians if they no longer have living parents.

Unfortunately, the convention is also silent regarding the obligations of any state to help people under the age of majority who have taken on the burdens of adulthood, such as parenting or supporting their family. So, too, this convention fails to incorporate notions found in the common law regarding "emancipated minors" such as teenagers who have already borne children and are therefore parents, imbued with adult parental responsibilities while still under the age of majority. In many legal systems, such children have the right if not the obligation to make decisions governing their own medical care, the health of their spouse if the spouse is incapacitated, or for their own children.

It is unclear whether by bearing children they are excluded from the protections offered under this convention or remain a child under international law, with perhaps greater support from the state due to the responsibilities that create their quasi-adult status. This question of when or whether there exists an adult status for a child who should have and enjoy the rights under this convention is especially important for nanomedicine decisions regarding personal health, treatment for sexually transmissible disease, HIV/AIDS prevention breakthroughs, LGBT reproductive health, abortion, prenatal care, and consent to treatment that impacts the health of a child whose parent is an adolescent. An individual's status as adult or child may determine the ability to consent to procedures and treatments such as fetal surgery, neonatal treatment of cystic fibrosis, and a variety of other treatments made possible by nanomedicines. Typically, individuals who are economically independent orphans or who are teenage parents are considered mature minors or emancipated minors under law [24] in many nations, but the convention does not address this problem.

Therefore it is possible that nanomedicines could provide life-transforming help to the children of teenage parents, but the law might not protect the parents' ability to give their consent for the operation or treatment in some circumstances. The ability to allow for informed consent and access to care for children without parental intervention is obviously altered by the absence of such principles in the convention. This absence of clear lines of responsibility will therefore indirectly alter the global health impact of nanomedicine for children and may therefore indirectly create a paradigm shift in the global interpretation of the scope and effect of this convention concerning the rights of the child. Nanotechnology, promising new robotics, also can be used as a tool to promote the goals of the convention, helping to end child labor. Nano-enabled products promising a mechanism to provide health care and improved quality of life by making accessible nanomedicines offer the rare opportunity to fill the important void in international law, protecting children and to ensure the well-being of children of all ages throughout civil society, offering cheaper and faster electronics to facilitate in-school and online education, and providing some really cool new toys.

3.2.2 Potential Consumer Exposures among Children

Gaps in public health protections for children under the UN International Convention on the Rights of the Child militate in favor of regulatory vigilance within nations to protect children from secondary or passive exposures to risk of harm. The US courts have been sensitive to this issue, as demonstrated by one case that gave very careful scrutiny to children's need to be protected from exposure to nanosilver. The court riveted its attention upon potential exposure among toddlers and infants (under six months of age) when it stopped premarket registration of nanosilver antimicrobial powder used in textiles.

In the case of *NRDC v. EPA*, a nongovernmental organization challenged the federal administration's risk assessment of dermal and oral exposure to nanosilver in a rodenticide that resulted in denying pre-registration of the product. The court agreed with the petitioner and denied the validity of parts of the Environmental Protection Agency (EPA) pre-registration procedures and therefore required the agency to rewrite its pre-registration standard. All subsequent or pending pre-registration applications to the EPA involving nano-enabled products were therefore placed on hold as an indirect consequence of this litigation. The litigation lasted three years.

The problems raised for industry, regulators, and public understanding of health impacts are spotlighted in this case because the court was forced by the petitioner to wander far beyond the realm of reasonable judicial expertise. The facts before the court as raised by the petitioner went so far beyond normal legal expertise that the court engaged a special expert to recalculate and then explain the risk assessment calculations made by the administrative agency (EPA) and challenged by the petitioner (the Natural Resources

Defense Council [NRDC]). The court's clearly stated desire to protect public health, especially the health of small hildren who are incapable of consenting to harmful exposures or understanding risk analysis, underscores that courts hesitate to accept risk when children are potentially in danger. Consistent with the protection of children under international law, the common law, and national consumer protection laws, the court erred on the side of precaution when confronted with confusing new complex scientific evidence.

The court's confusion about emerging science despite good intentions is evident because many of its sentences are gibberish. In essence, the court hesitated to approve agency pre-registration of a nanosilver-containing rodenticide for fabrics because it could not understand the science involved.

If the court could not understand the science, then it is reasonable that many ordinary citizens could not understand it either. Erring on the side of embedded precautionary principles therefore was appropriate, and the court consequently required the federal agency to rewrite parts of its standard for pre-registration. In its opinion in *NRDC v. EPA*, the court found a credible threat of harmful exposure from the pre-registered product with nanosilver because "the ubiquity of textiles" combined with the vulnerability of infants so "members cannot reasonably assure that the carpets at the daycare center, the jackets worn by a caretakers or the seat on the school bus" have no nanosilver treatment [25]. The agency had selected the age group of six months to three years old as the most vulnerable population exposed, but the NRDC successfully argued that a more vulnerable population consisted of infants who were less than six months old and that their needs were not addressed in the agency's pre-registration risk assessment procedures.

Factual questions before the court involved the methods used by the agency to determine an estimate of the daily dose for toddlers who might chew or suck on fabric that has nanosilver embedded in its threads or as a coating. The court's special experts recalculated the agency estimates in order to determine whether the agency had correctly chosen its exposure assumptions. The court discussed rounding the estimated doses to two significant digits instead of carrying the analysis further, because of the rule that "[w]hen measured quantities are added or subtracted, the numbe of decimal places in the result is the same as that in the quantity with the greatest uncertainty and hence the smallest number of decimal places" [26].

Furthermore, the court expressed its own opinion about the quality of the agency's calculations, inventing a new word in order to prove its point: "Having established a rule of decision of less than or equal to 1,000 EPA cannot unmake it" [27]. The court concluded that the agency had not done correct calculations, and vacated the agency decisions, and then to remedy this problem, it set forth procedures requiring the agency to redo its risk assessments concerning potential future nanosilver exposure for toddlers if the product in question were granted permission to be registered because "EPA's conclusion that short- and intermediate-term aggregate dermal and oral exposure to textiles and surface-coated with AGS-20 is not supported by substantial evidence."

The lesson learned from the long years of the *NRDC v. EPA* litigation is that if the regulatory framework becomes so complex that smart people who are judges cannot understand the reasoning, then it is impossible to expect the general public of consumers to understand the risks and benefits and balance them for their personal needs. The subtext of this court of appeals decision hesitating to approve use of nanosilver in fabrics indirectly reflects the asbestos litigation history. The impact of the asbestos litigation and its subsequent regulatory history is manifest in the court's reluctance to find a level of acceptable risk for unquantifiable and unknown risks involving the application of nanotechnology for consumer products. The court appears to be influenced in its opinion by this history, when confronted with the issues presented by nanosilver exposure to infants and toddlers. Given the asbestos litigation precedents, after the public had been exposed to consumer products that used asbestos, the court's hesitation is not surprising. Even though asbestos was used first and then litigated after the fact of harm, the risk assessment requirements under law that were written in the wake of the asbestos litigation require deep thought and scrutiny before products can be approved for consumer use, even when the end user is part of the supply chain. Ultimately, sound compliance programs that feature due diligence can address this evidentiary problem, which can be solved by creating a clear and transparent regulatory framework that will use such evidence for governing nanosafety.

3.3 Beauty Babies and Dieting: The Impact of Nanotechnology on Women's Health

Will the well-documented empirical gender-based health disparities between men and women be exacerbated or reduced following widespread use of nanotechnology?

The 2009 keystone WHO report *Women and Health: Today's Evidence, Tomorrow's Agenda* documented health disparities between men and women and found that women's health lags behind their male cohorts at five key stages of the life cycle:

1. Birth to 5 years
2. Adolescence (including implications of adolescent pregnancy)
3. Reproductive years
4. Postreproductive years (menopause and greater risk for cancer among sex-based target organs)
5. Advanced aging (65–80 years)

In an effort to understand systemic approaches that might reduce these documented health disparities, WHO later examined the role of constitutional law and local legislation ensuring the rights of women and children in light of their health outcomes is several nations. These early linkages between constituional language protecting health and a postiive health impact on the lives of women and children are discussed in the WHO expert report *Women and Children's Health: Evidence of Impact of Human Rights* [28]. Progress has been slow, according to the report, despite recognition of local impacts for a shared global problem. Many facets of women's uniquebiological, cultural, and caretaking needs have been long ignored in the health research context, especially when science relies on male models as a benchmark. There is an implicit difference between men and women regarding indirect commercial nanoparticle exposure for cosemetics and skin care, and therefore a possible impact regarding cumulative use cannot be captured for retrospective study but may confound our understanding of health outcomes. Gender-based differences have been identified in response to cellulose nanocrystals [29]. Transplacental transfer of toxins and impacts upon reproductive health are also an emerging concern that may be different for women compared to their male peers [30].

Like sand along the riverbanks, nanomaterial exposure has been accruing in human bodies that use cosmetics daily, thus also raising the question whether women have different exposures, biological responses to cumulative impact of exposure to some MNMs and whether those impacts will be expressed in any health outcomes. Whether for pleasure or as a requirement of their job, makeup is vital to women more often than to men and throughout their lifetime, but nanocosmetics that can be absorbed through the skin and migrate throughout the body raise important questions of concern. Strong but subtle exposure impacts women disproportionately compared to men, regardless of their age, ethnicity, demographics, or profession. Many personal care products are currently on the market, but remarkably little is known about their chemical makeup and exposure hazards. This potential cumulative effect may be more complex than many synergistic effects that epidemiology has previously attempted to measure. But this challenge can be viewed as an opportunity to deconstruct epidemiological approaches to womens' health and to change the methods used in existing paradigms.

Increasing appreciation of the unique needs of female patients and women's health care providers has created a need for innovative approaches to women's health. Nanotechnology offers an opportunity to discard skewed methods that have been used in the past, and thereby reduce women's health disparities, using a creative, new approach to solving these ancient puzzles that contribute to documented health disparaties. Nanotechnology's arrival in commerce and public health can revolutionize the field by breaking a detrimental cycle that historically has caused harm.

3.3.1 Reducing the Global Disease Burden of Maternal Mortality

The sound of freedom that resonates from civil and political rights rings hollow to a newborn who has a low birth weight because the baby's mother had no access to a clean workplace, good nutrition, or adequate prenatal care. Some people may argue, too, that political and civil rights are of little value to a baby who has lost a parent due to an occupational accident or whose parents are debilitated by occupational disease or to a baby who may suffer personal injury due to the effects of its parents' workplace exposure to mutagens or unchecked but foreseeable harms caused by the absence of sound best practices. International law is rife with references to maternal health or the needs of pregnant women and nursing mothers in a special capacity. A plethora of international documents implemented via expensive programs state that mothers are entitled to special care and assistance, but those documents fail to offer criteria for that care, and therefore maternal mortality during pregnancy and childbirth remains a plague for humanity.

Nanotechnology's revolution for commerce can revolutionize public health by transforming the delivery of care and medicine during pregnancy by shaping new, sorely needed special criteria for protecting the health of women whether at work, when driving and traveling, or at home. As WHO discussed in 2009 in *Women and Health: Today's Evidence, Tomorrow's Agenda* [31], differences in male versus female health outcomes using their indicators showed better health among men in five stages of the life cycle [32] and issues in reproductive health, especially in clinical strategies for the diagnosis and treatment of common gynecologic and medical disorders of women. Traditional patterns of gender-based health outcomes can either be avoided or be exacerbated by the advent of nanomedicine and nanotechnologies in consumer goods.

The European Cooperation in Science & Technology (COST) Science Café for genderSTE , "Networking the Way to Gender Equality in Science and Technology," November 30, 2012, Brussels, at the European Parliament assembled a wealth of female talent in order to tackle these questions, with a view to:

- Sharing ideas for future programming to advance gender equity under Article 15 of the Horizon 2020 final document of the European Union
- Examining specific areas that have been unexplored or problematic for achieving gender equity in the past.

Worldwide, experts agree there is an urgent need for capacity building concerning skills and strategies to:

- Discuss and implement new guidelines in office-based preventive medicine for women for prevention and early detection of cancer with clinical examinations, pap tests, diagnostic imaging, and human papillomavirus (HPV) and influenza vaccines

- Diagnose and treat common problems in women's health, including osteoporosis, menopause, breast masses and breast cancer, allergies, sports injuries, incontinence, anxiety, and depression
- Counsel patients about treatment options in contraception and therapeutic abortion
- Diagnose and treat common skin problems in women and discuss options in cosmetic dermatology
- Assess and treat common sports injuries and musculoskeletal complaints in women
- Better care for older patients presenting with osteoporosis, cerebrovascular disease, and chronic medical illnesses
- Prepare and distribute emergency kits with nano-enabled instruments, light, and medicines
- Redesign systems that will effectively address women's health

Unfortunately, nearly a decade after the pathbreaking WHO report, most of its observations remain true. Infant mortality and maternal death or illness during pregnancy and childbirth have baffled humanity since life began and probably touch a core value for each religion or political faction that values the right to life. Civil society has been studying this challenge for centuries. Therefore we already have many of the outcome metrics and tools to measure and explore this problem. And, in many nations, women are forced out of the workforce during pregnancy or childbearing in the first five years of the child's life or face difficult medical situations that are avoidable according to WHO and leading experts on pediatrics. This loss of economic strength, added with the physical demands of pregnancy, childbirth, and infant care, may have an impact on the widespread problems of maternal mortality and infant mortality, because mothers who do not work cannot afford the same level of nutrition, housing, or medical care compared to salaried peers. As WHO has continued to note, the impact of women's health disparities is a strong source of the global disease burden.

In response, the European Union launched its "Birthday Project" to reduce mortality during pregnancy and childbirth [33]. According to the website, hundreds of thousands of women and babies die around the world on the day of birth, and millions more are left with serious illness [34]. Thanks to global collaborations to reduce the magnitude of these problems, since 1990, maternal deaths have dropped worldwide by 44%, but that is almost three decades ago, with remarkably little progress to advance maternal and infant health.

Deaths and serious health effects for mothers and newborns are surprisingly high, especially given the amazing developments in nanomedicine and nano-enabled technologies that could be applied for every phase of pregnancy and delivery, from fetal surgery and transplacental treatments to light and affordable emergency kits to assist with sterile disposable instruments, lights, and portable energy generators for use during surprise childbirths. Implementing these benefits of nanotechnology will require uprooting embedded errors in old methods of creating and administering health care and creating new approaches to health during pregnancy and childbirth.

The question of how to reduce negative health outcomes is a transdisciplinary conundrum that can bring together nanotechnology researchers and public health systems, while supporting important economic development. WHO estimates that 303,000 women died in 2015 from preventable causes related to pregnancy and childbirth. Therefore the European Union Research and Innovation program Horizon 2020 has taken the lead to tackle the global problem of maternal mortality and infant mortality. Horizon 2020 quotes data from UNICEF: 5.9 million children per year die before their fifth birthday, of which 2.65 million are newborn babies. An excellent starting point to focus the best use of nanoscience and law protecting reproductive health therefore should confront the grand challenge of reducing health disparities based on gender or sex, especially those impacting infant mortality and maternal mortality [35].

Nanomedicine offers a wonderful opportunity to improve the quality of life and longevity of women. Birth and pregnancy offer a natural place to start.

3.3.2 International Convention on the Elimination of All Forms of Discrimination against Women: Friend or Foe of Pregnant Women?

When people care about reproductive health, they care about protecting posterity, thereby extending human rights to lives beyond their own. Universal norms, codified in international instruments, outlast lifetimes and reach across geographic borders and ethnicities and national jurisdictions to enrich all human society. Pregnancy underscores the complex reality that no one is either self-sufficient or completely dependent on others. Human existence is a confluence of interdependence, which makes even the highest-ranking leader dependent on human rights protection. Civil society has not fully replaced human childbearing for human reproduction, and therefore protecting posterity, especially during childbirth and pregnancy, is a shared health concern of people throughout the world. But the international law, like many domestic laws, reflects profound, long-term ambivalence about the role of women as mothers and the protections that should be afforded to their health during pregnancy, whether or not they engage in paying work. Nanotechnology, by increasing the ability to observe and better understand every phase of pregnancy, can be a crucial tool for improving the quality of life for mother and child and for those people who choose not to bear children at all.

Although critics complain about the UN system, there have been few if any calls for reform of the oversight mechanisms that investigate and prevent discrimination. Nonetheless, several features of the existing international law regarding reproductive health are, at best, ambiguious and worthy of careful scrutiny for reform. For example, the UN Convention on the Elimination of All Forms of Discrimination Against Women (CEDAW) [36], Part III, Article 11(2) a prohibits "sanctions, dismissal on the grounds of maternity leave" [37]. For pregnant women and parents who work, these important issues remain unresolved in the jurisprudence of reproductive health because there is little international consensus on the meaning of these terms within nations or across the world. CEDAW Article 11(2) is silent regarding leave for prenatal care, special accommodations for pregnancy during paid employment, and the time allowed for paid maternity leave after childbirth, despite stating that pregnant women should enjoy "special protections," a term of art that historically has included laws that kept women out of the workforce [38]. There is an open question of whether there exists a legal basis for employing special risk assessments or additional preventive strategies to protect fetal health when pregnant women are exposed to workplace hazards that may contribute to adverse pregnancy outcome(s). The international law of reproductive health is silent about this question even as increasingly large percentages of the workforce are female and increasingly larger percentages of reproductive age women seek paid employment outside their household. It is unclear from the CEDAW text what the standard of proof is that triggers a necessary special protection.

Equally uncertain is the proper response for politically loaded question of whether a special protection can deploy an approach that is inappropriate even if it protects, for example, the health of the unborn but is harmful to the body or employability of the mom. As noted by Dr. Morton Corn, Former Assistant Secretary of Labor for the Occupational Safety and Health Administration (OSHA), there are three major types of adverse effects that may be considered the endpoints for evaluation of reproductive hazards after occupational exposure: (i) subfertility/infertility, (ii) pregnancy loss, and (iii) disorders at birth or in infancy [39]. Nothing in the CEDAW text requires modification of a drastic method in favor of reasonable approaches, and nothing especially guarantees safety and health special protections in the workplace for mothers who work with embryotoxins in pharmaceutical manufacturing, dangerous substances or radiation, or new unquantified sources of risk, including nanotechnology [38a, 40].

It is unresolved and hotly contested whether the potential presence of a vulnerable unborn human fetus in a toxic and hazardous workplace gives rise to special exemptions from federal discrimination laws in nations or under international law. For example, in the EU, fetal presence is protected from workplace exposure by directives that provide paid leave to pregnant workers. In the United States, unpaid leave is the norm, with problematic results. US case law pertaining to fetal protection policies (a euphemism for policies that required women to be sterilized in order to keep their jobs) was once considered a valid legal approach to resolving difficult issues concerning reproductive health hazards in the workplace. Fetal protection policies, fetal vulnerbility policies, and similar exclusionary policies were not illegal before the US Supreme Court's decision in *IUAW v. Johnson Controls* (1990) [38a, 41]. This case was the subject of media attention because of its discussion of the implications surrounding fetal protection policies for women's choice and the right to life. The Supreme Court ruled that women could not be excluded from working in fetotoxic work environments by requiring that they show their employers a surgical scar after they have been voluntarily sterilized. But the case did not clarify the standards to be employed in order to make the workplace clean and safe and was silent about the universal concept under international law regarding special protections articulated in CEDAW.

Previously, in the case of *Oil Chemical and Atomic Workers (OCAW) v. American Cyanamid*, sterilization was a condition of employment, too. In OCAW, the court allowed the policy, claiming this approach was reasonable because a policy cannot constitute a hazard of employment: "An employee's decision to undergo sterilization in order to gain or retain employment grows out of societal factors which operate primarily outside the workplace . . . . The women involved were thus faced with a distressing choice. Some chose sterilization, some did not." OCAW incorrectly understated the employer's responsibility and control regarding working conditions under laws that expressly require employers to accept the responsibility for workplace safety and health; the case's view ultimately undermines health protection for all employees [40].

One sensitive area of the law that remains controversial is whether there are places where pregnant workers should not be allowed. It remains unknown to what extent nanotechnology will be studied in a manner that overcomes past exclusion of women from clinical trials and workplace settings during pregnancy.

This revolutionary opportunity to discard embedded prejudice in existing methods of researching and delivering health care using nanotechnology and new nanomedicines will have an important global health impact, as the opportunity to place a new research paradigm that can become a great equalizer, solving ancient health care issues. "Protecting Children from Environmental Health and Safety Risks" formed a vital part of the United States' occupational health justification for participating in globally harmonized chemical safety programs [42], and that section requires disclosure of potential hazards to the unborn , notably without defining the term "unborn."

Complete analysis of possible reproductive impairment for the purposes of Safety Data Sheets (SDS) in the Globally Harmonized System of Classification and Labeling of Chemicals (GHS) include disturbances in ovulation and spermatogenesis, male sexual dysfunction, pregnancy loss at all gestational stages, developmental defects of the fetus or infant, genetic defects, low birth weight, and other abnormalities in infancy and childhood.

Several key international efforts attempt to improve conditions for pregnant workers, for all women workers and their mates, and for the next generation despite the absence of support in the international law preventing discrimination against women. Efforts to provide meaningful implementation for job security and paid leave of absence for maternity leave immediately before and immediately after the birth or adoption of a child are consistent with International Labour Organization (ILO) Conventions C155 and C168, Article 5(4)(h) and ILO "Safe Maternity" policies. The importance of instituting these protections is underscored by developments in nanotechnology, whereby "the placenta is likely to come into contact with novel nanoparticles, either accidentally through exposure to these materials, or intentionally in the case of nanomedical applications" [43], but it is not clear whether CEDAW has incorporated these provisions by reference.

CEDAW also remains silent about the duration of special protections, for example, whether these protections reach postpartum or early childhood phases of the human birth process. Effects on breastfeeding, early childhood development, or learning disabilities may be included as potential outcomes to be studied, but there is no guidance under CEDAW to require or prioritize such studies and no penalty or stated remedy for failure to provide special protections. Thus, women cannot rely on CEDAW to support a demand to exercise the right to special protections or to set limits on methods for purported special protections (such as the old fetal protection policies that are illegal in the United States), even if those protections are unreasonably harsh or place them at risk for unquantified but potentially grave harms.

In these cases, nanotechnology applications and nanomedicines are both a sword and a shield: as transplacental therapies can improve the health of a fetus during pregnancy, so, too, unwanted migration of nanoparticles across the placenta or across cell membranes may be a source of harm. Nano-enabled devices, such as lightweight emergency medical equipment, nano-enabled heating systems in blankets and emergency lights, and nanomedicines may be a source of invaluable emergency care if distributed widely with instructions for use in the event of needed urgent care. CEDAW's silence regarding these long-standing issues of maternal health, however, does not provide any international legal argument for promoting the widespread use of these important advances in medical care.

The Grand Challenge to Apply Nanotechnology to Reduce Maternal Mortality and Improve Infant Health

"A unique, established strength of the nanotechnology enterprise lies in its interdisciplinary nature. A broad nanotechnology R&D portfolio invests at the frontiers and intersections of many fields including biology, chemistry, computer science, ecology, engineering, geology, materials science, medicine, physics, and the social sciences. Recently, NNI agencies have been exploring efforts focused on research at the convergence of nanotechnology, biotechnology, information technology, and cognitive sciences that leverage knowledge and approaches in each of these areas to solve problems of national and societal importance" (National Nanotechnology Initiative [NNI] 2016 Strategy, Draft for Public Comment, p. 7).

Four key goals offered by the OSTP [44] and the NNI mission:

1. Advance a world-class nanotechnology research and development program.
2. Foster the transfer of new technologies into products for commercial and public benefit.
3. Develop and sustain educational resources, a skilled workforce, and a dynamic infrastructure and toolset to advance nanotechnology.
4. Support responsible development of nanotechnology.

Nanotechnology applications to nanomedicine and nano-enabled devices, sensors, new foods for better nutrition, a cleaner environment, construction for sturdy housing, and clothing that is almost immune from insects and rodents are no exception to this fine use of public funds for the greater public good. Saving lives of babies and their mothers is an excellent vehicle to capture the public's imagination and one that will reap benefits on behalf of billions of stakeholders, some perhaps unborn, using lab-to-market strategies. It is difficult to posit a global health impact of nanotechnology that better fits the OSTP requirement to tackle a problem that "remains an important problem nationally and globally and whose conquest will represent responsible and sustainable use of nanotechnology and that captures the imagination of the general public." Therefore, the Work Health and Survival (WHS) project proposed the Maternity Outcomes Made Better (MOM) Project as a grand challenge to reduce maternal mortality and infant mortality that have plagued humanity for millennia.

3.4 New Promise for Aging Populations

Many government publications across the world have expressed concern over the growing population of people who are aging and who will soon outnumber younger cohorts because good science has been applied to increase longevity. Few of these publications offer solutions to the policy dilemma of financing life for the aging population (Fig. 3.5), but many opportunites to live healthier lives and to maintain productive work will be possible because of nanomedicine.

Figure 3.5 Retired man in the United States. Photo by Ilise L. Feitshans.

3.4.1 A "Revolving Door" for Health Care of the Aging

Aging brings profound changes in every body. People with chronic diseases, previously incurable, who can benefit from nanomedicine will ask public health systems to fulfil the promise of nano-enabled therapies to reverse degenerative diseases and will be able to enter the workforce. Whether to really enjoy retirement with sports, travel, and leisure activities, or as a source of return to gainful work, nanomedicines promise new opportunities among aging populations who might have been marginalized by family, colleagues, and friends, absent the benefits of nanomedicine.

Nanomedicine therefore will offer aging populations renewed health and thus expand exponentially the available labor market across all age groups, especially in the case of diseases that occur in later life. For example, nanotechnologies can create new devices that mimic normal biological processes. According to the chart in Fig. 3.6, from WHO's *World Report on Disability* (June 2011), the largest single demographic variable for the global burden of disease is aging. According to this chart, depending on income, disability will impact between 40% and 60% of the aging population, with a consistently higher impact among women compared to their male peers, even when the data have been stratified by income. This projection has two implications: (i) The long-term health disparities between men and women last throughout the life cycle, and (ii) there is an unmet need for research into health problems that impact only women, especially concerning cancers and degeneration of reproductive organs.

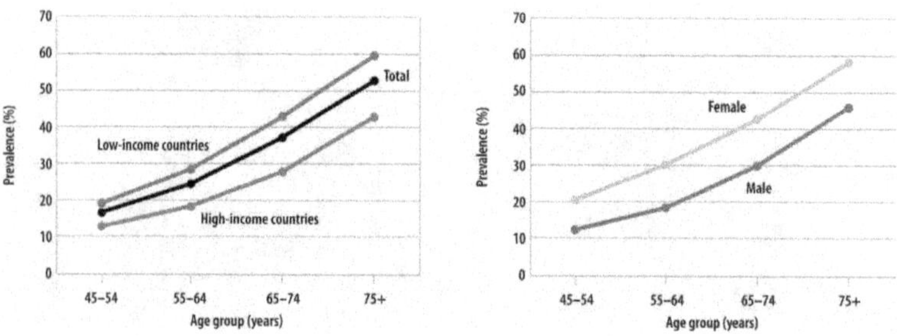

Figure 3.6 Scope of the aging population of people with disabilities (PWD) who may be patients or consumers of nanomedicine [45]. *Source*: WHO report.

Nanomedicines and nano-enabled medical devices for care will therefore have the global health impact of creating a "revolving door" approach to long-term disability care and chronic illness. Nanomedicines will offer treatment for people who will need intensive care before treatment and then may be able to rehabilitate and return to work afterward, regardless of their age. Nanomedicine and medical uses of nanotechnologies therefore present important law and public health policy challenges for civil society regarding healthy disabled people. Civil society must prepare to meet these challenges. Once nanomedicines give large populations of people with chronic illness renewed functional ability, health systems will require new health economics research.

The impact of nanomedicine on work, health, and the quality of life for older people will require needs assessment [46]. There is a dearth of empirical data about the quality of life of older workers. Studies in 2005 by the Swiss government pointed to problems such as discrimination based on age and prejudice about knowledge and skills as key obstacles to employment of older workers [47]. The situation of older nurses is one of the few examples of an older workforce that continues to work. The nursing shortage [48] has been attributed to the increasing dissatisfaction with nursing as a career, and therefore older nurses continue working because they are not replaced by younger staff [49].

These issues have been discussed since the early twenty-first century in occupational health literature [50]. Wages are low and that nursing is undervalued and less respected compared to allied health professions that are available with similar training, such as licensed practical nurses (LNPs), social workers, public health workers, lab assistants, or clinic support staff. For these reasons, nurses with experience have been encouraged to keep their jobs instead of retiring.

The number and proportion of nurses working at older ages is significantly higher than that of the general workforce: a large percentage of nurses are over 55 years of age [51]. A physically demanding occupation, nursing is often characterized by severe musculoskeletal injuries, particularly from lifting activities. Data suggest that overexertion injuries to health care workers are double the rates occurring in general industry, and for those in home health care, the rates are triple those of the general labor force [52]. But relatively limited data exist about the relationship of age to the likelihood of injury among nurses. Studies of the general labor force suggest that older workers have a lower overall risk of injury but a greater long-term disability when an injury does occur [53]. Another confounding variable that suggests older people will work longer is the so-called sandwich generation phenomenon. In particular, women in greater numbers than men find that they continue working because their parents require assistance and their children still need their income, too.

These working women, especially nurses, are working longer hours and continue full-time work beyond retirement age because several people in their families are sustained by their income. Women in this cohort commonly experience the social phenomenon of taking care of younger children, such as their own grandchildren, while a working mother is not home, and, due to social patterns of increased longevity, are also likely to have living parents for whom they also perform tasks of nursing or caretaking. Yet, this group of professionals continues working despite their age and despite the myth that older workers cannot perform the same tasks as they did previously. It remains to be seen whether this experience among nurses will be generalizable in comparison to other professions within the older working population, with implications for returning older workers to their jobs.

Key Areas of Nanomedicine Research Have a Greater Impact on the Health of Aging Populations:

Cancer research

Alzheimer disease

Parkinson disease

Bone regeneration

According to the UN Special Rapporteur on Older Persons, nongovernmental organizations within civil society and stakeholders within the health care industry will need big data to describe and protect ageless rights to health. Specifically, public health planning strategists will need better information regarding a positive or negative impact on health derived from an extended productive working life, with particular concern for the well-being and quality of life among older people. A longer, productive working life holds implications for employers, too, as the changing demographic of an older working population means that greater professional experience can easily be required as part of job qualifications in society [49].

The societal economic and bioethical question that will arise once again is whether older citizens should work indefinitely, and health status as both a rationale for working and as a reason to stop work will once again be a player in this millennial question. Nanotechnology will sharpen civil society's focus on the economic and moral aspects of these issues by offering new opportunities to improve health status, engage in social aspects of well-being, and prolong life.

The ability to work at gainful employment may become a rallying point for proving victory against disease among survivor support groups, who then can express their opinions as stakeholders in the health care system. Whether healthy older individuals need to work for financial or emotional reasons and, if so, whether they are taking the jobs to be filled by younger populations will require political judgments by the surrounding society at large. Legal issues about defining the scope of insurance coverage for a variety of replacement parts and medical testing and procedures will no longer raise theoretical questions, even in nations that provide basic coverage for health care as a constitutional right. Financing nanomedicines on a mass scale, while addressing individual needs that use nanotechnology for precision medicine by individuals, will ultimately confront the question of whether there is a maximum lifespan that society can afford. Older populations that benefit from nanomedicines are therefore likely to be the first constituency to confront what limits exist under law regarding their lifelong right to the full panoply of nano-enabled treatment regardless of age.

Conditions That May Be Ameliorated by Nanomedicine

Aging populations are more likely to have alterations of sleep patterns, decreased tolerance for shift work, and a reduced ability to adjust to temperature extremes. The impact of aging has a high rate of individual variation with little predictability, related to genetics and lifestyle, combined with the impact of working conditions—the same working conditions. Older workers frequently have one or more chronic medical diseases or disorders (see Fig. 3.2). The presence of several chronic conditions that are often linked in the disease profile is termed "comorbidity." The rates of hypertension, chronic pulmonary or cardiac disease, diabetes, obesity, cancer, neurological disorders, and renal and liver disease are increased among older people and may require some type of long-term medication with side effects that impact stamina, overall health, reaction time, and workplace safety. On average, there is weight gain with increased body fat but a decline in lean muscle mass, termed "sarcopenia." Men and women both begin to lose bone mass by age 30, with an accelerated rate of loss in women after menopause. In the peripheral nervous system there are age-related losses of vibratory, tactile, and thermal sensation, including changes in cognition. In addition, there may be a decrease in fluid intelligence (the ability to dynamically evaluate, accommodate, and respond to new events) compared to crystallized knowledge (that which is accumulated over a lifetime) [54].

3.4.2 Ageless Reproductive Health

Prolonging fertility and the ability to engage in sexual activity into the sixth and seventh decades of life was the hallmark of research regarding reproductive health for older people in the twentieth century. Although many of those strides continue along the same trajectory by applying nanotechnology to assisted reproductive technologies, the major contribution of nanotechnologies to reproductive health will be the ability to outlive diseases that target reproductive organs. Stems cells from an individual can be used to grow new organs, and nanoscale delivery systems can treat prostate cancer and cervical cancer efficiently.

Nanomedicine will also have a global health impact on the treatment and possible cure of latent, long-term sexually based diseases such as HIV/AIDS and venereal disease. The ability to bring drug delivery systems into cancer cells, for example, will radically prolong the survival curve in many patients.

References

1. US EPA, Interim Technical Guidance for Assessing Screening Level Environmental Fate and Transport of, and General Population, Consumer, and Environmental Exposure to Nanomaterials, 2010, http://www.epa.gov/oppt/exposure/pubs/nanomaterial.pdf.
2. Rinat Meir, Katerina Shamalov, Oshra Betzer, Menacham Motiei, Miryam Horovitz-Fried, Ronen Yahuda, Aron Popovtzer, Rachela Popovtzer, and Cyrille J. Cohen, Nanomedicine for cancer immunotherapy: tracking

cancer-specific T-cells in vivo with gold nanoparticles and CT imaging, *ACS Nano*, **9**(6):6363–6372, 2015.

3. Ilise Feitshans, *Forecasting Nano Law: Risk Management Protecting Public Health Under International Law*, Thesis deposited for Doctorate in International Relations, Geneva School of Diplomacy, Geneva, Switzerland, December 16, 2013.
4. Abby Lippman, Prenatal genetic testing and screening: constructing needs and reinforcing inequities, *American Journal of Law & Medicine*, **17**(1–2):15, 1991.
5. Troy Duster, *Backdoor to Eugenics*, Routledge Press, 1989.
6. Emory Medical Center, Pediatric Nanomedicine Center Links Health Care and Engineering, March 2011: First-of-its-kind research center includes physicians and scientists, press release.
7. Ilise Feitshans, Confronting AIDS in the workplace: balancing equal employment opportunity and occupational health, *Detroit College of Law Review*, **1990**(3), 1990.
8. L. B. Andrews and A. S. Jaeger, Confidentiality of genetic information in the workplace, *American Journal of Law & Medicine*, **17**(1–2), 1991.
9. Links from Nano World Cancer Day: http://www.etp-nanomedicine.eu/public/about-nanomedicine/nanomedicine-applications/nanomedicine-in-cancer/nanomedicine-in-cancer; http://www.etp-nanomedicine.eu/public/about-nanomedicine/nanomedicine-applications/nanomedicine-to-fight-diabetes; http://www.etp-nanomedicine.eu/public/about-nanomedicine/nanomedicine-applications/nanomedicine-for-the-eye; http://www.etp-nanomedicine.eu/public/about-nanomedicine/nanomedicine-applications/nanomedicine-and-tissue-engineering/marrying-nanomedicine-and-tissue-engineering; http://www.etp-nanomedicine.eu/public/about-nanomedicine/nanomedicine-applications/amr/nanomedicine-to-combat-infections-from-antimicrobial-resistant-bacteria; http://www.etp-nanomedicine.eu/public/about-nanomedicine/nanomedicine-applications/nanotechnology-and-alzheimer-disease/new-science-for-an-old-problem; http://www.etp-nanomedicine.eu/public/about-nanomedicine/nanomedicine-applications/nanomedicine-to-treat-arthritis/nanoparticles-for-diagnosis-and-therapy-of-arthritis; http://www.etp-nanomedicine.eu/public/about-nanomedicine/nanomedicine-translation/nanomedicine-pilot-programmes; http://www.etp-nanomedicine.eu/public/about-nanomedicine/nanomedicine-community/nanomedicine-related-national-platforms-initiatives.

10. *Using Nano to Treat Cancer: TED Talk for TV*, https://slice.mit.edu/.../using-nano-to-treat-cancer-ted-talk-for-tv.

11. Executive Office of the President of the United States, Office of Science and Technology Policy, National Nanotechnology Initiative 2016 Draft Strategy for Public Comment, Washington, D.C., US Government Printing Office 2016; see also press release, the BRAIN Project, www.whitehouse.gov/administration/eop/ostp/nstc.

12. Ilise Feitshans, A new grand challenge for nanotechnology: the MOM Project; public comments for NNI, *BAOJ Nanotechnology*, 2016.

13. (a) Ilise Feitshans, Foreshadowing future changes: implications of the AIDS pandemic for international law and policy of public health, *Michigan Journal of International Law*, **15**(3), 1994; (b) Ronald Bayer and David Kirp, *A Review of AIDS in the Industrialized Democracies: Passions, Politics and Policies*, Rutgers University Press; (c) National Research Council, National Academy of Science, The social impact of AIDS in the United States, *Michigan Journal of International Law*, **15**:807, 1994.

14. Raj Bawa, Nanoparticle based therapeutics in humans, *Nanotechnology Law and Business*, **5**(2):155, 2008.

15. Joint United Nations Program on HIV/AIDS, *2015 Progress Report on the Global Plan: Towards the Elimination of New HIV Infection among Children and Keeping Their Mothers Alive*, UNICEF, WHO, and UNAIDS, Geneva, Switzerland, 2015.

16. Louis Henkin, ed., *The International Bill of Rights*, Columbia University Press, 1981 (introduction).

17. Louis Henkin, Privacy and autonomy, *Columbia Law Review*, **74**:1410, 1974.

18. Ibid., p. 1419.

19. UN Charter, signed June 26, 1945, entered into force October 24, 1945. Center for the Study of Human Rights, *Twenty Five Human Rights Documents*, Columbia University, New York, 1994. Preamble: "To promote social progress and better standards of life in larger freedom . . . to employ international machinery for the promotion of economic and social advancement of all peoples."

20. UN Charter, Chapter I, Article 13: "1. The General Assembly shall initiate studies and make recommendations for the purpose of: (b) promoting international cooperation in economic, social, . . . and health fields, and assisting in the realization of human rights and fundamental freedoms for all".

21. Ilise Feitshans, *Law and Science, Perfect Together? Solving Key Challenges to Reproductive Health*, Keynote speech for the conference on Reproductive Health in a Pluralist Legal World Faculty of Law, University of Santiago de Compostela, March 30–31, 2017.

22. (a) Convention on the Rights of the Child, adopted by the UN General Assembly on November 20, 1989, entered into force on September 2, 1990, in accordance with Article 49(1), as reproduced on pp. 80–93; (b) Center for the Study of Human Rights (ed.), *Twenty Five Human Rights Documents*, Columbia University Center for the Study of Human Rights, 1981.

23. Henry Campbell Black, *Black's Law Dictionary, Revised Fourth Edition*, West, Saint Paul, MN, 1968, defines "competent" as "having sufficient ability and authority; possessing the requisite natural or legal qualifications . . . [possessing] mental capacity to understand the nature of his [sic] act." Similarly, many state laws follow this old pattern found in the laws of the state of New York, Public Health Law, Section 2780, McKinney's Laws of New York 1988: "Capacity to consent means an individual's ability, determined without regard to such individual's age, to understand and appreciate the nature and consequences of a proposed health care service, treatment or procedure, and to make an informed decision concerning such service, treatment or procedure."

24. (a) Fay A. Rozovsky, *Consent to Treatment*, 1987 edition; (b) Angela Roddey Holder, The minor's consent to treatment, in *Legal Issues: Pediatrics and Adolescent Medicine*, Yale University Press, 1977; (c) NYC Public Health Law, "Minors and Health Care: The Age of Consent," Section 2504 (4) (1988) states: Any minor who has been married or borne a child may consent to services for the child." This right of minors to act as adults is also extended to medical conditions where "an attempt to secure consent would result in the delay of treatment which would increase the risk to the person's life or health."

25. *NRDC v, EPA*, slip opinion p. 10.

26. Ibid., p. 22.

27. Ibid., p. 23.

28. Flavia Bustreo and Paul Hunt, *Women's and Children's Health: Evidence of Impact of Human Rights*, WHO Press, Geneva, Switzerland, May 2013.

29. Anna Shevedova, Elena R. Kisin, Naveena Yanamala, Mariana T. Farcas, Autumn L. Menas, Andrew Williams, Philip M. Fournier, Jeffrey S. Reynolds, Dmitriy W. Gutkin, Alexander Star, Richard S. Reiner, Sabina Halapanavar, and Valerie E. Kagan, Gender differences in murine

pulmonary responses elicited by cellulose nanocrystals, *Particle and Fibre Toxicology*, **13**:28, 2016.

30. Elpida-Nikki Emmanouil-Nikolousis, Nanomedicine and embryology: causative embryotoxic agents which can pass the placenta and induce birth defects, in Varvara Karagkiozaki and Stergios Logothetidis, eds., *Horizons in Clinical Nanomedicine*, Pan Stanford, Singapore, 2015.
31. Ilise Feitshans, Invited presentation, "Beauty, Babies and Dieting: The Impact of Nanotechnology Law on Reproductive Health and Women's Occupational Health Disparities, IDA background briefing for use by OSTP, Washington, D.C., December 2013.
32. World Health Organization, *Women and Health: Today's Evidence, Tomorrow's Agenda*, WHO Press, Geneva, Switzerland, 2009. ISBN: 9789241563857, http://whqlibdoc.who.int/publications/2009/9789241563857_eng.pdf.
33. European Union, *Birthday Project*, press release.
34. European Union Conference, *Report Together for the Next Generation: Research and Innovation for Maternal & Newborn Health*, European Union DG Research and Innovation, December 2015, https://ec.europa.eu/programmes/horizon2020/.
35. Ilise Feitshans, Public comment on Draft Strategy 2016 for OSTP and NNI: A New Grand Challenge; *The MOM Project; Eliminating or Reducing Women's Health Disparities Impacting Infant Mortality and Maternal Mortality during Pregnancy*, Work Health and Survival Project, Haddonfield, NJ, September 23, 2016. Posted on Researchgate Oct. 2016.
36. Convention on the Elimination of All Forms of Discrimination Against Women. Adopted and opened for signature, ratification, and accession by United Nations General Assembly resolution 34/180 on December 18, 1979. Entered into force on September 3, 1981. Reprinted in the Center for the Study of Human Rights, *Twenty Five Human Rights Documents*, pp. 48–56.
37. Ilise Feitshans, Job security for pregnant employees: the Model Employment Termination Act, *Annals of the American Academy of Political and Social Sciences*, **536**:119, 1994.
38. (a) Jack Levy, Ilise Feitshans, and John Kasdan, on behalf of the Industrial Hygiene Law Project, Brief Amicus Curiae, July 1990, in the Supreme Court of the United States *IUAW v. Johnson Controls* . US Supreme Court cases in this area span from a concern for limiting the hours of womens' work because of their need to be home, raising families, upheld in *Muller v. the State of Oregon*, 208 U.S. 412 (1908), to

a later Supreme Court decision banning forced sterilizations of women who are exposed to reproductive health hazards in the workplace in *IUAW v. Johnson Controls*, 499 U.S. 187 (1991); (b) Ilise Feitshans, Job security: prohibiting wrongful discharge of pregnant employees under the Model Employment Termination Act, *Annals of the American Academy of Political and Social Sciences*, **536**:119, 1994; (c) Mary Becker, Reproductive hazards after Johnson Controls, *Houston Law Review*, **31**:43, 1994.

39. Morton Corn, Exposure assessment challenges presented by investigations of reproductive effects in the semiconductor industry, in *Exposure Assessment for Epidemiology and Hazard Control*, American Conference of Government Industrial Hygienists (ACGIH).

40. See discussion of *IUAW v. Johnson Controls* and *OCAW v. American Cyanamid*. In these cases, women were compelled by their employers to be sterilized in order to obtain or retain employment in an arguably fetotoxic work environment. The Convention on the Elimination of All Forms of Discrimination Against Women remains silent, however, regarding whether its notion of "special protections" includes such extreme approaches to protection.

41. Ilise L. Feitshans, "The Demise of Exclusionary Policies and The Triumph of Public Health and Social Justice," Award winner Delta Omega National Competition, 1991, Department of Health Policy and Management, John Hopkins University School of Hygiene and Public Health, Baltimore, MD, presented at APHA, Atlanta, GA, 1991.

42. US Department of Labor, Occupational Safety and Health Administration (OSHA), 29 CFR 1910, 1915, and 1926, Hazard Communication Final Rule 77FR17574 at 17574.

43. Tina Buerki-Thurnherr, Ursula von Mandach, and Peter Winch, Knocking at the door of the unborn child: engineered nanoparticles at the human placental barrier, *Swiss Medical Weekly*, **142**:w13559, 2012.

44. The White House Office of Science and Technology Policy (OSTP) was created by the US Congress's National Science and Technology Policy, Organization, and Priorities Act of 1976. The OSTP's responsibilities include advising the US president in policy formulation and budget development on questions in which science and technology are important and fostering partnerships among federal, state, and local governments and the scientific communities in industry and academia.

45. World Health Organization, *World Report on Disability*, WHO report, June 2011.

46. Brigitta Danuser, Vieillir au travail en Suisse, presentation citing Carol Black/Waddell 2005: le travail est un pré-requis pour la santé, le travail est thérapeutique and DGAUM 2006: les objectifs de la médecine du travail résident dans la promotion, la préservation et la participation au rétablissement de la santé, de la capacité de travail et de l'*employabilité* des individus.

47. Accroissement de la participation des travailleurs àgés au marché du travail seco-admin.ch, bern Nov. 2005, pp. 49–50.

48. R. Z. Booth, The nursing shortage: a worldwide problem, *Revista Latino-Americana de Enfermagem*, **10**(3):392–400, 2002.

49. K. C. Fleming, J. M. Evans, and D. S. Chutka, Caregiver and clinician shortages in an aging nation, *Mayo Clinic Proceedings*, **78**(8):1026–1040, 2003.

50. OSHA, *OSHA Recommends Eliminating Manual Lifting of Residents When Feasible*, ergonomics guidelines announced for the nursing home industry, March 13, 2003, press release, http://www.osha.gov/pls/oshaweb/owadisp.show_document?p_table=NEWS_RELEASES&p_id=10129.

51. P. I. Buerhaus, D. O. Staiger, and D. I. Auerbach, Policy responses to an aging nurse workforce, *Nursing Economics*, **18**(6):278–286, 2000.

52. T. Galinsky, T. Waters, and B. Malit, Overexertion injuries in home health care workers and the need for ergonomics, *Home Health Care Services Quarterly*, **20**(3):57–73, 2001.

53. M. E. Personick and J. A. Windau, *Characteristics of Older Worker Injuries: 1992*, Bureau of Labor Statistics.

54. R. W. Keefover, Aging and cognition, Neuroimaging Clinics of North America, **16**:635, 1998.

Chapter 4

Integrating Nanotechnology into International Laws Protecting Health

4.1 An Abundance of International Laws: Resisting the Fad of Nanoregulation

The protection and promotion of the health and welfare of its citizens is considered to be one of the most important functions of the modern state [1].

—George Rosen

A vast and vibrant corpus of laws protects health. This concept of government obligation to protect its citizens is as old as the Great Wall of China, which was built thousands of years ago to keep out invaders and preserve the integrity of an empire. And this concept of government responsibility is met in the actions of thousands of diplomats and government workers around the world, civil servants who meet to plan and implement health policy and protect rights, as seen in Fig. 4.1. International laws reflect, and do not ignore, the societal need for a legislative response to hazards that exist in daily life. Legal tools exist for promoting the implementation of precautionary principles without civil society seeking to reinvent them. Strong international consensus among laws across a majority of nations and a parallel system of codified international norms demonstrate a universal desire to protect consumers, protect the

Global Health Impacts of Nanotechnology Law
Ilise L. Feitshans

ISBN 978-981-4774-84-0 (Hardcover), 978-1-351-13447-7 (eBook)
www.panstanford.com

environment throughout the life cycle of product use, and enhance occupational health protections for all societies.

Figure 4.1 Swiss representative to the United Nations Human Rights Council prepares to address a public session in the Palais des Nations Geneva, Switzerland. Photo by Dr. Ilise Feitshans.

Precautionary activities are among the fundamental responsibilities of governments, such as primary care [2]. Protecting public safety, defense, and national security and controlling toxic or hazardous substances are reflected among national laws and intergovernmental agreements designed to promote those goals. International laws governing health and safety for consumers, workers, and the general public therefore provide an important backdrop against which the efficiency of national laws can be compared and measured.

International laws provide more than a yardstick for measuring compliance. The large array of national local and international laws protecting health underscores the fundamental character of public health protections, which seem to be universal across all societies. Embedded with precautionary principles, many legal systems in the world universally provide the terms and conditions for promoting future development in society. Many laws clearly articulate and codify shared social values that safety and health protections (indirectly following scientific precautionary principles) are invaluable to preserving society.

The dilemma for policymakers concerns context: Nanosilver that is wonderful for destroying bacteria and HIV in condoms and destroys rodents that eat fabric also may harm children; nanogold that makes rapid communication for electronics and cell phones may bring a message that needlessly hurts someone; titanium dioxide that makes fluffy cream to place in a bowl so that a young child will be attracted to taking his or her needed medicine may also be the source of harm.

Addressing major unforeseen problems before the fact, therefore, requires balancing an admixture of quantifiable and unquantified variables in order to give their approval to major programs, sometimes making hard choices before the risks are known. As Fig. 4.2 illustrates, policy inputs for information can be an admixture of views from all constituencies and all ages; it is widely accepted that policy can be most responsive when it functions without artificial limits regarding the type of stakeholders involved. Policy choices must be made: sometimes hard choices dictate a response to the will of the people with a law that is popular but unsound, sometimes championing the best practices that meet their long-standing responsibility for protecting people, while, at the same time, balancing the urgent need to foster and develop new industries to stimulate a broken economy.

Figure 4.2 Emalyn Levy Feitshans speaks at the US Capitol. Photo by Dr. Ilise L. Feitshans.

Figure 4.3 Commercial center built with nanomaterials and selling products that apply nanotechnology. Photo by Dr. Ilise L. Feitshans.

From an economic standpoint, there can be little doubt that the development of nanotechnology is a shot of adrenalin that can stimulate new commerce and new jobs for an interdependent global economy and therefore a dynamic instrument of social change that can transform many aspects of daily life in civil society, beyond nanotechnology's financial implications. As Fig. 4.3 illustrates, the size and shape and availability of commercial centers and the contents they sell, including electronic transfers of information, are enhanced and accessible, thanks to nanotechnologies. This brings more jobs and more commercial activity every day, globally. Yet, diffuse use of sovereign power by having too many different systems governing nanotechnology can lead to confusion and conflicts of laws and ultimately block the flow of commerce that had brought the hope of nanotechnology's promises to civil society.

4.1.1 Health Protections under the United Nations Charter

Governmental functions protecting health under international law are as old as the treaty law of the World Health Organization (WHO) Constitution and the soft law of the Universal Declaration of Human Rights (UDHR) [3] treaties such as the International Covenant on Economic, Social and Cultural Rights (ICESCR) [4] and, perhaps, as old as civilization itself. Precautionary principles form the foundation of international rules about toxic and hazardous substance exposure. Few member states of the United Nations (UN) will state aloud that they violate such principles; if anything, rhetoric from every member state can be found claiming that these vital principles are upheld by their law and the actions that implement them when address in the general assembly, as seen in Fig. 4.4, And even though this simple consensus-based notion may seem to be old fashioned, the precautionary principles as applied to nanotechnologies in the context of human exposure to nanomaterials are alive and well.

Figure 4.4 Dignitary address the United Nations General Assembly, New York. Source: US government.

Precautionary principles and government responsibilities protecting health are not new, and therefore these notions are deeply embedded in many laws at every level of governance. These concepts are operationalized in national state and local laws, major international legislation, such as the Globally Harmonized System of Classification and Labeling of Chemicals (GHS), and international nongovernmental programs such as the International Organization for Standardization (ISO) working groups regarding nanotechnology [5] and work in close partnership with international governments and treaty organizations such as the Chemicals Committee and the Working Party on Chemicals, Pesticides and Biotechnology, Organization for Economic Co-operation and Development (OECD). All of these organizations include in their rationale for their findings references to the bedrock notions of applying precautionary principles—even though the actual text of these principles is elusive and never codified.

According to UN Charter Article 13 [6], "The contracting parties state their desire to promote: 'economic and social advancement' and 'better standards of life'." Putting these powerful words into programs and strategic plans has been the work of hundreds of thousands of people who have passed through the gates of the UN in Geneva, Switzerland, shown in Fig. 4.5. As a result of these deliberations, not every policy has been effective or inexpensive or equitable, but many important global health decisions have been successfully implemented at the regional and international levels, with the hope that national laws will be consistent with the best points in the international model. Big risks have been successfully addressed by governments and their legislatures in the past, enabling industry and commerce to flourish by promoting technology, while limiting the scope of liability by creating regulatory barriers to activities that cause avoidable harms. These laws use flexible frameworks for oversight with placeholders in final legislation for new methods. And historically, society has won great benefits by gambling with regulated risk.

By contrast to the slow development at an evolutionary pace for laws about health and international governance infrastructures, nanotechnology laws are sprouting like mushrooms in every nation! Governmental structures at all levels of society presently face a situation in which there is potential risk to public health, but insufficient data exists about actual risk in order to make key policy judgments. Consequently, the regulatory picture of the legal landscape presently looks patchy and disorganized—large gaps in the law where there is uncertainty about the magnitude of risk, and many different sources of law clustered around tangible, established practices for toxic or hazardous materials. None of these laws question the juridical basis for their enforcement of the state's power to protect health, even when mandating costly engineering controls, medical surveillance through employer-based occupational health services, or global sharing of chemical hazard information.

Figure 4.5 Main entrance, Palais des Nations of the United Nations, Geneva, Switzerland, home of expert meetings about precautionary principles and the law of health. Photo by Dr. Ilise L. Feitshans.

Recalling the International Labour Organization's (ILO) constitutional mandate in 1919 in the Treaty of Versailles, UN Charter Article 55 specifically notes the link between "creation of conditions of stability and well-being" for peace and "higher standards of living" first and "universal respect for, and observance of, human rights and fundamental freedoms." During the half century that followed the codification of these words, an elaborate apparatus of capacity building for health care created an infrastructure for implementing these values worldwide, at national, multinational, and international levels [7]. Subsequently, protection for the right to health was written into the fundamental constitutional principles of many nations [8]. This rhetoric equates "adequate" health with related basic rights. But it is difficult to patch together any text explaining "better standards of life" in detail. Like other international instruments that address health, its vague descriptions of protections for life, security of the person, without a benchmark for "well-being," have been used as the legal basis for thousands of programs and new treaties. For this reason, many theorists challenged the notion that health is among the universal human rights codified under international law. The debate has become dormant, however, once the infrastructure has grown to command an impressive role in controlling disease and ameliorating the quality of life worldwide.

4.1.2 International Covenant on Economic, Social, and Cultural Rights

Broader questions regarding the international legal right to health protections have been addressed, without detailed guidance, in the UN Charter, in ICESCR Articles 7 and 12 and subsequent standards by UN-based international organizations and various international conventions. Obligations that arise under the ICESCR stem from a principle of progressive implementation of the rights it defines, and the principle of nondiscrimination in the enjoyment of those rights. Article 1 establishes substantive obligations that the signatory states undertake to implement in their home legislatures. Using a unique enforcement system of international monitoring and compliance, the treaty tries to balance risks and preventive strategies for using global resources, through international assistance and technical cooperation [9].

4.1.2.1 Article 7 of the International Covenant on Economic, Social and Cultural Rights

Article 7 provides greater insight into the meaning of the right to just and favorable conditions of work discussed in other UN documents. "Favorable conditions of work" include terms of remuneration as well as "safe and healthy working conditions." The use of this phrase within the context of favorable conditions of work lends greater meaning to the UDHR's protections and demonstrates the clear nexus between other human rights principles and protection of health, as further amplified in ICESCR Article 12.

4.1.2.2 Article 12 (Right to Health: Promotion of Industrial Hygiene and Protection of Environmental Health)

Of all the UN-based international human rights documents, ICESCR Article 12 most clearly and deliberately addresses health. It is the clearest of all human rights instruments regarding the explicit right to protection for "industrial hygiene" and protections against "occupational disease." Further, Article 12's mandate to improve "industrial hygiene" is consistent with Article 7(b) of the ICESCR, regarding safe and healthful working conditions. Yet, even this express guarantee of occupational safety and health protections does not offer a detailed exposition of the meaning of these rights, nor does it list the possible approaches that could be applied for achieving the ICESCR's goals. Consistent with the principles articulated in many other international human rights documents, Article 12 employs WHO's constitutional notions of health. Article 12 states the following:

> *The States Parties to the present Covenant recognize the right of everyone to the enjoyment of the highest attainable standard of physical and mental health. 2. The steps to be taken by the States Parties to the present Covenant to achieve the full realization of this right shall include those necessary for: ...*
>
> *(b) The improvement of all aspects of environmental and industrial hygiene;*
>
> *The prevention, treatment and control of epidemic, endemic, occupational [10]*
>
> *and other diseases . . . .*

The plain meaning of this language in Article 12 treats occupational disease as a vector that has the potential global health impact of diminishing public health and increasing the "global disease burden." Under ICESCR Article 12, the states' parties recognize the right to physical and mental health proclaimed in the WHO Constitution as the basis for their commitment to international health programs. Under these terms, countries commit to four "steps" to be taken along a strategic path to achieve the "full realization" of this right. Significantly, Article 12 also pays direct attention to the impact of occupational disease on health and upon disease burden within society, thereby giving validity to occupational medicine as worthy of human rights protection [11]. The text from the Paris Accords on Climate Change also relies very heavily on this precedent in international regulatory history.

Under ICESCR Article 12, the states' parties recognize the right to physical and mental health coequally with all parameters of the WHO constitutional definition. The language of this article is consistent with the view that "health protection in the workplace is as a matter of great significance, but it is also inextricably linked to the task of protecting the population generally" [12]. Like many laws designed to operationalize precautionary principles, the main subject of the law remains undefined: standards for industrial hygiene and the quality of environmental health remain undefined. It is worth noting that this definition, like its predecessors in international law, avoided a list approach in favor of an open-ended definition that allows any condition to be considered an occupational disease for the purposes of prevention, treatment, compensation, and eradication [13]. According to Grad and Feitshans [11], Paragraph 1 of the Draft Covenant prepared under the auspices of the Commission on Human Rights decided to allow the WHO definition to fill this void. ICESCR drafters urged member states "to develop and strengthen occupational health institutions and to provide measures for preventing hazards in work places." Repeating a theme expressed in many international documents and originating in the WHO Constitution, the ICESCR states, "The right of everyone to the enjoyment of the highest attainable standard of physical and mental health." This goal remains as elusive as it is universal.

4.1.3 The World Health Organization Constitution: Codifying Precautionary Principles

The movement to codify health norms as legal principles had a defining moment at the end of World War II, when the entire world cared about attempting to set written legal limits upon behavior by governments and individuals. UN activity brought codification of international norms regarding the right to health into the positivist, plain language of several key international human rights instruments, with a spirit of hope for all humanity's survival.

Although merely an administrative agency, many view WHO as the paragon of rights-based health programming and respected references for health research and health policy (Fig. 4.6). The most widely accepted definition of "health" in the world is written in the preamble to the WHO Constitution [14]. This text has been quoted around the globe in constitutions, international treaties, and public health practical guides [15]. Programs to provide vaccination, preventive strategies, quarantine, and successful risk management of diseases relying on WHO's flexible legal framework have become the bedrock for a jurisprudence about human rights to the health, life, and security of a person, codified in subsequent major international human rights instruments. International agreements following this path of legal analysis facilitate transnational scientific collaboration, national and international inspection, and regulatory cooperation and also offer an opportunity for capacity building to implement national programs designed to improve well-being [16].

Figure 4.6 Logo from the WHO website. This logo, as used here, is only to assist readers to identify WHO and does not constitute or imply any endorsement of content.

The WHO Constitution's two-page definition of health begins with this famous text:

Health is a state of complete physical, mental and social well-being and not merely the absence of disease and infirmity.

This remarkably broad but flexible definition of health bespeaks the basic human need for health. The WHO constitutional definition is so encompassing, however, that it has been criticized as making virtually any human endeavor a matter of health jurisdiction. Few endeavors have no impact on human health. The drafters of the WHO Constitution envisioned scientific breakthroughs that would give insight into the disease process.

WHO's legislative drafters did not know the names and diagnoses of those diseases that had not yet become pandemic or their methods for treatment, but they accurately anticipated that there would be new health problems. The WHO Constitution offers an elastic framework that can be expanded to include new developments and reduced when a problem has successfully been diminished, as in the case of eradication of smallpox. The drafters of the WHO Constitution also understood the notion of latent disease, which is of increasing significance in the areas of reproductive health and human reproductive toxicology, whereby long-term exposures that may appear to be harmless at the outset may, after many years, take their toll due to a cumulative effect. In those situations, a person may enjoy many years of seemingly good health, while, in fact, the illness or disease is silently, slowly progressing. The structure the drafters created therefore is sufficiently precise to offer an irresistible promise of global health and sufficiently vague so that new interventions can be designed and implemented to prevent harm and thereby protect health.

Nanomedicine, for example, promises presymptomatic treatments or cures for conditions that may once have been completely disabling or caused death in humans that might not fit into a classic definition of diagnosis or treatment based on old models of recognized symptoms and harm [17]. For this reason, flexible language concerning health protection "in the absence of disease or infirmity" has a very important jurisdictional effect, for it gives rise to WHO's permission to research and implement protective measures regardless of whether people manifest illness or appear to be sick, and will therefore touch presymptomatic nanomedicine patients in disease-specific populations. This ultimate precautionary principle is fundamentally important for justifying any health programs.

WHO's definitions are constructed on the basis of solid fundamental principles rather than a towering paper list of specific conditions and diseases. Consequently, WHO's definition has endured for nearly a century. For example, this language has been used to enable WHO to pioneer efforts in the research, study, treatment, and avenues of prevention of HIV/AIDS. HIV/AIDS was not a known disease at the time the WHO Constitution was written. But once the disease became internationally recognized, the international medical community called upon WHO to create and implement international cooperative research and preventive programs, consistent with its Constitution; no one challenged the basis for WHO's authority to create interventions and inspire research for treatments. Instead, WHO's role as a leader in the global response to the pandemic was encouraged, and the agency had the statutory authority to go forward. The principle of the rights of sovereigns to engage in international relations embraces their obligation to use international relations as a tool to protect the health and well-being of their people. Quarantine, by prohibiting people with specific diseases to cross borders in prior centuries, is the first crude example of this right and duty.

Instituting even-handed needs-based quarantine was an early task of WHO in the late twentieth century. Vaccination as a preemptory strike against pandemics that lead to quarantine provides another example of international collaborative strategies for health protection, facilitated by WHO's constitutional text. To be effective, vaccination must occur to seemingly healthy or disease-free human beings [18]. WHO's efforts have also consistently applied to occupational exposures and related environmental health risks that may be compounded by synergy with human factors or contaminants. Scientists try to determine the precise quantities of toxins that cause cancer, acute toxicity, or other harms in the workplace, compared to the general environment, personal lifestyle, or genetic predispositions. The confluence of these factors may be very different among individuals or populations, but the patterns of disease transmission and illness know no borders.

Recently, WHO reinterpreted key definitions by emphasizing the importance of social context when defining disability [19]. WHO's revised global legal framework includes a commitment from the agency and from each WHO member state to improve capacity building for disease prevention, detection, and response. It provides standards for addressing national public health threats that have the potential to become global emergencies. The rationale for such activity is linked to preventive principles taught in public health schools from the first day: "Investment in these elements will strengthen not only global public health security but also the infrastructure needed to help broaden access to health care services and improve individual health outcomes, which would help break the cycles of poverty and political instability . . ." The advantage of this approach is its applicability to existing threats as well as to those that are new and unforeseen, such as natural disasters, industrial or chemical accidents, and other environmental changes, which might cross international borders [20]. WHO Constitutional language has strong links to the right to life and security of person, codified in major international human rights instruments such as the UN Charter, and subsequent international agreements [16]. This text can serve equally well to facilitate evaluation, prevention, and regulation of health risks from emerging nanotechnologies. Nanotoxicity and the implications of exposure for human health therefore are within the scope of WHO's mission to protect human health.

4.1.3.1 WHO's International Programme for Chemical Safety: A Precursor to the GHS

WHO's International Programme for Chemical Safety (IPCS) Harmonization Project enables governments and others to work toward the achievement of goals first outlined in Agenda 21, Chapter 19, in Rio in 1992 at the United Nations Conference on Environment and Development. According to Rio Declaration Principle 10, "States shall facilitate and encourage public awareness and participation by making information widely available" [21]. As elaborated in recommendations of the Intergovernmental Forum on Chemical Safety Bahia Declaration of 2000, reaffirmed by governments in the 2002 Johannesburg World Summit on Sustainable Development (WSSD) Plan of Implementation [22], the IPCS is consistent with the Rio principles, which were carried forward in the Strategic Approach to International Chemicals Management (SAICM) in Dubai in 2006 [23].

Implementing WHO's IPCS is one component of the very complex endeavors involving over 25 UN agencies, such as the United Nations Institute for Training and Research (UNITAR) [24] and hundreds of national government agencies in the GHS [25]. Regional governments such as the European Union (EU), the ASEAN nations, and trade organizations such as the ISO have teamed with various stakeholder partners to promote a unified system of identifying chemicals with the same set of labels worldwide. This approach ensures that handling and storage information travels with the chemical so that the same level of accurate right-to-know information is accessible to everyone involved in transport, storage, or industrial use of the substance. This approach avoids spills or other accidents and provides midstream commercial users with a layer of protection against liability.

4.1.3.2 WHO Beijing Declaration: Occupational Health for All

The Beijing Declaration, Occupational Health Occupational Health for All (1994), signed at the second meeting of the WHO Collaborative Centers and adopted in 1996, builds upon WHO's HFA2000 Plan for Action by addressing rapid changes in work organizations that impact human health worldwide. WHO, the ILO, the United Nations Development Programme (UNDP), and the nongovernmental organization (NGO) International Commission on Occupational Health (ICOH), in addition to 27 countries, adopted a proposal for action about target goals.

Human existence is an admixture of interdependence. Even the highest-ranking leader can face hazards as a worker that jeopardizes his or her right to health. Too often, policymakers in occupational health and the attendant allied health professions act as benevolent caretakers of the health of a small segment of the working population: speaking in the third person about working conditions, an aloof third-party observer of the flaws and strong points in labor or management structures, unattached to risk and rarely noticing their own health and safety. Following this strategy, WHO's Ninth General Programme of Work, for 1996–2001, stated, "Occupational health and safety at work is a fundamental human right and a worldwide social goal as the basis for his view that successful implementation of WHO's global strategies will depend upon: (1) data sharing and (2) research and collaboration in partnership with industry and (3) similar activities in partnership with occupational safety and health compliance professions, through professional associations" [26]. Research examining costs of the so-called disease burden in society reveals that it is in everyone's economic interest to be part of the activities that prevent work-related impairment. WHOs global strategy for promoting occupational health has been developed through a network of collaborating centers (CCs) that share "a common vision . . . to mitigate the adverse effects of occupational hazards and to meet emerging problems."

Pursuant to WHO's General Authority Mandating Action to Protect Worker Reproductive Health: Implications of the WHO Global Strategy for Health for All Plan of Action 1996–2001 [27], the director general of WHO was requested to implement an occupational-health-for-all strategy embracing occupational health care, small enterprises, migrant or informal sectors, and women, as part of the high-risk groups with special needs. Key priorities included developing and disseminating evidence-based prevention tools and raising awareness. By 2009 this mandate was expanded to include programs for the elimination of related disease [28], the creation of toolkits to improve assessment and management of physical risks in workplaces, prevention of the effects of noise and vibration, assessments of psychosocial risks in the workplace, self-contained learning units, and a project assessing the hazards from nanoparticles and communication of their risks. Specifically addressing occupational health for the very first time among WHO international instruments, this international instrument, although vague, applies concepts from valid international laws and the WHO Constitution in order to craft a platform for implementing precautionary principles protecting health at work. Point 9 of the declaration reaffirms each worker's "right to know the potential hazards in their risks in their work and workplace, including the development and use of appropriate mechanisms . . . in planning and decision-making concerning occupational health and other aspects of their own work. Workers should be empowered to improve working conditions by their own action, should be provided information and education, and should be given all the information, in order to produce an effective occupational health response through their participation, including the right to know information about health hazards from long term and acute exposures to substances in their workplace." Unfortunately, the scope of hazard information disclosure is not discussed in the declaration, and thus, risk managers must rely on text from other international documents and national laws to fill this void.

WHO hosts several bodies specializing in hazardous chemicals and waste activities, including:

- The **International Programme on Chemical Safety** (IPCS)
- **Inter-Organization Programme for the Sound Management of Chemicals** (IOMC)
- The **Intergovernmental Forum on Chemical Safety** (IFCS)
- The **Health and Environment Linkages Initiative** (HELI)

WHO joins UN Environment in providing the secretariat for the **Strategic Approach to International Chemicals Management** (SAICM).

4.1.3.3 WHO *Guidelines on Protecting Workers from Potential Risks of Manufactured Nanomaterials*

The purpose of WHO guidelines (Fig. 4.7) is as follows: "These guidelines aim to facilitate improvements in occupational health and safety of workers potentially exposed to nanomaterials in a broad range of manufacturing and social environments. The guidelines will incorporate elements of risk assessment and risk management and contextual issues. They will provide recommendations to improve occupational safety and protect the health of workers using nanomaterials in all countries and especially in low and medium-income countries" [29].

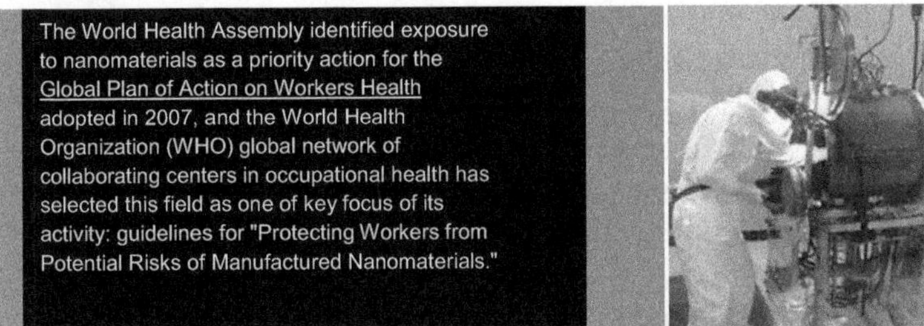

Figure 4.7 (Left) WHO's *Guidelines on Protecting Workers from Potential Risks of Manufactured Nanomaterials*. (Right) Working with engineered nanomaterials. Photo source: WHO.

Nanotechnology provides the perfect opportunity for WHO to pull together its discordant strands of programming in the global health tapestry and to use the coordinating effort in order to correct long-standing systemic problems in the access, public awareness, and delivery of services associated with workplace health.

Under the auspices of the "Global Plan of Action on Workers' Health" adopted in 2007 [30], WHO published its final report, *Guidelines on Protecting Workers from Potential Risks of Manufactured Nanomaterials* [29] on December 12, 2017. Previously, WHO had accepted and reviewed comments on a draft that it published in 2011: World Health Organization's *Proposal for Guidelines on Protecting Workers from Potential Risks of Manufactured Nanomaterials* [31]. The report's clearly stated reliance on precautionary principles throughout its text underscores the inextricable link between work, health, and the economic viability of any employer in society. By using highly accessible tools to implement best practices that already govern bulk materials, WHO in partnership with the private sector and nonprofit employers has created "a golden opportunity to educate the general public—the novice who is thrown a text and told, 'we need risk mitigation, give me a list for risk mitigation and have it on my desk tomorrow'" [32].

4.1.3.3.1 *WHO guidelines: a flexible definition of nanomaterials*

The guidelines offer the standard definition, "'Nanomaterials' refers to materials that have at least one dimension (height, width, or length) that is smaller than 100 nm (10^{-7} m)," but also take into account that manufactured nanomaterials (MNMs) may glom together into substances of much larger sizes [29]. The ability to include the larger group of MNMs without reaching into standards for a bulk form of the same substance is a conceptual breakthrough, showing that nanomaterials are now viewed in context and not studied in isolation anymore.

4.1.3.3.2 *Main message: disclose possible risks*

WHO developed these guidelines for the target audience of workers, policymakers, and professionals in the field of occupational health and safety in making decisions about protection against the potential risks of MNMs, health professionals, and decision makers at the local, national, or international level, who are responsible for the health and safety of workers exposed to MNMs. Additionally, the guidelines focus on low- and middle-income (LMI) countries where nanotechnology is an important means of economic progress. Countries such as Brazil and South Africa produce MNMs and have research laboratories that produce CNTs and produce nanosilver that is incorporated in milk packs, fabrics, and clothes, and MNMs are also produced for use in pharmaceuticals. The authors noted, "Despite the publication of a large number of scientific articles about nanotechnology by authors from LMI countries, only a few are about the potential toxicity of MNMs and very few report on safety or risk assessment" [33].

As stated at the start of the report, the guidelines project aims to avoid the catastrophic effects of uncontrolled exposure such as the asbestos industry experienced in the time before occupational exposure regulations existed [34]: "Recourse to precaution should be used to reduce or prevent exposure as far as possible. This was seen as an important underlying approach in the interest of protecting workers' health, especially given previous experience with asbestos" [29].

The big news for people who apply the final guidelines is one message: Use the GHS methods of classification and labeling of chemicals, and using the authorized Safety Data Sheets (SDS), disclose potential harm from workplace exposure to nanomaterials. The Guidelines Development Group (GDG) stated that its scope and purpose included prioritization and classification of hazards posed by MNMs [35]."The GDG recommends updating safety data sheets with MNM-specific hazard information or indicating which toxicological end-points did not have adequate testing available including respirable fibres and granular biopersistent particles' groups" [29]. The GDG report calls for worker exposure assessment using similar methods already in use for bulk materials and for the proposed specific occupational exposure limit (OEL) value of the MNMs, noting that the employer's decision about OEL should be at least as protective as a legally mandated OEL for the bulk form. Since MNM safety is evolving rapidly, the GDG proposes to update the guidelines in 2022.

The final report in 2017 stated that it followed precautionary principles because "while humans have long been exposed to unintentionally produced nanoparticles, such as those from combustion processes, the recent increase in MNM production demands greater investigation into the potential toxicity and adverse health effects of these materials following exposure. Since newly developed MNMs are not tested sufficiently for possible health hazards, it is generally recommended to take a precautionary approach until testing results are available. This means that MNMs should be considered as hazardous unless there is clear proof that they are not."

To underscore this point and operationalize these goals easily, the authors of the guidelines recommend using the GHS for all MNMs, which requires worker training and disclosure of potential hazards in paperwork that accompanies materials throughout their supply chain in global commerce. Offering a specific list of MNMs, the guidelines strongly recommend disclosure on the SDS for carbon nanofibers (CNFs), CNTs, MNMs, Si-based, and titanium dioxide (TiO_2). To achieve maximum precaution, the hierarchy of controls is an important guide. WHO stated that there is scientific consensus that MNMs do pose potential serious risks to human health, even though high-grade scientific evidence to quantify risks does not yet exist. Yet the reality that it is only a matter of time before scientists will have clear evidence of important factors leading to nanotoxicity looms like a shadow across the entire guidelines. Therefore, exposure must be reduced despite uncertainty about specific adverse health effects.

It would be wrong to claim that the report was based on poor-quality data. Rather, the report forecasts the importance of future risk assessments despite the immaturity of the state of the art. These steps are reasonable and can demonstrate due diligence despite weak evidence in the current state of the art because "the toxicity of MNMs may differ for physicochemical properties, including size, shape (i.e. size in a particular dimension), composition, surface characteristics, charge and rate of dissolution. There is currently a paucity of precise information about human exposure pathways for MNMs, their fate in the human body and their ability to induce unwanted biological effects such as generation of oxidative stress." The WHO guidelines can be used as a tool for establishing due diligence to prevent harm: every employer should have a program to address exposure to MNMs, including training for exposed workers. Due diligence, embodied in the ability to create a paper trail of evidence demonstrating an employer's good faith efforts to engage in reasonable protective efforts, can reduce harm and subsequent liability [36].

4.1.3.3.3 *Role of due diligence for crafting worker protections*

Due diligence, as suggested in the comments about the draft guidelines in 2012 from the Work Health and Survival Project (WHS), is the coherent strand that pulls the entire nanomaterials safety mechanism together. Due diligence is implicit in the process of following these guidelines: First, noting the premature, if not primitive, state of the art of understanding nanotoxicity and the responses for nanosafety, employers, workers, and policymakers who apply these guidelines have nonetheless shown a keen awareness of potential hazards that mandate SaferNano programming. Second, the process itself allows employers and policymakers to write compliance programs with a blank check regarding specific methods of assessment and precautionary measures to be implemented.

Although this approach in the final report offers more of a mandate for safety training than was available in the draft proposal, the political reality is that the absence of a clear method for training is open to criticism because it leaves unresolved the legal questions of voluntary compliance that have plagued products liability and workplace health for a century. The guidelines state, "The GDG considers training of workers and worker involvement in health and safety issues to be best practice but cannot recommend one form of training of workers over another, or one form of worker involvement over another, owing to the lack of studies available."

To assess worker exposure, the authors recommend control banding [37] in combination with the same or similar methods used for the proposed specific OEL value of the MNMs listed in its Annex 1, noting that the MNM OEL should be at least as protective as a legally mandated OEL for the bulk form of the material. Since these lists are guides and not law, there are no penalties for exceeding the proposed OEL. For dermal exposure assessment, there was insufficient evidence to recommend one method of dermal exposure assessment over another.

When exposures exceed the OELs, the report suggests a step-by-step approach for inhalation exposure:

1. First, assessment of the potential for exposure
2. Second, a basic exposure assessment
3. Third, a comprehensive exposure assessment following the standards set forth by the Organisation for Economic Co-operation and Development (OECD) or the European Committee for Standardization.

Given the role of several trade unions such as the International Union of Food, Agricultural, Hotel, Restaurant, Catering, Tobacco and Allied Workers' Associations (IUF), the European Trade Union Confederation (ETUC), and the Australian Council of Trade Unions and their literature available to the experts submitted, the final guidelines are surprisingly ambiguous regarding the nuts and bolts of compliance for nanosafety training despite available information cited by the GDG [38]. The guidelines text is silent about suggested methods for worker training and information disclosures that protect workers' health, despite citing an array of information about safety and health training programs [39] and the consensus cited in the report favoring such programs. This inadvertently leaves to chance whether employers who are the target audience for these guidelines will actually engage in programs that deploy due diligence [40]. Nonetheless, more important than the words used in its actual text, the release of these guidelines represents a milestone in operationalizing the role of due diligence and creating effective SaferNano compliance programs. Effective in-house occupational safety and health programs remain the best-available tool for managing best practices to control risk, serving as both as a weapon to protect against liability and as a tool for preventing harm [41].

4.1.3.3.4 *Bioethical concerns about future research of MNMs*

Although no long-term adverse health effects in humans have been observed, there may simply have not yet been enough time for harms to appear; also the guidelines' authors inaccurately cite ethical concerns about conducting studies on humans as a roadblock to a better understanding of nanotoxicity processes in humans. The report authors note with frustration that, except for a few materials where human studies are available, health recommendations must be based on extrapolation of the evidence from in vitro, animal, or other studies from fields (such as air pollution) where humans have been exposed to nanoparticles. Their complaint about these limits overlooks remarkable strides in computer modeling, especially using artificial organs made from stem cells using nano-enabled techniques for research and nanomedicine. These new tools created by nano-enabled technologies can provide accurate simulation of human experimentation without jeopardizing the health and lives of real people.

4.1.3.3.5 *Guidelines are not standards*

A specific list of MNMs was published in the guidelines. The authors rated the list as based on moderate-quality evidence, which means that potential risks are more clearly documented than other aspects of nanomaterials where risks are unknown and therefore evidence of risk is low. This does not mean, however, that the unknown risks are less important than obvious ones; it simply means that further research is required about nanotoxicity before the risks can be prioritized.

Unquantified risks that may be difficult to explain to workers, consumers, and the general public are nonetheless quite real. For example, the guidelines note that data from in vitro, animal, and human MNM inhalation studies are available for only a few MNMs. Therefore the report authors strongly recommend updating SDS for the GHS by including MNM-specific hazard information or indicating which toxicological end points did not have adequate testing available. WHO lacks the authority to require enforcement of standards. Therefore no penalties for exceeding these standards or time frames for abatement are included in the guidelines.

4.1.3.3.6 *Transparency to satisfy UN reform*

The WHO guidelines also reflect an effort to modernize the public health approach to occupational health, reformed in order to meet the needs of nanosafety. Stakeholder calls for transparency on the heels of UN reform also have influenced the guidelines. Responding to criticisms by the WHS about a lack of transparency in the draft guidelines [42], the revising team took great pains to explain in detail the methods used by the working group and to spell out the required steps in the review process that answered 11 key questions in the scope of the work. Experts were listed in several annexes. Following established WHO procedures, the Interventions for Healthy Environments Unit in the Department of Public Health, Environmental and Social Determinants of Health obtained planning approval in 2010 to develop guidelines and established a WHO Guideline Steering Group and a Guideline Development Group (GDG) [43]. The GDG was composed of leading experts and end users responsible for the process of developing the evidence-based recommendations. (Members of the WHO Guideline Steering Group and the GDG are listed in Tables A.2.1 and A.2.2 of Annex 2.) Funding for meetings and the costs of the methodologist were provided by the WHO Department of Public Health, Environmental and Social Determinants of Health. Experts participated in the GDG on an in-kind basis, and systematic reviews were conducted by volunteer teams. The project of creating workplace exposure guidelines started with a small team of experts, who prepared a background paper on the development of guidelines for protecting workers from potential risks of exposure to MNMs in 2010–2011. Calls for experts were made to join the GDG and the External Review Group and to identify volunteers to carry out systematic reviews using a Delphi process. To incorporate significant research undertaken in the area of MNM health and safety, teams of researchers were identified who could carry out systematic reviews of the pertinent literature [37] according to the process outlined in the *WHO Handbook for Guideline Development*. (The systematic review teams are listed in Table A.2.3 of Annex 2 [44].)

Key Questions Identified by the GDG

1. Risks of MNMs: Which specific MNMs and groups of MNMs are most relevant with respect to reducing risks to workers, and which should these guidelines now focus on, taking into account toxicological considerations and quantities produced and used.
2. Specific hazard classes: Which hazard class should be assigned to specific MNMs or groups of MNMs and how?
3. Forms and routes of exposure: For the specific MNMs and groups of MNMs identified, what are the forms and routes of exposure that are of concern for the worker protection?
4. Typical exposure situations: What are the typical exposure situations and industrial processes of concern for relevant, specific MNMs or groups of MNMs?
5. Exposure measurement and assessment: How will exposure be assessed, and are there alternatives to current exposure assessment techniques for MNMs that should be recommended in LMI countries?
6. OEL values: Which OEL or reference value should be used for specific MNMs or groups of MNMs?
7. Control banding: Can control banding be useful to ensure adequate controls for safe handling of MNMs?

8. Specific risk mitigation techniques: What risk mitigation techniques should be used for specific MNMs or groups of MNMs in specific exposure situations, and what are the criteria for evaluating the effectiveness of controls?
9. Training for workers to prevent risks from exposure: What training should be provided to workers who are at risk from exposure to specific MNMs or groups of MNMs?
10. Health surveillance to detect and prevent risks from exposure: What health surveillance approaches, if any, should be implemented for workers at risk from exposure to specific MNMs or groups of MNMs?
11. Involvement of workers and their representatives: How will workers and their representatives participate in the workplace risk assessment and management of handling MNMs?

As the WHS pointed out in its comments on the draft guidelines, the first team did not have women [42]. But women experts were represented in the final guidelines. Extensive effort was made also to bridge the gap between nanotechnology's unique and novel attributes and existing methodologies in order to better understand nanotoxicity. The GDG commissioned systematic reviews of the literature according to the process set out in the *WHO Handbook for Guideline Development*. All recommendations were made on the basis of consensus within the GDG. Eleven systematic reviews were generated, applying the Grading of Recommendations, Assessment, Development, and Evaluations (GRADE) systematic review process, which WHO considers valid for the development of any guidelines. As the team itself noted, this was potentially inappropriate, but the GRADE system was customized The GRADE systematic review process was developed for medical topics, where clinical trials could be conducted, and health topics, such as environmental and occupational health. Thus the development of these guidelines also provided invaluable experience on adapting the GRADE systematic review process to occupational health and other nonmedical topics. Systematic reviews to answer questions about risks of MNMs and worker training were used to inform section 5 of the guidelines (about best practices). Using the systematic review and GRADE, the guidelines recommendations are rated as "strong" or "conditional," depending on the quality of the scientific evidence, values and preferences, and costs related to the recommendation. The GDG met in Johannesburg, South Africa (2013); then Paris, France (2015); Brussels, Belgium (2015); and Dortmund, Germany (2016). Following procedures outlined in the *WHO Handbook for Guideline Development*, experts were recruited to synthesize research about MNM health and safety in order to create teams of researchers to conduct systematic reviews of the literature. The systematic review teams were listed in the final report, along with their affiliations.

The first step in the evidence search and retrieval procedure was to identify and define the type of evidence required to address the scoping questions. First, the systematic review teams reformulated the key questions posed in section 1.2 so that they could be answered by a systematic review [45]. Then they defined the best-available evidence to provide answers despite the scarcity of experimental studies directly assessing the impact of interventions on occupational health and safety; several distinct areas of evidence were required for each scoping question. Very few existing systematic reviews were found, because this type of assessment is new in the fields of toxicology, occupational health, and environmental exposure assessment.

Systematic reviews were commissioned for all questions with the aim of locating studies that could answer the pertinent questions. The systematic review process used for each question varied slightly but followed four PICO elements: **p**opulation, **i**ntervention, **c**omparator, and **o**utcome(s), which are used to assess the exposure or the intervention. The PICO approach guarantees that the systematic review process collects the evidence that is needed to answer the question at hand. The searches conducted for the systematic reviews included observational or experimental study of people or workplaces exposed to MNMs. Systematic review conclusions were based on the findings of the included studies. Following this protocol, systematic review teams determined the quality of evidence for each conclusion. GRADE allows the reviewer to systematically and transparently grade the quality of the body of evidence for the effectiveness of medical interventions. The quality of the body of evidence is then graded on the basis of five specific qualifiers, including risk of bias and inconsistency of results. This results in one of four quality ratings: high, moderate, low, or very low quality of evidence. The GDG broke new ground attempting to craft a use of the GRADE approach for environmental and occupational exposure [43, 46]. The rating process ranked a study design as high quality if it was considered the best for the question at hand. The reviewers did not use any qualifiers for upgrading the evidence, as is possible in the GRADE approach for nonrandomized intervention studies [46].

4.1.3.3.7 *The general process from evidence to recommendations*

After the systematic reviews had been conducted, the GDG developed recommendations based on the expert conclusions. When formulating its recommendations, the GDG used the balance between harms and benefits, values and preferences, and monetary costs and the quality of evidence in order to determine some relative value of the strength or weakness of the conclusions. Qualitative methods were used due to the absence of no numerical values for benefits and harms. Costs of an intervention, or the implementation costs of a recommendation, were considered and based on expert opinions, but cost–benefit or cost-effectiveness analyses were not performed. There is an explanation for each recommendation linked to the evidence. Proposed recommendations were discussed through face-to-face meetings until there was consensus, with no minority report. Thus, the guidelines are an important step toward protecting workers worldwide from these potential risks.

4.1.4 The International Labour Organization

The Preamble to the ILO Constitution of 1919 states, "Universal and lasting peace can be established only if it is based upon social justice" [47]. Since the founding of the Committee of Experts on the Application of Standards in 1926, "ILO standards provide the legislative framework and rationale for dynamic but feasible workplace health protection under law." ILO documents describe occupational safety and health as a multidisciplinary field devoted to the anticipation and control of workplace hazards that may impair health. Precautionary principles codified in these documents reflect fundamental tenets of industrial hygiene, codified in ICESCR Article 12 and repeated in the scientific literature about risk management for nanotechnology [48]. ILO C155 (Convention on Occupational Safety and Health) and ILO C161 (Convention on Occupational Health Services) provide a framework for governance infrastructures that can ensure the implementation of a coherent national policy that can generate robust data for training and updated information about injuries, illness, statistics, education and training tools, risk assessment data, and best practices.

Figure 4.8 The International Labour Organization (ILO).

By contrast to many state, federal, or local labor codes that segregate workers on the basis of their economic sector, the ILO does not draw such distinctions from the standpoint of the scope of its jurisdiction. Laborers or professionals, as shown in Figs. 4.9 and 4.10, disabled or able-bodied, from seafarers to domestic help, and a wide range of highly paid staff across many job categories embracing agriculture, civil employment, and the precarious contracts of informal employees can enjoy the protections enshrined in the ILO precepts, regardless of whether they are paid by in multinational corporations like the Better Work Partner of the Disney Corporation or small and middle enterprises. The ILO Constitution protects health and the right to information to ensure safety and health at work. Since its founding at the end of World War I, the ILO has offered the opportunity for trade unions, governments, and employers to achieve social justice and fair employment conditions as a fundamental tenet of creating world peace. The ILO founders were men who believed that fair working conditions prevent a race to the bottom that could lead to slavery and that informed free men would make good decisions as voters and consumers. These notions are exemplified in many of the sculptures at the ILO headquarters, such as the borze in Fig. 4.10 and in the reproduction of the Stele of

Hamarabi, the first recorded written laws, also on the ILO worksite (Fig. 4.10).

Figure 4.9 (Left) Laborer cutting wood in Switzerland (2015) and (right) signing of the Memorandum of Understanding (MoU) between the American Society of Safety Engineers (ASSE) and the ILO, San Antonio, 2009. Photos: Charoy family.

Figure 4.10 (Left) Statue honoring labor on the ILO grounds and (right) the ILO's main headquarters building before renovation in 2017, Geneva, Switzerland. Photos by Dr. Ilise L. Feitshans.

The ILO Constitution of 1919 makes the link between work health and the survival of civil society quite clear: "Whereas universal and lasting peace can be established only if it is based upon social justice; And whereas conditions of labour exist involving such injustice hardship and privation to large numbers of people as to produce unrest so great that the peace and harmony of the world are imperiled; and an improvement of those conditions is urgently required; as, for example, by the regulation of . . . protection of the worker against . . . disease and injury arising out of his employment [sic]."

ILO C161 and ILO C170 (Convention on Safety in the Use of Chemicals at Work) [49] complement the provisions of ILO C155, which requires member states to establish competent governmental institutions that regulate and inspect workplaces. These duties involve surveillance of the work environment to identify, assess, prevent, and control occupational health hazards at the source. ILO C170 [49] states, "It is essential to prevent or reduce the incidence of chemically induced illnesses and injuries at work by: (a) ensuring that all chemicals are evaluated to determine their hazards; (b) providing employers with a mechanism to obtain from suppliers information about the chemicals used at work so that they can implement effective programmes to protect workers from chemical hazards; (c) providing workers with information about the chemicals at their workplaces, and about appropriate preventive measures so that they can effectively participate in protective programmes; (d) establishing principles for such programmes to ensure that chemicals are used safely."

ILO C187 (Convention on Promotional Framework for Occupational Safety and Health) reaffirms ILO C155 by providing an infrastructure for managing the Occupational Safety and Health (OSH) Act, establishing a prevention culture and progressively enhance occupational health services. This flexible regulatory framework was written as a compromise to end heated debates about whether standards should be customized for specific worker populations or "one size fits all." The visionary flexibility of this text means that ILO C155 is consistent with the trend toward "personalized medicine" that will build upon genetic profiles using the application of nanomedicine.

Since its inception in 1919, the ILO has encouraged promotion of better working conditions. Consistent also with this heritage, ILO C155, Article 3(e) [50] offers the definition of health as "in relation to work, indicates not merely the absence of disease or infirmity; it also includes the physical and mental elements affecting health which are directly related to safety and hygiene at work." This is no typographical error, no stray text from elsewhere that was accidently printed because someone left it on the photocopy machine. Here, the ILO has deliberately applied the WHO constitutional language previously discussed, because it is so well recognized. This classic definition from WHO, as modified for occupational health, however, is deceptively simple and comprehensive at the same time: it assumes that there is a consensus about the term "work"—for example, is it always paid, and if so, how does the law separate out protections for volunteers or slaves, or does it treat them the same as paid employees?

This definition of work also relies heavily upon a strong assumption about the relationship between health and work. The text does not question the role of work in disease causation but simply states that the relationship exists. Therefore this definition bespeaks the complex interaction between dangerous workplace exposures, individual lifestyles, and environmental factors that impact the combined effects of working conditions. In addition, this approach is multidimensional because its concern for both physical and mental elements of health and well-being implicitly takes into account effects of occupational stress. These concepts are important to people throughout society because of the ubiquitous nature of nanotechnology.

The heart of ILO C155 about national occupational safety and health laws concerns the creation of effective national, regional, and workplace mechanisms for implementation and compliance with other ILO standards. ILO C155 fosters the creation, implementation and periodic evaluation of occupational safety and health standards among member states of the ILO. For example, Article 4.1 states ILO C155's goal of fostering the development of a "coherent national policy" concerning occupational safety and health protections. To this end, ILO C155 obligates ratifying member states to promote research, statistical monitoring of hazardous exposures (such as medical surveillance measures, not unlike technical standards in member states), and worker education and training. ILO C155 uses broad terminology to provide a regulatory framework. Consultation with representative organizations and employers is required before exemptions will be granted, and any exclusion for categories of workers requires reporting on efforts to achieve "any progress towards wider application" pursuant to Article 2.3. Some have argued that if a broader economic goal supplants the interest in occupational health, then a government could justify reducing occupational health programming under the terms of this convention, but this rhetoric has not gained traction, because occupational health programming saves resources and promotes growth by protecting the employer's economic health.

ILO C155 also fosters education for "representative organizations" and worker participation in the development and enforcement of occupational safety and health regulations internally and on regional, national, and international levels. This language foreshadows the role of worker training and the consumer's right to know that was operationalized by the GHS a decade later. Under the terms of ILO C155, "National competent authorities" who are the authorized administrative agencies undertake to ensure that they will create and implement an inspection system to enforce national laws and regulations on occupational safety and health. If nanotechnology is determined to be part of the occupational health protection system, then nanotechnology workplaces will be embraced in this law without anyone writing anything new. Much of ILO C155 reads like a checklist for a sound occupational health compliance program, except that the requirements apply to governments instead of employers. Governments that ratify this convention basically promise to include these components of risk management systems into their laws, or else they are not in compliance with the international labor standards.

Oversight of such activities by governments is included in the role of the Committee of Experts on the Application of Standards, which meets at the International Labour Conference (ILC) in Geneva, Switzerland, every June. Some of the features that are required for the design of a complete occupational health national regulatory scheme include inspection, development of national safety and health standards, and protection of worker rights to information and training and to complain. These principles for information dissemination in the convention represent an international codification of the so-called right to know that Dr. Alice Hamilton and her peers worked hard to develop at the ILO in the 1930s. National inspection services may also complement their activities with an advisory role on voluntary initiatives. Thus, a nationwide system should be in place in each nation consisting of a national system of recording and notification of occupational accidents and diseases, regularly updated for preventive purposes. Collecting statistical data about the incidence and prevalence of injuries and ill health due to accidents and exposures allows administrators to develop a clear picture of priorities for intervention. Occupational injury and disease compensation and rehabilitation systems can also use this information to offer experience/rating incentives for reduced injury.

ILO C155 is an international template for the so-called right to know that is granted to workers and communities in civil society. This host of rights includes the right to be informed about hazards, safe handling, and use of dangerous materials; access to working safety equipment free of charge; the right to be involved in the management and supervision of occupational health and safety measures at the workplace; the right to be organized in a representative group that can select delegates to occupational health and safety committees; the right to regularly scheduled updates concerning information and training on hazards/risks associated with their work and the measures to prevent them; the right to complain with impunity about unsafe circumstances; and the right to refuse hazardous work and not be required to return, in case of imminent serious danger to their health and life, without retaliation, with representation. In parallel, responsibilities of workers regarding such information require that workers follow safety and health rules when using protective equipment; participate in safety and health training and awareness-raising activities; cooperate with their employer to implement safety and health measures; and inform their direct supervisors if they withdraw from an imminent and serious danger, stating their reasons. The same core values of the right to know are applied to communities to protect consumers and the environment, as well as individuals whose time and exposure is controlled by their employers within industry.

The ILO/WHO Committee on Occupational Heath, created to advance the purposes of ILO C155 and the chemical safety convention, adopted a comprehensive definition of the aim of occupational health programming: "Occupational health and safety should aim at the promotion and maintenance of the highest degree of physical, mental and social well-being of workers in all occupations" [51]. In 2007, the World Health Assembly (WHA) endorsed the WHO Global Plan of Action on Workers' Health (GPA) (2008–2017) to follow up the WHO Global Strategy on Occupational Health for All endorsed by the WHA in 1996. WHO has developed a global work plan in collaboration with WHO's network of CCs on the basis of the objectives of the GPA for 2009–2012. The main objectives of the GPA apply remarkably well for instituting precautionary measures for workplace exposures involving nanotechnology [52]. Once again, international law text applies a definition that is practically a photocopy of the WHO constitutional definition of "health." The committee has oversight authority to review policies and to promote implementation of these goals without regard to age, sex, nationality, occupation, type of employment, and size or location of the workplace.

A key tool for implementation of the goals expressed in ILO conventions is ratification of the conventions by member states under the ILO Constitution. According to Dixon, "Ratification is the process whereby a state finally confirms that it intends to be bound by a treaty that it has previously signed, consent not being effective until such ratification" [53]. Economists might disagree, noting that the economic imperative for having consistent labels and consistent requirements for material safety during handling outweigh ordinary political concerns, and therefore shift the political will of states in favor of ratification. The universal need for valid information in a predictable format is a common thread across all industries and part of the global safety net that ties together corporations across borders and ties employers to occupational safety and health programs despite governmental monitoring and surveillance.

Alston views the ILO as an international model for procedural requirements, which, in his opinion, "legitimize the declaration of new norms" (1984) [54]. Such features of ILO procedures include preparation of a preliminary survey of relevant laws among member states, followed by its governing body's decision on whether to place the item on the agenda of the annual International Labour Conference (ILC), followed by a questionnaire from the ILO Secretariat to participating member states. After the draft has been referred to a technical committee, a draft instrument is circulated to member states and the appropriate worker and employer representatives; a revised draft instrument is then prepared and submitted to the technical committee, discussed by plenary and drafting committee, and adopted after voting by the ILC. This approach allows for maximum discussion and communication between regulated entities and governments.

These procedures, initiated in 1926 at the inception of the Committee of Experts on the Application of Conventions and Recommendations, have continued vibrancy in the international system. For example, the ILO's model forms the blueprint in Convention on the Elimination of All Forms of Discrimination against Women (CEDAW), as discussed in detail in Chapter 3: Article 18 sets forth a mandatory reporting mechanism before an international committee also described within the provisions of the convention. Except in the case of complaints, mandatory reports regarding activities toward implementation and compliance are heard by the committee at the end of the first year following ratification and then at least every four years. Additional reporting procedures for monitoring the application of ILO standards and conventions include but are not limited to direct contact missions [55], commissions of inquiry to investigate particular cases of egregious violations of ILO conventions and constitutional provisions, and regularly scheduled periodic oversight through reporting to conference meetings and reporting to the governing body and the administrative tribunal.

Reporting mechanisms are slow but invaluable; these constitute an important component of a much larger process of mobilizing world opinion toward positive change regarding labor issues.

In addition to generating rules for international labor standards through its conventions, the ILO offers codes of practice. These expressions of the best thinking of collective experts regarding safety protections have served as the blueprint for occupational safety laws and regulations in such areas as dock work, transfer of technology to developing nations, civil engineering, and heavy industries. These model codes, which are sometimes applied with minor modification as draft legislation, share the values expressed in several ILO conventions pertaining to occupational safety and health: C62 Safety Provisions (Building) (1937), C77 Medical Examination of Young Persons (Industry) (1946), C78 Medical Examination of Young Persons (Non-Industrial Occupations) (1946), C119 Guarding of Machinery (1963), C120 Hygiene (Commerce and Offices) (1964), and C152 Occupational Safety and Health (Dock Work) Convention (1979). Precedent exists, therefore, to create a code of practice for the safe handling of MNMs and nanosafety, even though no such strategy exists in the ILO agenda.

4.1.5 The United Nations Environment Programme

The United Nations Environment Programme (UNEP) has a mandate that strives to balance benefits from chemicals and the recognized potential to adversely impact human health and the environment if not managed properly. UNEP's mission reflects the international consensus about its working assumption: "Chemicals are an integral part of everyday life with over 100,000 different substances in use. Industries producing and using these substances have an enormous impact on employment, trade and economic growth worldwide. There is hardly any industry where chemical substances are not implicated and there is no single economic sector where chemicals do not play an important role" [56].

UNEP is, therefore, a repository for data concerning health-related effects ranging from acute poisoning to long-term effects, such as cancers, birth defects, neurological disorders, and hormone disruption. UNEP examines effects on ecosystems, eutrophication of water bodies, and stratospheric ozone depletion. UNEP is a focal point for integration knowledge across disciplines, regardless of whether people are exposed through occupation, activities in daily life through intake of contaminated drinking water, ingestion of contaminated food (e.g., fish contaminated with mercury, dichlorodiphenyltrichloroethane and/or polychlorinated biphenyls), inhalation of polluted air (outdoor as well as indoor), or direct skin contact. UNEP serves the international community and civil society as an informational resource for many key technical areas of scientific interest that involve data repository and capacity building. UNEP is also an international governmental focal point for coordinating across groups of stakeholders, bridging the channels for communication across governments, NGOs, civil society, and opinion leaders in industry. The Chemicals and Waste subprogram assists countries and regions in managing, within a life cycle approach, chemical substances and waste that have potential to cause adverse impacts on the environment and human health, including:

- Persistent, bioaccumulative, and toxic substances (PBTs)
- Chemicals that are carcinogens or mutagens or that adversely affect the reproductive, endocrine, immune, or nervous system

- Chemicals that have immediate hazards (acutely toxic, explosives, corrosives)
- Chemicals of global concern, such as persistent organic pollutants (POPs), greenhouse gases, and ozone-depleting substances (ODS)
- Health care wastes
- E-waste

Figure 4.11 A meeting of governments and stakeholders at a conference center to deliberate the UNEP agenda, Geneva, Switzerland. Photo source: United Nations Environment Programme.

One key successful conduit for these efforts is Strategic Approach to International Chemicals Management (SICAM), a coalition of several UN agencies, private sector actors, and NGOs that meets regularly in Geneva, Switzerland, to iron out old problems and deliberate strategies for emerging issues. Nanotechnology is included among the emerging issues. The Inter-Organization Programme for the Sound Management of Chemicals (IOMC) (Fig. 4.12), established in 1995, has the purpose of promoting coordinated policies and activities among the participating organizations, jointly or separately, to achieve the sound management of chemical safety protecting human health and the environment.

Figure 4.12 The Inter-Organization Programme for the Sound Management of Chemicals (IOMC). The logo appears here for information purposes and does not imply any endorsement of this text or its contents.

Given the complexity of the scientific and technical questions to be translated into policy and the remarkably large number of international agencies that have jurisdiction over these issues, the integration of several key UN agencies under the umbrella of one interagency authority to tackle chemical safety issues without duplication is justified. According to UN agreements, the need for the IOMC is underscored by global growth patterns for production, impacting the work and health of future generations. The IOMC, housed in UNEP, is predicated on the precautionary principle that chemicals are major contributors to national economies but also require sound management throughout their life cycle in order to avoid expensive impacts upon human health and the environment.

IOMC participating organizations include UNEP, the ILO, the Food and Agriculture Organization (FAO), WHO, the United Nations Industrial Development Organization (UNIDO), the United Nations Institute for Training and Research (UNITAR), the OECD, the World Bank, and the UNDP. The agencies are tasked under international legal agreements with a division of responsibilities as follows: UNIDO undertakes activities on chemicals through its Energy and Environment Programme; the FAO has activities in the field of pesticides, a major chemicals sector that will be reconfigured as less but more potent pesticide materials will be used by applying nanotechnology; UNITAR provides training for governments and stakeholders; the ILO oversees and provides technical assistance for the application of standards for safe and healthful working conditions and prevention of major chemicals accidents; and the UNDP facilitates amelioration of chemicals management from the start of commercial programming and in tandem with the OECD [56].

In parallel, UNEP and its IOMC partners study the potential risk that hazardous wastes at the end of their life cycles might contaminate otherwise nonhazardous wastes. Chemical processes of concern include sludge from wastewater treatment plants, waste oils, and waste batteries. Addressing this perennial issue requires the legal authority vested in the IOMC to engage governments, industries, civil society, and other stakeholders to seek consensus regarding creative solutions to these intractable problems. UNEP's efforts to improve capacity to manage chemicals and waste soundly throughout their life cycles are realized by developing policy instruments, developing regulatory frameworks, and providing scientific and technical knowledge and tools needed to ensure a successful transition among countries toward sound management of chemicals and waste in order to minimize the impact on the environment and human well-being. The UN Environment Chemicals and Waste Branch is the focal point of UN Environment activities on chemical issues and the main catalytic force in the UN system for concerted global action on the environmentally sound management of hazardous chemicals. UNEP's Chemicals and Waste Branch works directly with countries to build national capacity for the clean production, use, and disposal of chemicals and promotes and disseminates state-of-the-art information on chemical safety. In response to mandates from the UN Environment Governing Council, it facilitates global action, including the development of international policy frameworks, guidelines, and programs.

4.2 WTO Limitations Regarding Safety and Health for Trade Agreements

An ability to grant patents and then protect the integrity of its patents can turn "big science" into big money. The World Trade Organization (WTO) (Fig. 4.13) is a voluntary organization of member states, designed to negotiate global trade policies. The WTO's main functions relate to trade negotiations and the enforcement of negotiated multilateral trade rules (including dispute settlement) regarding the implementation of WTO commitments and multilateral negotiations and trade and tariff data relating to exports. The WTO and the World Intellectual Property Organization (WIPO) have worked together to create an intellectual property regime that prevents fraud and piracy, while, at the same time, promoting research and development of expensive technologies for trade by all nations. The everyday activity of industry and commerce at the WTO was inherited from the General Agreement on Tariffs and Trade (GATT). On the basis of the work of the Bretton Woods institutions that interact to foster global commerce since the end of World War II, the WTO was founded in

1994 as the result of rethinking its predecessor organization, GATT. GATT had a clear constituency framework and mission: reducing the barriers to global free trade. Although GATT remains famous for its efforts to clarify customs procedures and reduce tariffs [57], its provisions regarding the primacy of safety and health concerns have often been viewed as an interesting loophole in the otherwise unbroken chain of trade rights and responsibilities. Many GATT precedents regarding food quality or safety of chemicals used in industry may apply to nanotechnology. Several OECD projects [58] contribute to global efforts to enhance its regulatory oversight of nanoscale materials health standards for groups of nanomaterials that have a similar chemical composition compared to those covered by existing recommendations [59].

Figure 4.13 World Trade Organization (WTO). The logo appears here for information purposes and does not imply any endorsement of this text or its contents. Source: WHO.

Free trade is not without legal limits. WTO limits upon global trade have been carefully crafted within the fabric of the WTO's treaty and its precisely written exceptions. For example, the "public order" exception may be invoked when genuine and sufficiently serious threat is posed to one of the fundamental interests of society. This important exception to trade in favor of protecting human health is reaffirmed in Article 8, Principles ("1.Members may, in formulating or amending their laws and regulations, adopt measures necessary to protect public health and nutrition") and Article 14, General Exceptions ("Subject to the requirement that such measures are not applied in a manner which would constitute a means of arbitrary or unjustifiable discrimination between countries where like conditions prevail, or a disguised restriction on trade in services, nothing in this Agreement shall be construed to prevent the adoption or enforcement by any Member of measures:

(a)necessary to protect public morals or to maintain public order; [60] (b) necessary to protect human, animal or plant life or health;" . . . (iii) safety").

Too little attention has been given to this exception to the overall WTO course of business, but that may change with the ubiquitous use of nanotechnology by industry in a wide variety of consumer products. Unlike previous technology leaps through the development of "big science" in genetics, atomic energy, and astrophysics, many people who are not impressed by new scientific developments will find that their money purse and their shopping carts are nonetheless touched by nanotechnology—in addition to toys and exotic new foods that appear in grocery stores. If so, the nanotechnology "revolution" may herald trade limits based on the role of safety and health, and then, the right to health will take center stage in the WTO. The WTO is also a forum for the views of stakeholders, including but not limited to private firms, business organizations, farmers, consumers, NGOs, competitors, and trading partners, in order to engage in dispute resolution for trade problems. According to the WTO "Agreement on Trade-Related Aspects of Intellectual Property Rights" (TRIPS), Article 24, International Negotiations; Exceptions: "2. The Council for TRIPS shall keep under review the application of the provisions of this Section. Any matter affecting the compliance with the obligations under these provisions may be drawn to the attention of the Council . . ."

In the WTO, therefore, nanotechnology poses a two-edged sword: both as an area of safety and health concern under GATT limitations on trade and as an area of rapidly emerging intellectual property that seeks protection under TRIPS. From the standpoint of emerging risks in nanotechnology, the real challenge for the WTO when striking the balance between these competing values is found in the plain meaning and text in its own treaties and mandates. An invaluable opportunity exists to use the WTO to resolve this fundamental conflict, however, if the WTO proactively serves as a forum for discussion of the emerging risks to human health of applying nanotechnology to products in international trade or commerce, plant safety, or food.

4.3 OECD: Nanotechnology Protections across Industrialized Nations

The OECD in Paris, France, was among the first multinational treaty-based governmental organizations to address emerging risks and benefits of MNMs. The OECD has established a widely respected Working Party on Manufactured Nanomaterials (WPMN) that is engaged in a variety of projects to further our understanding of the properties and potential risks of nanomaterials for the following purposes: (i) development of a database on environmental health and safety, (ii) research strategies on MNMs, and (iii) safety testing of a representative set of MNMs. The OECD regularly updates its Research into Safety of Manufactured Nanomaterials database, which is a global resource that details research projects that address environmental, human health, and safety issues of MNMs. This database helps identify research gaps and assists researchers in future collaborative efforts. The database also assists the projects of the OECD WPMN as a resource of research information [61].

Figure 4.14 Organization for Economic Co-operation and Development (OECD). The logo appears here for information purposes and does not imply any endorsement of this text or its contents. Source: OECD.

The OECD WPMN approach also represents a shift in organizational priorities away from trade and economics into the deep inner workings of scientific research and the debates around its development resulting in innovation throughout the supply chain. Additional OECD products include Good Laboratory Practices (GLPs) and ongoing collaboration for information sharing to develop international standards for best practices for nanotechnology. One striking feature of this working party is that it offers a rare mix of stakeholders: it blends voluntary trade and industry-based associations, such as the ISO, with administrative agencies, such as the US Environmental Protection Agency (EPA) and ad hoc groups of scholars and experts from scientific institutions. This group may deal with complex scientific issues, but its existence is a fascinating development from the standpoints of the law. Such partnerships would have been unthinkable in the late twentieth century, when corporations resisted any effort at international regulations. By contrast, such organizations are at the forefront of funding and activity when discussing the revolutionary commercial applications of nanotechnology. The OECD WPMN Safety Testing of a Representative Set of Manufactured Nanomaterials project, in particular, is designed to find and help address important data gaps. The WPMN has identified a representative list of manufactured nanoscale materials for environmental health and safety testing, including fullerenes (C_{60}), single-walled carbon nanotubes (SWCNTs), multiwalled carbon nanotubes (MWCNTs). Additionally, the list for testing includes silver nanoparticles, iron nanoparticles, carbon black, titanium dioxide, aluminum oxide, cerium oxide, zinc oxide, silicon dioxide, polystyrene, dendrimers, and nanoclays. Stakeholders are nonetheless skeptical about the reliability of OECD data as a basis for risk governance and liability conclusions. A study commissioned by the Center for International Environmental Law (CIEL), the European Citizen's Organization for Standardization (ECOS), and the Öko-Institut suggests that much of the information made available by the Sponsorship Testing Programme of the OECD "is of little to no value for the regulatory risk assessment of nanomaterials" [62].

The study was published by the Institute of Occupational Medicine (IOM) based in Singapore. The IOM screened the 11,500 pages of raw data of the OECD dossiers on 11 nanomaterials and analyzed all characterization and toxicity data on 3 specific nanomaterials: fullerenes, SWCNTs, and zinc oxide. "EU policy makers and industry are using the existence of the data to dispel concerns about the potential health and environmental risks of manufactured nanomaterials," said David Azoulay, senior attorney for CIEL. "The fact that data exists about a nanomaterial does not mean that the information is reliable to assess the hazards or risks of the material."

This problem of methodology echoes the issues of scientific certainty that undermined the validity of early "threshold limit values" and scientific consensus standards produced by industrial hygiene organizations in the mid- to late twentieth century. In those cases, the matter of reliability and validation was settled by regulators only after years of protracted litigation [63]. The OECD WPMN published the dossiers in 2015, but the team was unable to draw conclusions on the data quality. The stakeholder groups objecting to the quality and use of the dossiers fear that despite missing analysis, the European Chemicals Agency (ECHA) and the European Commission's Joint Research Centre have presented the dossiers as containing information on nanospecific human health and environmental impacts.

Guidelines on Protecting Workers from Potential Risks of Manufactured Nanomaterials provides an important contribution to global efforts toward writing clear, reliable nanotechnology regulations and standards that will be consistently reproducible in the future, thereby advancing scientific certainty regarding exposure and its risks. By May 31, 2016, the OECD working party had published 58 reports, which have recently been subjected to scrutiny by nongovernmental organizations in civil society; the Center for International Environmental Law (CIEL), whose name also means "sky" in French; and its partner, the Institute of Occupational Medicine. Industry federations and individual companies have taken this a step further, emphasizing that there is enough information available to discard most concerns about potential health or environmental risks of MNMs. "Our study shows these claims that there is sufficient data available on nanomaterials are not only false, but dangerously so," said Doreen Fedrigo, senior policy officer of ECOS. "The lack of nano-specific information in the dossiers means that the results of the tests cannot be used as evidence of no 'nano-effect' of the tested material. This information is crucial for regulators and producers who need to know the hazard profile of these materials. Analysing the dossiers has shown that legislation detailing nano-specific information requirements is crucial for the regulatory risk assessment of nanomaterials." Yet, these same dossiers were not merely endorsed but relied upon by WHO [64] in its final guidelines for reducing exposure to MNMs.

The report *Analysis of OECD WPMN Dossiers Regarding the Availability of Data to Evaluate and Regulate Risk*" [65] provides recommendations for governance of nanomaterials in commerce. "Based on our analysis, serious gaps in current dossiers must be filled in with characterisation information, preparation protocols, and exposure data," said Andreas Hermann of the Öko-Institut. "Using these dossiers as they are and ignoring these recommendations would mean making decisions on the safety of nanomaterials based on faulty and incomplete data. Our health and environment requires more from producers and regulators."

The report indirectly represents a call for a new regulatory approach to risk governance, consistent with the revolution that has been heralded about every aspect of nanotechnology in civil society. It therefore remains to be seen whether traditional paradigms for risk management and standard setting, which were inadequate to address many complex risks created by use transport and disposal of toxic substances, can answer this vital need to protect work, health, and survival.

References

1. George Rosen, *A History of Public Health Monographs on Medical History*, New York, p. 17, 1958.
2. World Health Organization Constitution. The WHO constitutional language is repeated in the International Convention on Population and Development (IPCD), Cairo, 1994, in several national constitutions in the African (Banjul) Charter. The Alma-Ata Declaration, Article 1, reaffirms that "health . . . is a fundamental human right . . ."
3. Universal Declaration of Human Rights (1948): "Everyone has the right to life, right to work, to free choice of employment, to just and favourable conditions of work and to protection against unemployment."
4. International Covenant on Economic, Social and Cultural Rights (ICESCR), United Nations, adopted 1966, entered into force 1976, Articles 6, 7, 12: "Recognize the right of everyone to enjoyment of just and favorable conditions of work which ensure in particular, safe and healthy working conditions; the right to the highest attainable standards of physical and mental health, in particular, the improvement of all aspects of environmental and industrial hygiene; the prevention,

treatment and control of epidemic, endemic, occupational and other diseases."

5. http://www.iso.org discussing ISO TC 229, a series of consensus documents regarding the methods for measuring and examining nanomaterials used in commerce.
6. UN Charter, signed June 26, 1945, entered into force October 24, 1945. Center for the Study of Human Rights, *Twenty Five Human Rights Documents*, Columbia University, New York, 1994.
7. Ilise Feitshans, "United Nations Procedures Protecting International Human Rights," Lawline.com. Continuing Legal Education course 2012.
8. Constitutional rights to health protections exist in Canada, France, and many nations.
9. Oscar Garibaldi, Obligations arising from the International Covenant on Economic Social and Cultural Rights, in Hurst Hannum and Dana Fischer, eds., *United States Ratification of the International Covenants on Human Rights American Society of International Law*, Transnational, Washington, D.C.; Irvington-on-Hudson, New York, p. 164, 1993.
10. L. Parmeggianni, ed., UN ILO *Encyclopaedia of Occupational Safety and Health*, p. 1488, Occupational diseases "cover all pathological conditions induced by prolonged work, e.g., by excessive exertion, or exposure to harmful factors inherent in materials, equipment or the working environment."
11. Frank P. Grad and Ilise Feitshans, Article 12: right to health, in Hurst Hannum and Dana Fischer, eds., *United States Ratification of the International Covenants on Human Rights American Society of International Law*.
12. Ibid., p. 224.
13. Occupational Safety and Health Series 74, List of Occupational Diseases 2010 Geneva, Switzerland, p. 7, 2010. According to the ILO Protocol of 2002, "Occupational disease covers any disease contracted as a result of an exposure to risk factors arising from work activity." Note the ILO list definition is narrower than the encyclopedia because "conditions induced by prolonged work" may include illnesses that are not recognized as a 'disease'." This difference makes the ILO list a worrisome touchstone inviting litigation, once newly discovered diseases or new negative reactions harming health status are associated with new technologies.
14. WHO Constitution basic documents, in several languages, 1948.

15. Flavia Bustreo and Paul Hunt, *Women's and Children's Health: Evidence of Impact of Human Rights*, WHO Press, Geneva, Switzerland, May 2013.

16. The ICPD, Cairo 1994, adapts language from the WHO Constitution for its definition of "reproductive health."

17. Patrick Hunziker, *Nanomedicine: The Use of Nano-Scale Science for the Benefit of the Patient Nanomedicine*, European Foundation for Clinical Nanomedicine, 2010.

18. Gary Greenberg, Occ_Env_Med_L, April 2, 1999, **48**(12):241–243, citing to the *Mortality and Morbidity Weekly Report* (MMWR) from the Centers for Disease Control and Prevention (CDC) list of "Ten Great Public Health Achievements: United States; 1900–1999," which states, "Vaccination, which has resulted in the eradication of smallpox; elimination of poliomyelitis in the Americas; and control of measles, rubella, tetanus, diphtheria, Haemophilus influenzae type b, and other infectious diseases in the United States and other parts of the world."

19. WHO, *World Report on Disability*, WHO, Geneva, Switzerland, June 2012.

20. Allison L. Greenspan, James M. Hughes, David L. Heymann, and Guénaël Rodier, *Global Public Health Security Policy Review*, **13**(10), October 2007.

21. Fifty-Ninth World Health Assembly, WHA59.15, Agenda 19, May 27, 2006, collaboration within the United Nations system and with other intergovernmental organizations, including the United Nations Strategic Approach to International Chemicals Management.

22. Facts on aligning the Hazard Communication Standard to the GHS International Programme on Chemical Safety Harmonization Project Strategic Plan 2005–2009: harmonization of approaches to the assessment of risk from exposure to chemicals, www.who.int/entity/ipcs/methods/harmonization/strategic_plan_rev.pdf.

23. WHO, International Programme on Chemical Safety. WHO/IPCS meeting on strengthening global collaboration in chemical risk assessment in conjunction with the 9th meeting of the Harmonization Steering Committee, www.who.int/ipcs.

24. In 2008–2009, UNITAR supported national GHS implementation and capacity-building projects in Vietnam, Jamaica, and Uruguay. In 2005–2007, UNITAR supported projects in Cambodia, Indonesia, Laos, Nigeria, Senegal, Slovenia, Thailand, Gambia, and the Philippines. Meetings and workshops were supported in Malaysia, Singapore, and

the ASEAN OSHNET. Regional activities exist in ASEAN, SADC, ECOWAS, CEE, the eight Arab states, and other subregions. The projects are executed by UNITAR, in the context of the UNITAR/ILO Global GHS Capacity Building Programme, with funding from the Government of Switzerland, the European Union, and others (Unitar.org/cwm/ghs).

25. The IPCS promotes consistency among hazard and risk assessment products within the global system for classification of hazards, with a view to facilitating the GHS at the national level.
26. M. I. Mikheev, *WHOs Occupational Health Programme*, conference papers Session III, "Health a Changing Challenge," Safety and Health at Work Conference, London, England, March 4–6, 1997, citing WHO Global Strategy, Occupational Health for All, Resolution of the Forty-Ninth World Health Assembly, May 25, 1996 (offset document WHA49.12).
27. General Authority Mandating Action to Protect Worker Reproductive Health: Implications of the WHO Global Strategy for Health for All Plan of Action 1996–2001, WHA 49.12, May 19, 1996, reprinted in *International Journal of Occupational Medicine and Environmental Health*, **10**(2):113–139, 1997.
28. WHO, Facilitating Projects Guide Work Plan 2009–2012, WHO's global network of collaborating centers in occupational health.
29. WHO *Guidelines on Protecting Workers from Potential Risks of Manufactured Nanomaterials*, World Health Organization Department of Public Health, Environmental and Social Determinants of Health, Geneva, 2017; Cluster of Climate and Other Determinants of Health, www.who.int/phe, license: CC BY-NC-SA 3.0 IGO.
30. WHO Global Plan of Action on Workers' Health (2008–2017): Baseline for Implementation Global Country Survey 2008/2009 Executive Summary and Survey Findings Geneva, April 2013.
31. WHO draft proposal for *Guidelines on Protecting Workers from Potential Risks of Manufactured Nanomaterials* (WHO/NANOH), Background paper 2011, comments accepted until March 31 2012
32. The Work, Health and Survival Project (WHS), including the International Safety Resources Association (ISRA), Fullerton California, Earth Focus Foundation, Geneva, Switzerland, Digital 2000 Productions, Stafford, Texas, USA; Donald H. Ewert, IH, VP Field Services NanoTox, Inc., and Director Field Services AssuredNano, Dr. Gustav Grob, International Sustainable Energy Organization (ISEO) Geneva, Switzerland. Comments presented by Ilise Feitshans, WHO draft guidelines on *Protecting Workers from Potential Risks of*

Manufactured Nanomaterials, WHO/NANOH, background paper, 2011, comments until March 31, 2012.

33. F. Boccuni, D. Gagliardi, R. Ferrante, B. M. Rondinone, and S. Iavicoli, Measurement techniques of exposure to nanomaterials in the workplace for low- and medium-income countries: A systematic review, *International Journal of Hygiene and Environmental Health*, **220**:1089–1097, 2017. doi: 10.1016/j.ijheh.2017.06.003.
34. H. Nagai, Y. Okazaki, S. H. Chew, N. Misawa, Y. Yamashita, S. Akatsuka, et al., Diameter and rigidity of multiwalled carbon nanotubes are critical factors in mesothelial injury and carcinogenesis, *Proceedings of the National Academy of the* United States *of America*, **108**:E1330–E1338, 2011. doi: 10.1073/pnas.1110013108.
35. Scope of the guidelines and key questions identified by the GDG: Which specific MNMs and groups of MNMs are most relevant with respect to reducing risks to workers, and which should these guidelines now focus on, taking into account toxicological considerations and quantities produced and used? Which hazard class should be assigned to specific MNMs or groups of MNMs and how?
36. Ilise Feitshans, *Designing an Effective OSHA Compliance Program*, Thomson Reuters (available on Westlaw.com), 2013.
37. A. Eastlake, R. Zumwalde, and C. Geraci, Can control banding be useful for the safe handling of nanomaterials? A systematic review, *Journal of Nanoparticle Research*, **18**:169, 2016. doi: 10.1007/s11051-016-3476-0.
38. The guidelines cited the following sources: IUF: Small particles, big risks: IUF, international NGOs release recommendation on the use of nanotech in foods (http://www.iuf.org/w/?q=node/4073); ETUC: Second resolution on nanotechnologies and nanomaterials, 2010 (https://www.etuc.org/documents/etuc-2nd-resolution-nanotechnologies-and-nanomaterials#. WZrnez4jG70); Canadian Union of Public Employees: Fact sheet: Nanomaterials, 2016 (https://cupe.ca/fact-sheetnanomaterials); Australian Council of Trade Unions (http://www.actu.org.au/media/149927/actu_ factsheet_ohs_-nanotech_090409.pdf).
39. Y. von Mering and C. Schumacher, *What Training Should Be Provided to Workers Who Are at Risk from Exposure to the Specific Nanomaterials or Groups of Nanomaterials?* World Health Organization, Geneva, 2017. Licence: CC BY-NC-SA 3.0 IGO.
40. K. Kulinowski and B. Lippy, *Training Workers on Risks of Nanotechnology*, US Department of Health and Human Services/National Institutes

of Health, National Institute of Environmental Health Sciences, Washington, DC, 2011.

41. N. Lee, C. H. Lim, T. Kim, E. K. Son, G. S. Chung, C. J. Rho, et al., *Which Hazard Category Should Specific Nanomaterials or Groups of Nanomaterials Be Assigned to and How?* World Health Organization, Geneva, 2017. Licence: CC BY-NC-SA 3.0 IGO.

42. Ilise Feitshans on behalf of the International Safety Resources Association (ISRA), several individual stakeholders, and the Earth Focus Foundation, stakeholder comments to NANOH regarding the WHO draft proposal for *Guidelines on Protecting Workers from Potential Risks of Manufactured Nanomaterials*, WHO/NANOH, background paper 2011, filed March 31, 2012.

43. World Health Organization, *WHO Handbook for Guideline Development*, 2nd ed., World Health Organization, Geneva, 2014.

44. Individuals and partners involved in guideline development, including the WHO Guideline Steering Group Guideline Development Group (GDG), systematic review teams, External Review Group gave written statements about of conflicts of interest as part of the WHO protocol for guidelines.

45. B. Honnert and M. Grzebyk, Manufactured nano-objects: An occupational survey in five industries in France, *Annals of Occupational Hygiene*; **58**:121–135, 2014. doi: 10.1093/annhyg/met058.

46. S. Lewin, C. Glenton, H. Munthe-Kaas, B. Carlsen, C. J. Colvin, M. Gülmezoglu, et al. Using qualitative evidence in decision making for health and social interventions: an approach to assess confidence in findings from qualitative evidence syntheses (GRADE-CERQual), *PLOS Medicine*, **12**:e1001895, 2015. doi:0.1371/journal.pmed.1001895.

47. ILO Constitution, 1919, "Declaration of Philadelphia," 1944.

48. Editorial, *The Synergist*, February 2011.

49. C170 Chemicals Convention, 1990 Convention concerning Safety in the use of Chemicals at Work (Date of coming into force: 04:11:1993.) Convention C170, *International Labour Conventions and Recommendations 1977–1995*, Volume iii, International Labour Office, Geneva, Switzerland, p. 337, 1996.

50. C155 and all ILO conventions are available at ilo.org.

51. Workers' Health: Global Plan of Action, Sixtieth World Health Assembly (Wha60.26), Agenda 12.13, May 23, 2007.

52. Ibid. "Having considered the draft global plan of action on workers' health;1Recalling resolution WHA49.12 which endorsed the global

strategy for occupational health for all; Recalling and recognizing the recommendations of the World Summit on Sustainable Development (Johannesburg, South Africa, 2002) on strengthening WHO action on occupational health and linking it to public health;2Recalling the Promotional Framework for Occupational Safety and Health Convention, 2006,and the other international instruments in the area of one program occupational safety and health adopted by the General Conference of ILO;(International Labour Conference, Ninety-fifth Session, Geneva 2006. Provisional Record 20A) Considering that the health of workers is determined not only by occupational hazards, but also by social and individual factors, and access to health services; Mindful that interventions exist for primary prevention of occupational hazards and for developing healthy workplaces; Concerned that there are major gaps between and within countries in the exposure of workers and local communities to occupational hazards and in their access to occupational health services; Stressing that the health of workers is an essential prerequisite for productivity and economic development, . . . 2. URGES Member States: to devise, in collaboration with (1) workers, employers and their organizations, national policies and plans for implementation of the global plan of action on workers' health . . ."

53. Martin Dixon, *Textbook on International Law*, 6th edition, Oxford University Press, p. 65, 2007.
54. P. Alston, Conjuring up new human rights: a proposal for quality control, *American Journal of International Law*, **78**:607–621, 1981.
55. K. T. Samson, The impact of ILO direct contact missions, *International Labour Review*, 1984.
56. www.unep.org
57. Archived at Stanford University, California, USA. The purpose of GATT included but was not limited to Article IV, Increasing Participation of Developing Countries: "(a)the strengthening of their domestic services capacity and its efficiency and competitiveness, inter alia through access to technology on a commercial basis; (b)the improvement of their access to distribution channels and information networks; and (c)the liberalization of market access in sectors and modes of supply of export interest to them."
58. OECD Working Party for Manufactured Nanomaterials (WPMN), *OECD Emission Assessment for Identification of Sources of Release of Airborne Manufactured Nanomaterials in the Workplace: Compilation of Existing Guidance*, ENV/JM/MONO (2009)16, http://www.oecd.org/dataoecd/15/60/43289645.pdf; *OECD Preliminary Analysis of Exposure*

Measurement and Exposure Mitigation in Occupational Settings: Manufactured Nanomaterials, ENV/JM/MONO(2009)6, http://www.oecd.org/dataoecd/36/36/4259402.pdf; *OECD Comparison of Guidance on Selection of Skin Protective Equipment and Respirators for Use in the Workplace: Manufactured Nanomaterials*, ENV/JM/MONO(2009)17, http://www.oecd.org/dataoecd/15/56/43289781.pdf.

59. OECD WPMN, *List of Manufactured Nanomaterials and List of Endpoints for Phase One of the OECD Testing Programme*, ENV/JM/MONO(2008)13/REV, http://www.olis.oecd.org/olis/2008doc.nsf/LinkTo/NT000034C6/$FILE/JT03248749.
60. Ilise Feitshans, *China in the WTO: The Future of Regulation Protecting the Safety and Health of Workers Using Nanotechnology*, on file at the Geneva School of Diplomacy, Geneva, Switzerland, 2009.
61. http://www.oecd.org/document/26/0,3343,en_2649_37015404_42464730_1_1_1_1,00.html.
62. Joint press release by ECOS, CIEL, and Öko-Institut: "1,500-page OECD Dossiers on 11 Nanomaterials are of 'Little to No Value' in Assessing Risks," February 23, 2017.
63. Occupational Safety and Health Administration (OSHA). "PEL Project," under the Reagan administration, discussed in Ilise Feitshans, *Designing an Effective OSHA Compliance Program*, Thomson Reuters (available on Westlaw.com), 2013.
64. Harmonized tiered approach to measure and assess the potential exposure to airborne emissions of engineered nano-objects and their agglomerates and aggregates at workplaces. Series on the Safety of Manufactured Nanomaterials No. 55. Environment Directorate Joint Meeting of the Chemicals Committee and the Working Party on Chemicals, Pesticides and Biotechnology. ENV/JM/MONO(2015)19. Organisation for Economic Co-operation and Development, Paris, 2015, http://www.oecd.org/officialdocuments/publicdisplaydocumentpdf/?cote=env/jm/mono(2015)19&doclanguage=en, accessed August 31, 2017; http://www.oecd.org/science/nanosafety/publications-series-safety-manufactured-nanomaterials.htm; http://www.iso.org/iso/home/store/catalogue_tc/catalogue_tc_browse.htm?commid=381983&published=on&includesc=true, accessed May 15, 2017.
65. Michael Riediker, Yu Ting, and Rob Aiken, *Analysis of OECD WPMN Dossiers Regarding the Availability of Data to Evaluate and Regulate Risk*, report 200-00310, Institute of Occupational Medicine, SaferNano, Singapore, December 2016.

Chapter 5

Nanotechnology's Opportunity to Revolutionize Health for All

International public health laws affirm and operationalize precautionary principles across the globe. Basic international public health treaties such as the World Health Organization (WHO) Constitution and the conventions and treaties that form international governments are flexible enough to underscore the role of precautionary principles embracing developments in nanotechnology applications that meet basic human needs that impact health: food, clothing, shelter, and use of energy. The importance of the precautionary principles as a bedrock value when crafting laws and rules protecting health was underscored by WHO in the new *Guidelines on Protecting Workers from Potential Risks of Manufactured Nanomaterials* published on December 12, 2017. Each of these basic human needs is taken for granted in daily life and therefore only rarely discussed, as if they are variables that impact human health. Yet, from a practical standpoint, either the absence or the overabundance of food, causing either malnutrition or obesity, can impair well-being, with negative outcomes for physical or mental health and social adjustments. Clothing must, of course, be appropriate to the climate and to the activities involved in order to effectively protect health, regardless of whether it is coats, bathing suits, or temperature-controlled lightweight protective clothing for first responders, whose ill-health also has consequences for people

Global Health Impacts of Nanotechnology Law
Ilise L. Feitshans

ISBN 978-981-4774-84-0 (Hardcover), 978-1-351-13447-7 (eBook)
www.panstanford.com

whom they try to rescue. Housing, including office construction and institutional buildings, has implications beyond climate control in heat or cold; using materials such as silica or asbestos for insulation and concrete can cause negative respiratory health impacts but gives wonderful new products, such as 3D printing of housing components in Singapore, and promises to provide shelter and thereby curb homelessness throughout the world. Accumulated chemicals in furniture and carpeting may impact people who have multiple-chemical sensitivity (MCS). Additionally, architecture and neuroscience come together to teach civil society that there are human health effects from the use of space as a source of healing and comfort and the use of color and sound insulation to create healthy buildings. These parameters for human existence are packaged together in energy use, with health impacts for communities in which people live and thrive.

Nanotechnology applications touch each and every one of these aspects of daily life and thus raise the law and policy question of when existing laws and embedded social, cultural, and ethical paradigms should automatically apply to new developments in these essential areas of human existence. Not only does the void at the interface between existing precautionary principles and new technologies raise the question of whether to traverse traditional legal jurisdictional barriers, but also the revolutionary opportunities raised by nanotechnologies enable society to alter existing harmful paradigms in order to create positive social change that will improve health worldwide. For example, the development of nanomaterials for antibacterial containers in transport and secure packaging has expanded significantly the global food supply [1]. Commercial industrial food that transports that food supply has liberated housewives from labor-intensive hours of cooking, baking, shopping, and food preparation so that they can spend more time with their families and paying jobs.

Hard choices regarding this development requires trade-offs regarding the acceptable risk of harm from nanotoxicity: the same engineered nanomaterials that effectively kill rodents, pests, and microorganisms, preventing them from denaturing food, also open the door for unquantified risk from nanosilver toxicity in the general environment [2]. Also, the greater availability of food raises distribution problems and brings the threat of obesity as a pandemic for the first time in history, with long-term issues about the environmental release and potential bioaccumulation of nanoparticles in human bodies and the atmosphere and the biopersistence of nanoscale pesticides in food [3]. The safety of novel foods and food ingredients is regulated by the European Union (EU) Novel Foods Regulation 258/97, which indirectly touches nanotechnology applications and nano-enabled products. In parallel, the European Commission's (EC) Recommendation 97/618 provides a framework for scientific assessment of genuinely novel foods. Engineered nanoparticles intended for direct food additive use, such as titanium dioxide and nano-enabled substances in food packaging, containers, or adhesives and inks for labels may migrate to contaminate food. For example, if nanosilver were incorporated into edible coatings for fruits and vegetable materials, such as asparagus [4] to preserve shelf life, research should address risks from ingestion and accumulation across time. Safety data and the underlying methodologies used to study risk that is revealed by safety data should be easily accessible to the public, but industry associated with safeguarding commercial interests in a competitive market is slow to participate. For this reason the niche market of kosher food, which has a private, highly trusted system for self-regulation, is of great interest as a possible model for voluntary nano-enabled product regulations regardless of the scope and effect of food regulations from a wide range of governments in different jurisdictions.

5.1 Food for Thought

Food represents a great slice of the global economy for demonstrating the pervasive role of nanotechnology in daily life. Everyone needs food. The practitioners of business intelligence who survey consumers may wonder whether food choices are driven by tastes and aromas, psychological impact of advertising, morality and religious teachings, geography, genetic allergies, diabetes, growing season, or price. But it is a universally valid, working assumption that everyone needs food. And everyone who eats has an opinion about food, which impacts the choices each person makes, so commercial food industries work hard to design foods that are attractive to buyers. According to the joint report by the World Health Organization (WHO) and the Food and Agriculture Organization (FAO) of the United Nations (UN), "Nanotechnology promises to revolutionize the whole food chain—from production to processing, storage, and development of innovative materials, products and applications" [5]. By applying nanotechnology, food products can appear creamier without additional fat, and intense flavors and bright colors can appear with less additives [6]. These alterations in traditional industrial food processing are small and subtle, so "Some companies may not even know whether nanomaterials are present in their products" [7].

The EU that governs nearly 30 countries requires labeling of foods containing nanomaterials. The European Food Safety Authority has published guidance for assessing nanomaterials in food and animal feed. Nanotechnology has featured on agendas for both the UK Committee on Toxicity of Chemicals in Food, Consumer Products and the Environment (COT) and the Advisory Committee on Novel Foods and Processes (ACNFP), although consumers may be unaware of nano-applications in foods. The US Food and Drug Administration (FDA) draft guidance also suggests manufacturers should consult the FDA regarding use of nanotechnology for a food substance already in the market and food contact materials [8]. As part of the research program, the UK Food Standards Agency (FSA) commissioned studies to assess new and potential applications of nanotechnology for food packaging in the United Kingdom. The European Parliament has removed foods from the scope of the pan-European regulatory system that governs chemical exposures, called Registration, Evaluation, Authorization and Restriction of Chemical Substances (REACH), implemented by the inspectorate EU administrative agency housed in Finland, although food additives will continue to be included in the registration and inspection process under this law. The UN has two administrative agencies that have been at the forefront of the public discourse since their joint report appeared at the beginning of the 21st century: The FAO and WHO convened an expert meeting on the potential safety aspects of the application of nanotechnologies in the food and agricultural sectors [9]. Yet, use of nanoparticles as antimicrobials in food production is increasing despite the gaps in knowledge about risks to the environment and humans.

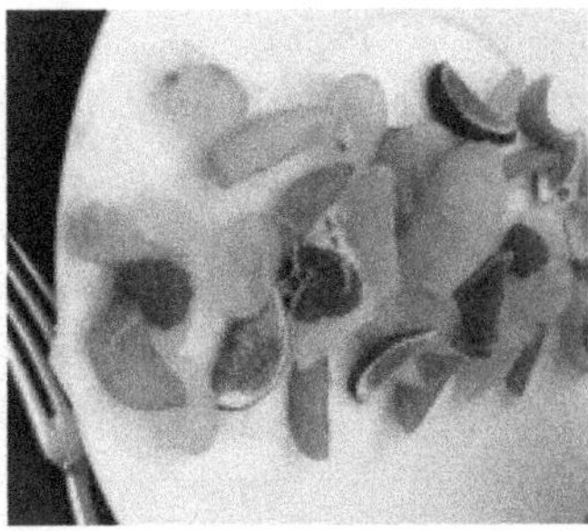

Figure 5.1 Fresh fruits locally grown but shipped by refrigerator trucks and packaged in plastic benefit from nanotechnologies in agriculture and nano-enabled containers. Photo by Dr. Ilise L. Feitshans.

5.1.1 Nanotechnology Impacts Our Daily Bread

New foods, lighter packages that keep food fresher longer, and packaging that changes color when food goes bad [10] sounds exciting but nonetheless raises important questions about safety and environmental health impacts on human health [11]. Consumer attitudes to nanotechnology are shaped, in part, by inaccurate information about nanotechnology for food [12]. According to the *New York Times*, a nongovernment organization (NGO) tested 10 varieties of powdered donuts for the presence of nanoparticles. Results from an independent lab revealed the presence of titanium dioxide of less than 10 nm, which used to brighten white substances such as creams, frosting, and foam. The impact of using titanium dioxide nanoparticles in food remains unknown and may have differing impacts, depending on context. But the importance of these claims remains unclear. For example, if its use encourages overeating then it is a negative impact when populations become obese. But what a wonderful positive impact if its use can encourage people to eat treats that can be used to coat medicines. Little is known about the behavior at the end of the product life cycle, the so-called fate of nanoscale materials in the human gastrointestinal tract. WHO suggests that "the biopersistence of dry or wet ENM (engineered nanomaterials) means their lack of digestibility, a factor that can induce adverse biological effects because they can form foreign bodies" [13]. Therefore it is feared that nanoparticles will bind with proteins or other compounds, acting as carriers of substances into an organism. This same property of polymers that is a benefit, protecting paint or prolonging fresh food, may threaten health in human digestion, depending on functionality, fate, and context.

This aspect of nanomaterials could be wonderful for preserving food, if the materials that bind are not forbidden by religious law or special dietary restrictions. For example, silver, known since ancient times for its antibacterial properties, has a unique property at the nanoscale: "antiseptic efficacy increases as particle size decreases" [14]. Widely used in food containers, refrigerators, and storage bags, WHO and FAO express concern: "Nanosilver as an antimicrobial, antiodorant, and a (proclaimed) health supplement and . . . an additive to prepare antibacterial wheat flour" may preserve the food by inhibiting the growth of microorganisms [15]. The report also notes that research is needed on (i) the potential cumulative effect of repeated exposure to these nanoparticles and (ii) whether different nanoscale food additives will interact with products such as cosmetics or workplace exposures to nanotechnology products to produce undesirable results, harming human health. Research to better understand and apply information regarding chemical identification and characterization of nanomaterials and their environmental fate, potential ecological effects, and environmental technology applications may also change the shape of regulatory governance [16].

Understanding the nature of nanostructures in food allows better selection of raw materials and enhanced food quality through processing [16]. For this reason, many food scientists would claim that they already embrace nanotechnology. Food proteins are globular particles with dimensions in the size range 1–100 nm, that is, true nanoparticles. Many polysaccharides (carbohydrates) and lipids (fats) are linear polymers less than 1 nm in thickness, that is, 1D nanostructures. Setting jellies, which maintain particles in suspension and prevent emulsions from separating into distinct oil and water phases, involves creating molecular networks—both 2D and 3D nanostructures. When starch is boiled to make custard, small 3D crystalline structures only tens of nanometers in thickness are melted. Creation of foams (e.g., the head on a glass of beer) or emulsions (sauces, creams, yogurts, butter, margarine) requires the generation of gas bubbles, or droplets of fat or oil, in a liquid medium.

Fluffy Titanium Dioxide: Friend or Foe? A possible problem with regard to nanoparticles and food is the potential and inadvertent introduction of "unnatural," persistent nanoparticles into food through their use in food packaging or food containers. The most cited and discussed example is the use of titanium dioxide (TiO_2). Because of its extreme whiteness and brightness, and its high refractive index, TiO_2 is widely used as a white pigment, primarily in surface coatings. In terms of food packaging, TiO_2 is used as a filler particle in paper and increasingly in plastics because of its resistance to UV radiation. The pigment is also used directly within foods, for example, in confectionery products such as sugar coatings, where it is used to produce a pure-white base onto which other colors can be added, thus enhancing the clarity of their colors, or as a clouding agent in dry beverage mixes. All of these applications utilize the whiteness of the pigment. However, nanoparticles of TiO_2 are transparent but retain their resistance to UV radiation. This has led to their use in transparent sunscreens and lotions. The transparency of the nanoparticles allows applications as filler particles in transparent films and plastic containers. Ingested TiO_2 nanoparticles might penetrate into tissue and cells, which could influence the accumulation and storage of these compounds with attendant risk of toxicity.

The Magic Question: Maximizing the Benefits and Reducing Risk

Will TiO_2 nanoparticles impact the separation between the pigment, the film or container, and the food?

5.1.2 Impact on Religious Food

Nanotechnologies touch each phase of the commercial cultivation and commercialization of food. New developments in nanotechnologies as applied to religious storage, transport, preparation, and service for food create a "traffic jam" at the intersection of science, law, commerce, and theology by using state-of-the-art nanotechnology to detect and remove impurities from food [17].

For example, kosher food can become easier to access, thanks to nanotechnologies, if the principle stakeholders in commercial kosher food production pay careful attention to these developments. At the same time, cheaper, faster, and long-lasting foods in packaging, which may contain nanomaterials from forbidden foods or where permitted materials might migrate places within packaging, mean that commercial kosher food is at a crossroads, requiring new vigilance about methods for agricultural planning and food process design. From a much larger perspective, the ease or difficulty of accessing the information for kosher food producers and consumers in society and commercial industry will determine whether science fosters religious practice. Whether nanotechnology will enable orthodox communities, as families and as consumers, to more easily eat food that is kosher than before or whether knowledge about nanoparticles simply creates the burden to be more vigilant against impurities may prove to be a great challenge for many religions in the world, and this dilemma will last, from the perspective of religious people, until the end of the world.

Kosher foods provide a useful case study of the impact of nanotechnology on daily life for religious people who express their religious beliefs through their choices about food. In particular, kosher food requires careful attention to preparation and handling during food production, a phase of daily food consumption that is increasingly in the hands of major commercial food industries that use nanotechnologies to grow and process food as well as nano-enabled products for food transport and storage containers. In the past, these aspects of religious life were almost exclusively in the hands of local retailers, housewives, and their families. The nano-enabled contribution to this ancient process could render purer foods for kosher consumers or, in the alternative, undermine the ability to protect kosher food from impurities derived from industrial application of nanoparticles to human food, depending on lab-to-market choices that are made at the design level of commercialization.

Three aspects of kosher food are intriguing for their health impacts as objects of applied nanotechnologies. First, kosher food represents a large multibillion-dollar portion of secular commercial markets that influences buyers beyond the religious community, primarily because of its unique approach to providing documented information about ingredients in packaged food and its special attention to every phase of the food preparation, storage and service process, including the disposal of some food products. Thus, commercial kosher food offers a real-life example of cradle-to-grave product management. Second, the ability of technology in general and nanotechnology in particular to liberate housewives from detailed food management regarding the storage of foods that are not allowed to contact each other has created available time for women who can now work outside the home or pursue leisure activities, calmly assured that the time-consuming and labor-intensive facets of religiously based food storage and preparation have met the religious standard. This includes lightweight and inexpensive nano-enabled plastics for plates, bowls, cups, flatware, and serving-ware that do not require management to maintain the separation required of fine porcelain and metals under religious law.

Third, and of great global health significance, the kosher experience in marketing commercial products for a multibillion-dollar industry is made possible by a self-regulated voluntary private system of inspection that tests foods in private laboratories and offers a seal of assurance to consumers who trust that seal and rely upon it when comparing products. Often the absence or presence of the Orthodox Union (OU) seal is outcome-determinative when selecting foods for purchase and consuming those products. This system can serve as a model for a wide array of nano-enabled products because its popular success has filled the void in laws across cultures in dozens of nations by providing a service that is trusted worldwide by stakeholders in the food industry and by consumers throughout the supply chain from food wholesaler to end user.

The parts of this system may be tied by a cultural unity, but they nonetheless offer a method for approaching regulation of risk, in this case the risk of eating something forbidden or of eating permitted foods in a forbidden way (in the wrong serving vessel, for example, or mixed with the wrong category of foods during the same meal). Observant Jews recite blessings over food before eating it, but these blessings do not render food kosher. The key criterion for determining whether food or other objects are kosher and therefore acceptable under religious laws is determined by the religious distinction between unclean and pure activities described in the Torah. Clean food is not merely from the appropriate species; it has been slaughtered according to religious law, with some parts discarded; cut using utensils that have been religiously cleaned; separated from other foods during storage and transport; and prepared according to religious laws, while maintaining the religious separations required by law. The religious importance of maintaining this separation is part of a belief that unclean or impure food enhances the undesirable or impure characteristics of human beings. Some religious scholars argue that there is unity between the food that reaches the body and the soul. The conscious choice of food either elevates or degrades so that eating, for example, birds of prey, causes a degenerative behaviour and impedes the spiritual quest for holiness. The ability to distinguish between right and wrong, good and evil, pure and defiled, the sacred and the profane, which is fundamental to holiness, is easily clouded by ingesting impurities under this view. Therefore, self-control regarding food and diet requires controlling basic, primal instincts, and therefore, following such rules elevates the simple act of eating into a religious ritual that can be sacred if done properly by respecting the rules.

Hirsch classifies foods according to their sources, which, in turn, determine how they affect the person who eats them [18]. Strong links between diet, health of the body, and health of the soul are discussed by Dr. I. Grunfeld in the *Jewish Dietary Laws*. Citing the words of the 13th-century Kabbalist, Rabbi Menachem Recanati (Taamei Hamitzvos), Grunfeld explains, "The body is the intermediary between the inner soul and the outside world. It matters greatly whether this intermediary is a willing and pliable servant or not. Just as the craftsman must have the finest tools in order to produce precision work, so must the human soul be housed in a pure body to succeed in its task. Forbidden foods contaminate the soul so that holiness will not flow within the body" [18, 19].

As Julien Bauer discussed in detail, in the modern, secular paradigm of processed foods, it is difficult to know what ingredients are in commercial food and how they were processed, even in the case of fresh vegetables [20]. Thus, without guidance from a trusted intermediary authority, confirming that food conforms to the laws is almost impossible for a manufacturer or consumer to be sure that the food is clean and pure under religious law. Industrially prepared kosher food, approved by established theological authorities, therefore has been for theologians, a miracle in itself. Mass marketing of these products to people beyond the scope of the religious community has transformed religious life for many religious people and also has brought kosher food to billions of nonkosher tables. People who might not take the time to prepare and store properly kosher food will gladly eat it if there is no extra time required to do so. And, in large markets such as the United

States, where the kosher certification has been obtained by major food manufacturers, kosher food does not cost more. For example, a bottle of ketchup costs the same whether it has the kosher symbol or not. Thus, voluntary compliance with industry-wide standards for food content has proven to be cost efficient because it actually expands the market of consumers willing to trust the kosher label and therefore make purchasing decisions, globally.

This huge commercial development was foreshadowed but not adequately discussed by Bauer [20], who accurately described discussed the emerging influence of kosher consumers on markets but could not have imagined the multibillion-dollar slice of the food economy represented by nanotechnology applications to kosher food.

Technology that controls the large-scale production of industrial kosher food also demonstrated that innovation can go hand-in-hand with human rights concepts such as religious freedom and women's rights to make choices about their lives beyond the religiously mandated lifestyle that requires women to control the food in a home. Previously, housewives procured the ingredients, checked their kosher status, handled their storage, and prepared the meals, including breads, cakes, pastries, and preserved foods, from scratch until commercial foods became available. Commercial kosher food thus offers an excellent example of how perennial societal issues tied to gender in the division of daily labor have been overcome through the use of new technology that has replaced underpaid and undervalued human labor. It is indeed possible to call from a cell phone anywhere in the world to a grocery store in the United States and have a ready-made kosher meal delivered within less than an hour without the head of the household being female or Jewish or lifting a finger if it is a voice-activated telephone. And, the meal that is delivered has a rabbi's seal, unthinkable in the homes of a grandmother in the centuries gone by.

Big business has also helped millions of women maintain their role as protectresses of the food for their families, while making possible paying work. The importance of lifting this burden from the shoulders of women cannot be understated. Food is kosher if it is prepared in accordance with Jewish law, a mandate that is the original life cycle approach to food from the grains in the field to disposal, an early commercial approach that lives today in modern commerce. Sound application of kosher principles of food processing offers what is really a cradle-to-grave approach to the life cycle of the products: Indeed, even if the food items are are legally acceptable as kosher and therefore edible in the first place, the ongoing process requirements of kosher laws require examination of the product at each step of the serving and storage process, causing consumers and the industries that address their markets to care very deeply about how each animal is slaughtered or vegetables are stored. This separation includes not only refraining from eating or serving forbidden foods but also maintaining the separation among the utensils, pots, and pans with which they are cooked, the plates and flatware from which they are eaten, the dishwashers or dishpans in which they are cleaned, the sponges with which they are cleaned, and the towels with which they are dried. A bagel, for example, can be kosher for year-round use but is not kosher for Passover, the holiday where all leavened bread is forbidden to be eaten. Similarly, there is no such thing as kosher-style food. Kosher is dichotomous: something is either really kosher or contaminated and not edible by people who keep kosher. Thus, using nanotechnologies to protect the kosher aspects of approved foods can also reduce food waste by assisting in the process of separating food. Cheaper containers, throwaway utensils, and nanosensors in food processing can be used to maintain this strict separation, thereby preventing food that was once kosher from becoming forbidden.

Many books, websites, and handy brochures in grocery stores throughout the United States explain that the Torah forbids to "boil a kid in its mother's milk," but the application of this rule is merely the tip of the iceberg in a mountain of complex regulations. A discussion of the methods for applying these laws fills volumes that have been the subject of debate among sages for centuries. It is widely agreed that this edict prohibits eating meat and dairy together, not merely in the same cheeseburger or sandwich, but not even during the same meal—from soup to dessert. This prohibition includes not eating milk and poultry together. It is, however, permissible to eat fish and dairy together, and it is quite common (lox and cream cheese, for example, in the United States is a mainstream breakfast food also consumed among non-Jewish populations) [21]. Dairy products may be eaten in the same meals as meat and eggs. The few types of mammals and birds that may be eaten must be slaughtered in accordance with Jewish law. And even permitted species cannot have died of natural causes or been killed by other animals. In addition, the animal must have no disease or flaws in the organs at the time of slaughter.

These restrictions do not apply to fish, only to the flocks and herds. But kosher consumers are mindful, too, that only some types of fish are allowed to be eaten. Therefore food ingredients derived from forbidden fish, such as shellfish and fish without scales, contaminate kosher food, rendering it forbidden. Nanotechnology can create nanosensors and devices to remain vigilant about these laws during food processing. Ritual slaughter is performed by a certified and pious man, well-trained in kashrut. The method of slaughter is a quick, deep stroke across the throat with a sharp blade without nicks or unevenness. Although the Swiss government opposes this approach and renders it illegal under Swiss law, in theory kosher methods for animal slaughter ensure rapid, complete draining of the blood and are believed to be humane and healthier compared to other methods. The question then arises, Is the prohibition against mixing milk and meat that gives rise to separate sets of metal flatware and silverware extend to the nanoscale when nanosilver particles are used in containers to prevent bacterial growth in foods [4]? If so, there might be separate meat and milk vegetables based on the method of transport and storage, even though vegetables are inherently kosher. Kosher households have at least two sets of pots, pans, and dishes, one for meat and one for dairy, so that during meal preparation and actual food service, the two separated types of food are never crossed. Although there is no prohibition against eating any specific fruits or vegetables, these foods too confront special rules for preservation and processing. Fruits and vegetables are kosher if they are without bugs or worms. If they are in contact with contamination, it means that vegetables, especially green leafy products (such as lettuce, romaine, kale, turnip greens, cabbage, and spinach), must be washed very carefully.

The industrial response to this problem of kosher food is to sell prepackaged parts of the vegetables, triple-washed and guaranteed clean. This saves the preparer time, avoids food waste, and gives the peace of mind that comes with the rabbi's stamp to guarantee the food is clean and safe for use. This religious vigilance has a profound impact on daily life, and therefore the rise of a multibillion-dollar kosher food industry is a natural consequence of food management in an era of agribusiness and industrially processed commercial foods. But what about a kosher transport of food, knowing about the possible migration of nanoparticles or the possible use of nanoparticles to manipulate flavour and color or enhance nutritional features?

Figure 5.2 Fresh fruit and foods fit for a feast. Photos by Dr. Ilise L. Feitshans.

It is this latter notion that even fruits and vegetables shipped in commerce could come in contact with nanoparticles that would contaminate them (or conversely could be coated with nanoparticles specifically designed to protect those foods against contamination) that will puzzle food chemists, nutritionists, and theologians in the years ahead. At the same time, any system that can be developed by applying nanotechnology to create an enhanced barrier to contamination will assure kosher consumers of a purity in their purchases that was never known to humans before but that, some rabbis will surely pronounce, was foretold by the separations decreed in the Torah [22]. Similar rules apply to certain fabrics, and there are specialists to determine whether clothing is kosher in accordance with religious law.

Reliable packaging means that food can be stored while important divisions separating meat from dairy can be maintained because food can be inspected at the production site, sealed, and marked with the stamp of the religious leaders on site who monitor the implementation of ritual laws with ongoing surveillance. Conversely, catastrophic health impacts on the diet of religious people could be anticipated if kosher food cannot be integrated into the new commercial food system driven by nanotechnology applications and new foods created using nanotechnologies. For example, titanium dioxide used to make fluffy vegetable-based whipped cream from soybeans might not be a subject of concern from the standpoint of cleanliness to be judged by the OU, but the laboratories might be concerned if there was a problem about the origins of the source from which the chemical is derived. It could not be extracted, for example, from forbidden foods such as shell fish or mixed with either animal fats or milk solids in order for the vegetable-based whipped cream to remain pure.

Kosher consumers are highly educated about the ingredients and processing of commercial food, and their religious choices are expressed using purchasing power. The Union of Orthodox Rabbis, founded in the nineteenth century, studies the chemical makeup of food that is commercially produced and offers a certification. According to their "web rebbe," the OU keeps abreast of all developments in food technology. Deferring to secular authorities regarding the quality and cleanliness of food processing, the OU is a kosher-certifying agency. Health aspects of food production are beyond its expertise because the expertise of the OU is deliberately limited to the domain of kosher supervision, and therefore evaluation of the health status of a product is beyond the scope of the OU's mandate to confirm that the recipe comprises kosher ingredients, without forbidden foods or forbidden mixtures. The commercial value of OU approval is so great that major food producers view acceptance by the rabbiate union quite seriously. For example, Kraft Foods in the United States exported internationally seeks the rabbinal laboratory approval for mundane products such as Smuckers jam, peanut butter, cranberry juice and sauces, and most cheese and pasta products, including the famous wacky mac macaroni and cheese. Heinz ketchup (used worldwide by McDonald's and many international franchises) and Mars candies (M&Ms, Mars bars, Snickers) also send their food for approval by these religious laboratories. If the U with a circle around it appears somewhere on the label, then the product recipe has been subjected to the private sector laboratory scrutiny to determine whether its conforms to the religious laws and certified that the ingredients are religiously permitted and clean. Even a cheeseburger may have kosher ketchup (but that will not render the meal kosher)!

By contrast, traditional Ashkenazic foods like knishes, bagels, blintzes, and matzah ball soup can all be nonkosher if prepared without regard to Jewish law. When a restaurant calls itself "kosher style," it usually means that the restaurant serves traditional Jewish foods, but it is unlikely that the food is kosher and therefore is not accessible to kosher consumers, even if the ingredients were kosher when the merchant purchased them. Of the "beasts of the earth," eating any animal that has cloven hooves and chews its cud is forbidden [23]. The camel, the rock badger, the hare, and the pig are not kosher.

Figure 5.3 Examples of powerful animals that cannot be eaten under kosher rules include lions, tigers, and bears. Source: Pexels, Creative Commons Zero (CCO) license.

This private sector apparatus provides oversight and surveillance of the processing, packaging, storage, and transport of kosher foods to meet this need that cannot be filled by most governments. The system of private oversight for publicly accessible foods is an instructive example of voluntary compliance with law that might prove to be a useful model across various aspects of commercialized nanotechnologies. For example, most parts of cattle, sheep, goats, deer, and bison are kosher so long as they were healthy when they were killed according to the rules for slaughter and their meat was not mixed with milk or any forbidden foods, but part of the animal will never be kosher according to the law. Because separating these parts of the animal is expensive and complex, many butchers simply sell the hind quarters to nonkosher butchers, and even in this regard nanotechnologies can enable a butcher to approach this delicate question with much more precision, thereby enjoying the economic benefit of reducing food waste. According to Chetan Sharma et al.

[24], "In order to meet effective food packaging requirements, advanced nanomaterial augmented polymers will help to amplify the benefits associated with existing polymers, with enhanced safety, besides addressing environmental concerns. . . . Undefined toxicity, scarcity of supportive clinical trials data and risk assessment studies limit the application of nanomaterial in the food packaging sector."

Industry-wide extensive use of private consultation to a laboratory for certification therefore provides a model for monitoring commercial food for other qualities when applying nanotechnology. Surprisingly, this inspection apparatus has not reduced the scope of marketing; on the contrary, oversight is the bedrock for building a thriving sector of the food economy. The existing role of an inspection apparatus and large-scale production for kosher food enables religious people to enjoy a modicum of comfort about their purchases and then, instead, to devote their time to religious endeavors, while maintaining a mainstream existence that was not possible before. For food served on a mass scale, thousands of meals per day in universities, nursing homes, hospitals, and prepared foods in major supermarket chains, packaged kosher food increasingly provides the major portion of daily foods consumed by religious families. In this regard, kosher food and other special religious products are at an important crossroads regarding the permeation of nanotechnologies into food processing for commercial markets. If there is no integration of theologians who otherwise control commercial food in the development of nanotechnology applications for food, it will be impossible for any housewife or any other consumer to be assured that food purchased or served is truly kosher. Yet, if kosher food is integrated into the design of nano-enabled food packaging and processing, it will be easier than ever for consumers to be kosher.

5.2 Clothing and Being Beautiful

5.2.1 Instant Clothing, New Textiles, New Dyes Creating Lighter, Stronger Fabrics

Nanotechnology has the important health impact of creating "instant clothing" that is so cheap and flexible that it can be purchased and worn with the ease of fast food. No one need die of cold in winter, because textiles that last longer with strong colors are cheaper to produce than the labor-intensive and slow processes of textile manufacture in previous centuries. The swift process for production and low price of production mean that inexpensive clothing is easily replaced with little cost. Nano-enabled textiles for clothing can therefore be available as uniforms or team T-shirts without complicated tailoring. And, Imperial College London researchers have developed a liquid clothing spray that hardens on the body and turns into a reusable garment [25]. The small fibers in a spray are mixed with polymers and a solvent that keeps the fabric in liquid form in the can. The solvent evaporates instantly when sprayed upon a surface, creating a smooth clothing material that can be washed and reworn, with implications for medical dressings or upholstery for furniture as well as shelter and keeping people warm.

Reducing the negative environmental health impacts of textile coloring and dyes for textiles has also benefited from nanoscience [26]. Textile dye waste can produce industrial wastes, resulting in water contamination. For example, methylene blue (MB) dye causes water pollution that has been linked to illness [27]. Dyes and pigments have wide application as colorants in the textile, pharmaceutical, food, cosmetics, plastics, paint, ink, photographic, and paper industries [28]. A smaller particle size means more surface area, and thus the adsorption capacity is high, and therefore textile dyeing and finishing industries have a shared interest with the general public of stakeholders who wish to reduce pollution by researching further the application of nanocomposites as adsorbents and due to accurate controls of their size, composition, morphology, and stability. Naturally available clays have been used as adsorbents in developing countries for wastewater treatment [29]. Like so many new products, MB requires balancing benefits and risks: MB has very useful clinical applications in diagnosis for carbon monoxide poisoning, methemoglobinemia, schizophrenia, cyanide poisoning, and herpes infections, but it is also believed to have negative health effects [26].

Wastewater treatment after using textile dyes may benefit from nanoscience, which can deconstruct the components of these dyes, using a novel nanocomposite that combines inorganic clay and organic papaya seed [26]. Nanoclays have properties that are highly absorbent. Nanocomposites therefore hold promise as catalysts with a novel source for remediation of the hazardous compounds persisting in soil water for clean-up of the environment.

5.2.2 Runway Fashion

Fashion designers may consider clothing to be works of art, but textiles are a print and fiber media with a large available surface area [30]. Manipulating nanoscale materials to create new combinations of natural fibers, while creating functionality and preserving appearance and comfort, is an important goal, as shown in the prototypes of lighter and stronger clothing in Fig. 5.4. Cotton, for example, is composed of cellulose molecules. But, new discoveries regarding self-assembly techniques, driven by electrostatic interactions, can modify the surface of cotton. Electrostatically driven assembly techniques can provide conformal coatings over the surface of the natural fibers. Several applications for metal nanoparticles over the surface of natural fibers such as cotton, a truly flexible and portable nanobiocomposite material, can be envisioned. For example, when silver nanoparticles are deposited, antibacterial properties are imparted onto cotton without reducing comfort. This has implications for first-responder clothing and uniforms (Fig. 5.5).

Figure 5.4 Athens Fashion Club merges science and runway beauty. Designs include wedding gowns, evening gowns, and costumes for theater. Photo by Dr. Ilise L. Feitshans.

Figure 5.5 Nanocoated fabrics are stronger and lighter and can reflect light in emergencies. Nanofashion featured prototypes for military camouflage at the NANOTEXNOLOGY conference and nanomedicine training, University of Aristotle, Thessaloniki, Greece. Photos by Dr. Ilise L. Feitshans.

An additional benefit of coating nanoparticles on top of cotton or textile fibers is the enhanced ability to control color by manipulating the interaction of light with the coated material. By tailoring the surface shape and size of the nanoparticles, nanotechnology offers designers a controllable color palette for any fabric. For example, if a designer desires to have a golden finish, gold nanoparticles can be used to create a shiny metallic-yellow color with antibacterial properties, as seen in Figs. 5.4 and 5.5. These new fashions run a diverse gamut, from clothes on government officials and the military that protect them to the high-fashion glitz of the runway. The ability to control color in fabrics without dyes can reduce environmental pollution and opens new portals for creativity in fashion (Fig. 5.6). According to materials science researcher Juan P. Hinestroza, whose new vision for nanocoated fabrics is illustrated in Fig. 5.6, "The ability to control color in fabrics in a tunable fashion, without the use of dyes, while adding functionality is a makeover for the old faithful, cotton; it is a match between fiber science and apparel design made possible by nanotechnology" [30].

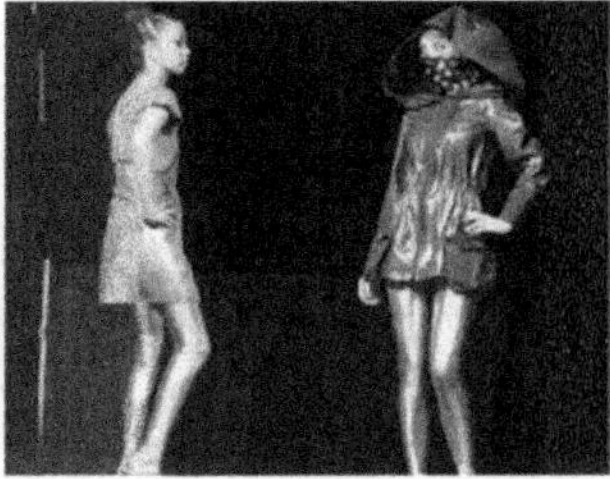

Figure 5.6 (Left) President Donald J. Trump and First Lady, Melania Trump, June 26, 2017, wearing nano-enabled textiles in clothing. Official White House photo by Shealah Craighead. (Right) A dress and jacket containing Ag and Pd nanoparticles with antibacterial and air-purifying qualities, designed by Cornell Fashion Design student Olivia Ong. Photo by Peter Moran. Reprinted from Ref. [30], Copyright (2007), with permission from Elsevier.

5.2.3 Cosmetics

Nanosilver and nanogold are used in cosmetics, perfumes, sunscreens, and a wide variety of personal care products, including underarm deodorants for antibacterial protection. Cosmetic manufacturers use nanoscale versions of ingredients to provide better protection against ultraviolet light, deeper skin penetration, long-lasting effects, and increased color and finish quality. Some cosmetic products, such as sunscreens, use mineral-based materials, and their performance depends on their particle size. In sunscreens, titanium dioxide and zinc oxide, in the size range of 20 nm, are used as efficient filters against the sun's rays. The question of whether rapid assimilation of nanomaterials is good or bad varies not by substance but by context. For example, if nanoscale titanium dioxide in personal care products reduces biological roles of bacteria after less than an hour of exposure, what happens next?

Close contact with the skin and internal organs that is common for cosmetics has traditionally been closely scrutinized by regulatory oversight for over a century in the United States and Europe. Cosmetics represents an area where quantities in use are usually small and therefore involve quantities that have been considered "generally recognized as safe" (GRAS) under law. Nanocrystals, made of tens of thousands of atoms that combine into a "cluster," can pass through the skin but may not constitute a size large enough to trigger regulatory oversight. Thus, populations at risk may not be studied to capture the data about health impacts whether good or bad. Typical sizes of these aggregates are between 10 and 400 nm, thus possibly exceeding the scope of regulation if conventional definitions of nanomaterials as 100 nm or less are adopted under law, providing a real example of the underinclusive character of numerical cut-offs rather compared to functional criteria when drafting nanosafety and risk governance protections. Will those particles form new clusters, or will they break down bacteria in target organs or wastewater treatment or groundwater?

Old questions about the composition of cosmetics and their impact on the skin and their toxicity if absorbed through the skin or digested have been revisited by regulators because of new concerns about the novel character and unquantified risk information surrounding nanomaterials in cosmetics [31]. Nanoscale forms of traditional materials that were previously recognized as safe in small quantities pose a puzzle for regulators, scientists, and consumers alike. Each of these stakeholder constituencies possesses an interest in finding the safest context for use with the least risk of harmful exposure and therefore requires accessible risk communication and careful medical surveillance.

5.3 Shelter from the Storms

Housing construction involves labor, land and material costs, maintenance costs, and interaction with laws of zoning, too. Nanotechnology offers the promise of relieving some of the burden from these ancient costs by reducing and tracking energy consumption, using lighter and stronger materials that can be rapidly applied, offering stronger paints and coatings inside and outside of buildings, and inventing new methods for transferring energy within the housing structure for lighting, heating, cooling, and gracious daily living [32]. Early applications of nanotechnology in architecture and construction have addressed energy efficiency, but nano-enabled products also hold importance for sustainability and low maintenance [33]. Energy-saving windowpanes apply several metal oxides or nitrides on a glass surface, with more precise control of the thickness. Such windowpanes are colorless and transparent, with unusual properties regarding energy and sunlight control levels. They also reflect a large percentage of infrared radiation and almost eliminate all the infrared radiation with a wavelength of 3000 nm, so heat transition through the windows is prevented. This function prevents sunlight and exterior heat entering inside in summer and reduces the need for cooling devices. Nanotechnology applications and nano-enabled products thereby improve energy efficiency, a long-standing challenge for engineers and architects, by offering new low-energy lighting [32]:

- Light-emitting diodes (LEDs) as a lighting technique with high energy efficiency for specific uses in buildings (facade lighting, guiding lights, and color lights)
- Color solar cells with transparency and aesthetic (decorative) characteristics, like glass facades as a replacement for silicon-based solar cells
- Core/shell nanostructures and catalyzers for more efficient fuel cells in electricity and heat section using natural gas

Using these materials and new methods of electronic printing made possible by nanotechnologies, it will soon be possible to build entire housing complexes with the speed and materials previously required for a few houses. Thus nanotechnology offers the prospect of low-cost housing without displacing construction workers but by expanding the construction industry to affordable units for people who might otherwise have nowhere to live. Instead of being homeless, people will have shelter from the storms [32].

5.3.1 Self-Cleaning Concrete

Nanotechnology offers multiple solutions to improve the beauty and durability of exterior surfaces of a building [33]. Carbon nanotubes (CNTs) are used in concrete; copper nanoparticles improve oxidation and the ability for welding steel; and nanosilica is used in faster, cheaper, and lighter materials for construction. Using nanomaterials, surface properties can be customized to meet local design needs concerning water and climate, which is important because moisture and ultraviolet rays are among factors that promote decay and can destroy construction materials. Nanotechnology offers the possibility of making ultraviolet-resistant and antidecay coatings for wooden materials. Nano-insulation materials may surpass glass wool for insulation properties. Thermal insulators for housing facades reduce operation costs for energy and maintenance, extending the possibility of thermal protection and insulation for three modes of heat transfer: radiation, advection, and convection. Being thinner while providing greater insulation against moisture and fungus, the early nano-enabled insulation products can be washed with soap and water. Nanosilica components of concrete also reduce the impact of pollution on structural materials and respond to changes in temperature and light. Water-repulsive nano-enabled building products in roofing and tiles, for example, are expected to require less cleaning staff and less risk of contact with respirable dusts that can cause silicosis.

5.3.2 Consumer and Workplace Paints and Coatings

Paintings and coatings for cars, housing, office buildings, consumer products, and interior use by homeowners are a common source of human exposure to nano-enabled materials and an area that deserves close scrutiny from the standpoint of risk assessment for exposure, long-term half- life of toxic or hazardous features of materials, and the cumulative effect of consumer exposure, workplace exposure, and the combined effects of these exposures together. From the standpoint of epidemiology and prevention, these products offer a unified stakeholder problem because manufacturers of consumer products are both users and consumers of nano-enabled paints and coatings that cover many end-user products. This area of stakeholder concern therefore requires more consideration than has been given to it to date, and will be an area of possible litigation and liability if regulators and manufacturers use the monolithic approach to exposure. Many people and environmental health may fall through this gap unless funding gives attention to such multidimensional modeling to predict exposure scenarios and subsequent risk of harm.

5.3.3 3D Printing for Housing Components

Public housing may benefit from nanotechnology innovations. As shown in Fig. 5.9, creative uses of form made possible by new, lighter, and stronger building materials mean attractive, affordable housing may be a reality for millions of future residents. New materials for traditional housing approaches are merely the beginning, however, when nanotechnology is concerned. It has been reported that Singapore is developing a novel method for low-income housing construction, applying nanotechnology and 3D printing.

Figure 5.7 Public housing near Geneva, Switzerland. Photo by Dr. Ilise L. Feitshans.

The Singapore Centre for 3D Printing, established with $107.7 million in government and industry funding, is reportedly working with a company to test the feasibility of 3D printing public housing units story by story, off-site, before assembling them at their destination. Using concrete 3D printers, the center has plans to build a test-bed prototype, designed to decrease dependence on foreign labor, typically used in the construction industry [34]. Structural components of buildings are those most likely to be printed, while others will be made by traditional methods. Professor Chua Chee Kai, executive director of the Singapore Centre for 3D Printing, says, "The idea is to print them maybe a unit at a time. So if you have a 10 storey building, you will probably do one storey at a time. These will be transported to the construction site where they will be stacked up like lego" [35].

5.4 Energy

5.4.1 Smarter, Driverless Cars

The car called "nano" and the changes in paints, coatings, upholstery, tires, and gasoline configurations are all nano-enabled applications that have been designed in order to reduce efficiency and increase the safety of motor vehicles. Aside from the reduced liability of motor vehicle accidents, which is a huge cost in terms of litigation, replacement parts, and global disease burden, many nano-enabled features of new cars and trucks are lauded for their fuel efficiency. Lighter cars, smoother tires that can carry a load longer without wearing down, and engine parts that require less fuel are part of a green package for energy-saving approaches to transportation. These changes are designed to transform the calculus of risk regarding driving, from the standpoint of both environmental health and public health, because there are so many cars in use around the world and driving is nonetheless a major killer [36]. The calculus of risk regarding transportation and driving changes dramatically when contemplating the prospect of driverless cars compared to ordinary cars with the potential for human error. Applying nanotechnology to create sensors and wireless cameras that communicate within a system, several major car manufacturers plan to have autonomous cars on the road in a commercial setting, such as a ride-sharing program, by 2021 [37]. Ford is reportedly preparing to launch self-driving cars with level 4 autonomy. Such vehicles won't have a brake pedal, an accelerator, or a steering wheel and will be able to operate in a predetermined geographical area without human intervention. Carmakers will rely on branding to fill the gap between experience and risk assessment when applying these new forms of nanotechnology.

A wide variety of sensors and cameras will substitute human response mechanisms to changes in driving conditions. Cameras can be strategically placed to detect close-range objects and monitor lane markers. Rear-facing transmitters will help change lanes by detecting vehicles that approach from behind. A trifocal camera at the top of a windscreen can help to identify sudden movements that impact driving conditions, such as pedestrians in a crosswalk or debris. Radar transmitters can provide 360° view of the car's surroundings. Ultrasonic sensors on the car's exterior can detect objects, and a multiple-beam laser scanner at the front of the car will increase automatic vision within 150 m. Nano-enabled electronics will facilitate 3D digital maps and a high-performance Global Positioning System (GPS) so that the car will select the quickest, most efficient route. Volvo offers customers to "Think about safer, less congested roads - and an effortless and more convenient commute for you. Your Volvo will know exactly where it is and what's ahead, allowing you to hand over control of the driving" [38].

Legal issues in this branch of nanotechnology have been faster to appear on the horizon than in other aspects of nanotechnology law; a swift sideswipe has already brought a near wreck in Silicon Valley, where there is a legal battle involving the Google self-driving car spinoff Waymo and Uber Technologies, Inc. When one company fired a key employee, swift action by the second employer allegedly resulted in leaking trade secrets to the competition [39]. The plaintiff company reviewed its internal compliance system and allegedly discovered that the former employee had downloaded 14,000 confidential files before leaving, including technical details of its proprietary laser technology that helps self-driving cars (Fig. 5.8) perceive obstacles when navigating.

Figure 5.8 Fusion photo of prototype driverless cars. Montage by Dr. Ilise L. Feitshans.

While one side attempts to link the lost information to willful acts by executives, the other side seeks to shield all information by claiming old-fashion attorney–client privilege, a concept that enables lawyers to prepare a defense without subjecting their notes and impressions to the scrutiny of courts of law.

5.4.2 Reduced Energy Needs Favoring Environmental Health

Many people believe that nanotechnology can offer a new era that will save energy by reducing energy consumption and also by creating new methods to transfer and store solar energy (Fig. 5.9). A smaller particle size means more surface area, and thus the capacity to absorb or react as a catalyst is greater than in traditional materials. Nanomaterials are also highly reactive due to their high surface area-to-mass ratio, providing more area by weight for chemical reactions to occur. Studies have revealed that because of this increased reactivity, some nanoscale particles may be potentially explosive and/ or photoactive. For example, some nanomaterials—such as nanoscale titanium dioxide and silicon dioxide—may explode if they are finely dispersed in air and they come in contact with a sufficiently strong ignition source.

Figure 5.9 New approaches to lighting buildings and public spaces: (Left, middle) The University of Freiburg, Germany, library outside and inside maximizes use of stronger steel and glass to provide an open-air environment that is highly energy efficient. (Right) Long-lasting, reduced-energy lighting in public spaces in Haddonfield, New Jersey. Photo credits: Dr. Ilise L. Feitshans.

Figure 5.10 Logo for Hellenic–Japanese alliance in organic photovoltaics (PV) is one of several initiatives showcased at the annual Nanotexnology meetings in Thessaloniki, Greece. The logo appears here for illustration and does not imply endorsement or affiliation with these products or their commercial entities.

Nanomaterials can address many unmet needs in energy applications; they can be used to improve energy densities and efficiencies, increase rates of energy transfer, and increase stabilities, for example. Nanotechnology also has the potential to facilitate hitherto underdeveloped or unknown commercial applications for energy (smart clothing, flexible electronics, covert inks, energy-scavenging/energy-converting devices), as well as reducing the cost of new energy technologies and enable their mass production. Within the field of nanotechnology, new material development will play an important role toward a sustainable energy future.

The "Nanotexnology Workshop on Renewable Energy & Storage" examines innovative materials, devices, and concepts for energy harvesting, conversion, storage, functionality, distribution, smart grids, and EU policies on renewable energy. Decentralized renewable energy installations and battery storage systems for domestic power supply will be combined in intelligent networks, providing balanced power to the public grid, while the devices and communication technology needed for the integration of these elements, intelligent transformers, and virtual power plants will be discussed. Despite challenges related to integrating renewable energies in isolated or weakly interconnected systems and in developing self-sustainable energy systems on islands, smart specialization strategies in energy (e.g., solar, bioenergy, sustainable buildings, and smart grids) and at the regional level and interregional cooperation will be possible in the future.

- Next-generation photovoltaics (PVs), PV systems
- Bioenergy, fuel cells, and hydrogen generation
- Smart energy consumption
- Energy storage and batteries and storage for smart cities and smart islands
- Sustainable buildings
- Energy and energy autonomy in agriculture water treatment and desalination
- Smart grid evolution and grid integration

These approaches have been less controversial that fracking [40], which potentially destroys land and can make solid land unstable despite the governmental claims that fracking is a safe [41] and economical approach to financing small businesses by leasing land. Fracking has been promoted as a source of natural funding to maintain family farming [42].

An important question of balancing risks regarding nanotechnology will therefore emerge in the field of energy because the environment is also at risk from exposure to nanomaterials through release into the water, air, and soil during the manufacture, use, or disposal of these materials. Antibacterial nanomaterials are a two-edged sword: the materials could interfere with beneficial bacteria in sewage and wastewater treatment plants or be used to clean up water and hazardous waste if directed properly. For example, materials scientists and bioelectrochemical engineers may have created an innovative, cost-competitive electrode material for cleaning pollutants in wastewater by using bacteria-coated nanofibers. The researchers created electrospun carbon nanofiber electrodes and coated them with a conductive polymer called poly(3,4-ethylenedioxythiophene) (PEDOT) to compete with carbon cloth electrodes on the market in order to digest pollutants.

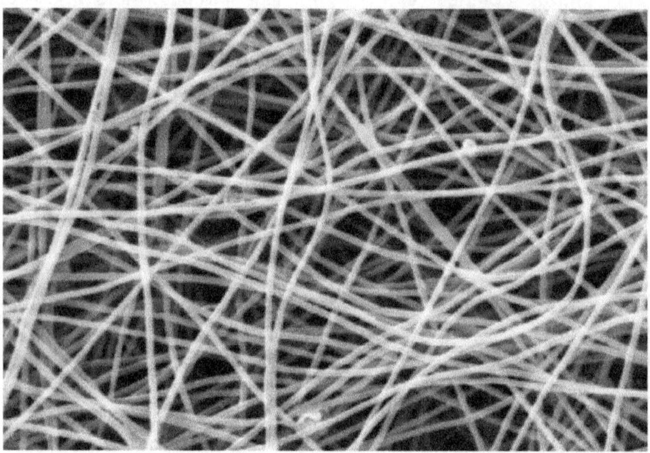

Figure 5.11 Carbon nanotube fibers. Source: US government.

Nanofibers create a favorable surface for bacteria, which digests pollutants from the wastewater and produces electricity according to research. When the PEDOT coating is applied, an electrically active layer of bacteria, called *Geobacter sulfurreducens*, grows to create electricity and transfer electrons for wastewater treatment, at the same time producing electrical energy. Carbon nanofiber electrodes are highly porous, have an extensive surface area, and are biocompatible with bacteria. The bacteria at the electrode could capture and degrade pollutants from the wastewater that flows by it, thereby having clean-up systems that use less land. "Electrodes are expensive to make now, and this material could bring the price of electrodes way down, making it easier to clean up polluted water," said co-lead author Juan Guzman [43].

Figure 5.12 Screen shot of the European Union website entry portal to details about EU-funded nanotechnology research activities, with focus on energy.

An increased surface area gives a promising future to solar energy and to bacteria-coated nanotubes for wastewater treatment. New uses for materials with a high surface area include catalysts, drug delivery molecules, solar energy generation, energy storage, and cosmetics.

References

1. http://www.csrees.usda.gov/nanotechnology.cfm.
2. http://www.efsa.europa.eu/en/topics/topic/nanotechnology.htm.
3. Joint FAO/WHO Expert Committee on Food Additives.
4. J. S. An, M. Zhang, S. Wang, and J. Tan, Physical, chemical and microbiological changes in stored green asparagus spears as affected by coating of silver nanoparticles: PVP, *LWT Food Science and Technology*, **41**:1100–1107, 2008.
5. UN Food and Agriculture Organization (FAO) and the World Health Organization (WHO) expert meeting on the application of

nanotechnologies in the food and agriculture sectors: *Potential Food Safety Implications* meeting report, p. 10, Rome, 2010.

6. FAO, State of the art on the initiatives and activities relevant to risk assessment and risk management of nanotechnologies in the food and agriculture sectors. CIEL, Friends of Earth Europe, and several NGOs, and stakeholders' response to the *Communication on the Second Regulatory Review on Nanomaterials*, Brussels, October 23, 2012.
7. Stephanie Strom, *Study Looks at Particles Used in Food*. February 5, 2013. A version of this article appeared in print on February 6, 2013, on page B5 of the New York edition: *Study Looks at Particles Used in Food*.
8. http://www.fda.gov/ScienceResearch/SpecialTopics/Nanotechnology/.
9. http://www.fao.org/docrep/012/i1434e/i1434e00.pdf.
10. The EU seminar on *Polymer Nanomaterials for Food Packaging: Characterization Needs, Safety & Environmental Issues* hosted by St. Mary's University College, London, on September 1–2, 2010, jointly under COST Action FA0904 (Eco-sustainable Food Packaging based on Polymer Nanomaterials) and FP7 NaPolyNet (Setting up Clusters of Research-Intensive Clusters across the EU on Characterization of Polymer Nanostructures).
11. V. J. Morris, Emerging roles of engineered nanomaterials in the food industry, *Trends in Biotechnology*, **29**(10):509–516, 2010.
12. http://www.who.int/foodsafety/biotech/nano/en/index.html.
13. FAO/WHO expert meeting on the application of nanotechnologies in the food and agriculture sectors: *Potential Food Safety Implications* meeting report, p. 25, Rome, 2010.
14. Marina E. Quadros and Linsey C. Marr, Environmental and human health effects of aersolized silver nanoparticles, *Journal of Air and Waste Management Association*, **60**:770–781, 2010.
15. FAO/WHO expert meeting on the application of nanotechnologies in the food and agriculture sectors: *Potential Food Safety Implications* meeting report, p. 12, Rome, 2010.
16. EU Scientific Committee on Food SCF/CS/ADD/EMU/199 Final, February 21, 2003. Opinion of the scientific committee on food on carrageenan, http://www.cybercolloids.net/sites/default/files/EU-carrageenan-opinion.pdf.
17. Ilise Feitshans , *Food for Thought: Nanoparticles and Kashruit, Crowded Traffic at the Intersection of Science, Law and Theology*, lecture for the

Womens' Group at Beth Jacob Synagogue, Geneva, Switzerland, 2013 and updated 2015.

18. Samson Raphael Hirsch, *Horeb: A Philosophy of Jewish Laws and Observances* (translated by Isidor Grunfeld), Soncino Press, London, 1962 (reprinted 2002). See also Samson Raphael Hirsch, *Commentary on the Pentateuch* (translated by Dr Isaac Levy, introduction by Isidor Grunfeld), Judaica Press, 1962 (reissued 1989).

19. I. Grunfeld, *The Jewish Dietary Laws*, Soncino Press, London, 1972.

20. Julien Bauer, *La Nouririture Cacher*, Presses Universitaires de France, Paris, 1996 Series "Que Sais Je?"

21. Tracey Rich, Judaism 101 notes that the religious commentaries on the Torah called the Talmud prohibits cooking meat and fish together or serving them on the same plates. This restriction has implications for commercial foods that may use genetic materials from one species extracted to protect shelf life or enhance flavor or color.

22. Prohibition of mixing foods "with unclean things" (Lev. 11:34).

23. Leviticus 11:3; Deuteronomy 14:6.

24. Chetan Sharma, Romika Himan, Namita Rokana, and Harsh Panwar, Nanotechnology: An untapped resource for food packaging, *Frontiers in Microbiology*, September 2017, https://doi.org/10.3389/fmicb.2017.01735.

25. Myles Burke, Spray-on clothing in a can to be launched, press release by the Imperial College London, *Telegraph*, 2010, http://www.telegraph.co.uk/news/newstopics/howaboutthat/8007246/Spray-on-clothing-in-a-can-to-be-launched.html.

26. A. A. Olajire and A. J. Olajide, Kinetic study of decolorization of methylene blue with sodium sulphite in aqueous media: Influence of transition metal ions, *Journal of Physical Chemistry and Biophysics*, **4**:136, 2014, doi:10.4172/2161-0398.1000136.

27. N. S. Verma, A. Gupta, M. Dubey, S. Mahajan, and R. Sharma, Resistance status of some pathogenic bacteria isolated from water of Yamuna river in Agra, *Asian Journal of Experimental Biological Sciences*, **2**(4):697–703, 2011. ISSN 0975-5845.

28. H. B. Singh and K. A. Bharati, *Handbook of Natural Dyes and Pigments*, Woodhead, New Delhi, 2014.

29. T. Ngulube, J. R. Gumbo, V. Masindi, and A. Maity, An update on synthetic dyes adsorption onto clay based minerals: a state-of-art review, *Journal of Environmental Management*, **191**:35–57, 2017.

30. Juan P. Hinestroza, Can nanotechnology be fashionable? *Materials Today*, **10**(9):64, 2007, doi:10.1016/S1369-7021(07)70219-5.
31. Silpa Raj, Shoma Jose, U. S. Sumod, and M. Sabitha, Nanotechnology in cosmetics: opportunities and challenges, *Journal of Pharmacy and Bioallied Sciences*, **4**(3):186–193, 2012. doi: 10.4103/0975-7406.99016.
32. F. H. S. Javad, R. Zeynali, F. Shahsavari, and Z. M. Alamouti, Study of nanotechnology application in construction industry (case study: houses in north of Iran), *Current World Environment*, **10**(Special Issue May), 2015, http://www.cwejournal.org/?p=11691.
33. S. Leydecker, M. Kölbel, and S. Peters, *Nano Materials in Architecture, Interior Architecture and Design*, 1st ed., Berlin, Springer Science+Business Media, 2008.
34. *Singapore Makes Plans to 3D Print Public Housing*, press release, Singapore Center for 3D Printing, 2016.
35. Clare Scott, *Singapore Developing Plans for 3D Printed Public Housing*, 3DPrint.com, Government section, February 10, 2016.
36. Ilise Feitshans, "Your Turn to Drive," PowerPoint presentation to the American Society of Safety Engineers (ASSE), Nigeria Chapter, explaining the United Nations Decade of Roadway Safety, March 2014. See also Richard A. Goodman, Richard E. Hoffman, Wilfredo Lopez, Gene W. Matthews, Mark A. Rothstein, and Karen L. Foster (eds.), *Law in Public Health Practice*, Oxford University Press, 2000.
37. Cadie Thompson, Ford CEO reveals a major fear about self-driving cars, *Guardian*, Feb. 13, 2017.
38. www.volvo.com.
39. Aarian Marshall, The Uber v. Waymo court showdown looms. Here's what you need to know, *Guardian*, July 13, 2017.
40. Bryan Henderson, Natural gas drilling coming to NC? *Charlotte Observer*, July 27, 2010.
41. Michele Rogers, et al., Marcellus shale: what local government officials need to know, *Marcellus Minutes*, April 8, 2009, http://marcellusminutes.com/marcellus-shale-what-local-government-officials-need-to-know/.
42. Theodore A. Feitshans, *Evaluating Oil and Gas Lease Proposals*, Powerpoint presentation, NC State University, 2010.
43. Juan Guzman et al., Performance of electro-spun carbon nanofiber electrodes with conductive poly(3,4-ethylenedioxythiophene) coatings in bioelectrochemical systems, *Journal of Power Sources*, **356**:331–337, 2017.

Chapter 6

Stakeholders, One and All

6.1 Public–Private Partnership Promoting Risk Management

Stakeholders engaged in expressing their views about nanotechnology need:

- Robust and reliable methods for measurement and detection of nanoparticles
- Accessible and trustworthy information about nanomaterials and nano-enabled products in commerce that explains the key issues concerning toxicity for humans and the environment
- Clear pathways for communication of their views within superstructures for risk governance

Stakeholders in civil society act as both a source of customary law and the secret weapon for effective implementation of law by exercising their political will. Competing interests using their opportunity to be heard may be clearly within the purview of the role of government legislators gathering information, and then it is the job of the legislators or regulators within the governing structure to serve several interests by striking a balance between the needs and desires of various stakeholders. Therefore it is important to examine the role of various stakeholders when studying the implementation of scientific precautionary principles and forecasting emerging

Global Health Impacts of Nanotechnology Law
Ilise L. Feitshans

ISBN 978-981-4774-84-0 (Hardcover), 978-1-351-13447-7 (eBook)
www.panstanford.com

laws to regulate nanotechnology in civil society. The need to involve stakeholders was immediately understood at the National Nanotechnology Initiative (NNI): "Development of a healthy global marketplace for nanotechnology products and ideas will require the establishment of consumer confidence, common approaches to nanotechnology environmental, health, and safety issues, efficient and effective regulatory schemes" [1].

Across the world, hundreds of federal, state, and international bodies of law are bottomed upon the philosophy that work-related illnesses are an avoidable aspect of industrialization and that consumer protection against risky products is a cornerstone of public health.

A natural step for the regulation of the risks of nanotechnology in order to provide society with maximum benefits therefore requires a partnership with industry, multinational corporations, governments, academia and research institutions, and stakeholder representatives. Murashov and Howard [2] offer an admixture of risk management and political discussion brought together to shape best practices in the event there is no consensus about regulations. Their approach would generate knowledge about the nature and extent of worker risk, utilize that knowledge to develop risk control strategies to protect nanotechnology workers, and provide an evidence base for a possible nanotechnology program standard later [2]. To fill this void, they proposed a national partnership between the administrative enforcement agency, the research agency, nanotech manufacturers and downstream users, workers, researchers, and safety and health practitioners. "A National Nanotechnology Partnership would generate knowledge about the nature and extent of worker risk, utilize that knowledge to develop risk control strategies to protect nanotechnology workers now, and provide an evidence base for a possible nanotechnology program standard later." Their six-prong approach to the management of occupational health risks combining qualitative assessment, the ability to adapt strategies and refine requirements, an appropriate level of precaution, global applicability, the ability to elicit voluntary cooperation by multinational companies, and stakeholder involvement is the glue that brings together a new approach to voluntary compliance by employers whether governmental or corporations.

6.1.1 The Wide Range of Stakeholders in Civil Society

Civil society includes institutions and voluntary-based organized groups that are independent from the government, such as nongovernmental organizations (NGOs), universities, private research centers, the so-called think tanks, and trade associations and professional societies. In theory, the term "civil society" should include the entire gamut of stakeholders, but especially in Europe, the concept tends to address the proactive efforts by NGOs in a watchdog capacity. In the European context, civil society activists may not overtly seek political power but powerfully influence geopolitics nonetheless. "Civil society organizations can help to develop the other values of democratic life: tolerance, moderation, compromise, and respect for opposing points of view. Without this deeper culture of accommodation, democracy cannot be stable. These values cannot simply be taught; they must also be experienced through practice" [3].

The definition of "stakeholders" includes a large and diverse collection of interest groups that possess the resources, information, and desire to make their views known by having their say. Government regulatory administrations around the world, stakeholder NGOs, and the diverse professional and trade associations that work with them in several different capacities are excellent conduits for information dissemination and knowledge sharing to achieve these goals of promoting commitment to long-term policy goals. Stakeholders can also serve as social partners that can support implementation efforts through social media, outreach, and knowledge sharing and include trade associations and professional societies as well as governments. In the 21st century, such groups represent many types of interests, from more nations and with more refined knowledge of the political system than in centuries before. This is a trend toward a democratization of international regulation, which was previously an exclusive governmental domain, ruled by kings endowed with divine powers, from the time of the Treaty of Westphalia to the time of the Treaty of Versailles.

Additionally, some major opinion leaders such as former government officials, heads of state, and corporate leaders can have a voice through stakeholder organizations that influence policy. NGOs that do not represent official entities can provide an opportunity for people who work at high levels to express their opinions without the label of their employer and therefore without confusing their private views with official employer policy. This is especially important for government officials who may disagree with political posturing at the highest levels of their administration and who wish to have a conduit for communication that allows them make their opinions known in order to enrich discourse among the general public. Organizations such as the Work Health and Survival Project (WHS), for example, enable officials to address policy problems under their sheltering umbrella so that their ideas will gain currency among other stakeholders.

By contrast to Europe, the US definition of "stakeholder" follows a two-tier approach: naming specific industries or people who may be impacted by proposed regulations, while also reassuring the general public that everyone is invited to participate, express their views, and attend. Typically, the stakeholder has some rights under statutory law, or a constitution or treaty whose rights are somehow touched by the proposed changes wrought by law, in order for a party to be considered a stakeholder. The party seeking to be heard must show a direct nexus between its interests and the problems before the decision makers in order to have the right to be heard in the precise context. Stakeholder comments in those contexts typically begin, therefore, with a "Statement of Interest." If prepared candidly, this statement offers clues for readers regarding the speaker's perspective and therefore serves as a safeguard for transparency and procedural fairness, as well as providing a voice for people who are not present at the deliberations. Several new approaches to developing stakeholder views and presenting them in various media are coming to the fore in the 21st century. For example, at the World Trade Organization (WTO), small groups of countries have clustered together around their common interest during agriculture negotiations. The growth of multistate coalitions reflects the broader geopolitical climate that demands multinational cooperation in order to stabilize the bargaining power of any single nation within international organizations. The European Union, offering its views at many international organization meetings, considers itself a stakeholder. So, too, policymaking integration exists for the Association of South East Asian Nations (ASEAN) as a stakeholder. Consequently, more voices speak as stakeholders in more places of regulatory governance than ever before.

6.1.2 Voluntary Compliance by Multinational Corporations

"A product must not be made available to the public without disclosure of those dangers that the application of reasonable foresight would reveal. Nor may a manufacturer rely unquestioningly on others to sound the hue and cry concerning a danger in its product. Rather, each manufacturer must bear the burden of showing that its own conduct was proportionate to the scope of its duty" [4].

The concept of voluntary compliance by corporations under laws addressing big-ticket items such as environmental health, workplace safety, products liability, and health care delivery poses a dilemma for policymakers. Voluntary compliance programming requires very careful balancing of practical constraints upon limited enforcement resources in light of the greater public interest expressed in the agency's statutory mission mandating inspection and enforcement of the law. Historically, the general public has viewed voluntary compliance, at best, with distrust and, at worst, as a euphemism for uninspected and unchecked hazards, because of fear that wealthy and powerful multinational corporations will use their influence to turn government regulators into captive agencies who will use the illusion of compliance as an excuse not to inspect and not to enforce laws that prevent harm to the public health [5].

Yet, regulatory models in complex aspects of science and law rely increasingly upon voluntary compliance among regulated entities in order to achieve compliance with the law. Although litigation is a very effective weapon against violations [6], the reality that only a handful of violations can logistically be subject to enforcement means that many important violations are ignored; the odds against getting caught for a violation undermines effective regulation. Therefore tools that encourage corporate voluntary compliance with law are a "necessary evil," needed in order to achieve statutory goals that protect public health.

At the same time, corporate actors complain that government regulators behave in an adversarial posture, using their power of sanctions, fines, and criminal penalties to litigate against violations of law without attempting to solve problems through negotiation and discourse in a manner that can reduce violations with law, promote corporate compliance, and avoid litigation expense [5]. The relationship between government and the corporations that are subject to regulation is triangulated by the role of stakeholder voices from civil society. Voluntary, nonstate, not-for-profit organizations formed by people with shared values have a potentially catalytic ability to promote capacity building and shape policy by "doing global health differently" [7].

Stakeholder organizations are not a monolith: many stakeholder organizations, especially charities, are also employers and producers of goods or services who are also subject to regulation, even though their efforts speak on behalf of a section of the general public. These organizations can include trade associations such as the Nanotechnologies Industries Association (NIA) or organizations representing beverage manufacturing or the construction industry, organizations promoting ideas from specific nations, or professional associations such as the International Commission of Occupational Safety and Health (IOSH) and the International Organization for Standardization (ISO) in Geneva, Switzerland. This triangular conversation is extremely important for any regulatory framework that is applied to nanotechnologies, whether new or used, and retrofitted to novel substances. Implementing any framework about nanotechnology protections of public health therefore requires recognizing stakeholder groups and supporting their interventions as vital partners and is a first step toward building coalitions, introducing novel policy alternatives, offering positive incentives for voluntary compliance by corporations (whether or not for profit), and ultimately transforming data into moral arguments to steer policy.

Creative approaches to this dilemma can produce effective, positive incentives that result in corporate compliance with the law [8]. Positive incentives for compliance with law offer the glimmer of hope for meeting society's need for implementation of existing laws and ethical standards in corporations. With over a half a century of civil and criminal enforcement authority over eight statutes and a multibillion-dollar budget, one of the most influential regulatory players in the traditional toxic substance regulatory framework has been the US Environmental Protection Agency (EPA), founded under an Executive Order from President Nixon in the 1970s. The EPA's 30-year-old voluntary compliance policies, which are the antecedents of many later regulations, trust in the power of self-policing another tool of voluntary compliance with law. The EPA's program for reduced penalties for corporate polluters who can demonstrate that the violation was discovered as part of a self-policing internal audit has the goal "to enhance protection of human health and the environment by encouraging regulated entities to voluntarily discover, and disclose and correct violations of environmental requirements" [9]. Citing reduced costs associated with inspections, prosecutions, and the overall shortage of governmental staff to provide adequate oversight that would be distributed fairly across all the possible manufacturers and end users who might be polluters, the EPA instituted a self-policing policy among the key players in its regulated industries nearly at the end of the 20th century [5]. Under the EPA's long-standing approach, corporate polluters are less likely to face stern discipline, in exchange for increased compliance using voluntary mechanisms, and offenders face reduced penalties if they have compliance programs in place designed to prevent, detect, and reduce harm following an unlawful incident [10]. The EPA's streamlined view emphasizes self-policing [11].

Criteria adapted from existing codes of practice, such as the 1991 USA Criminal Sentencing Guidelines for Corporations, are flexible to accommodate different types and sizes of businesses. The reward for demonstrated due diligence by self-policing, self-disclosure, and prompt self-correction is significantly reduced penalties. Prompt, voluntary written disclosure of violation can prevent the EPA from recommending criminal prosecution, so long as corporate officials do not conceal the violation or willfully ignore it. So long as the corporate polluter acts quickly without delay when offering information about the violation, prompt self-disclosure is rewarded with incentives that can include elimination or substantially reduced civil penalties for the violation or not recommending the case for criminal prosecution. Although it seems unfair to give polluters a reward for admitting that they have caused pollution, it is equally important that accurate information be available for rapid clean-up and that limited resources from the agency be spent on other matters instead of expensive and time-consuming adversarial investigation. Litigation obviously encourages hiding information, and despite the reduced penalties, sanctions under law remain in place. Therefore, from a practical standpoint, society cannot afford to avoid an approach based on voluntary compliance as one more tool for stakeholders to use for enforcement.

Positive Incentives for Voluntary Compliance

An alternative model to the "punishment and deterrence" model for governing organizational activity:

- Creating reasonable and workable alternatives to existing regulatory mechanisms, thereby saving administrative enforcement and monitoring resources that could never enforce every rule or inspect every violation.
- Implementing these strategies by recognizing positive achievements toward compliance, giving greater currency to so-called positive incentives for compliance, and encouraging stakeholder NGOs to serve as watchdogs.
- Note: Under Swiss law, there is an affirmative obligation to prove voluntary compliance with the law. The Swiss approach presumes that companies will come forward with information even if it is self-incriminating, and therefore the penalties for failure to do so are much more severe in the event of nondisclosure when violations are discovered by the government or third parties.

Skeptics correctly express concern that voluntary compliance and self-reporting open the door to abuse by unethical corporate actors, but the practical reality is that such partnerships are needed because of the limited resources available for watchdogs and enforcement of important laws that operationalize precautionary principles into the daily work of public health. This is applauded by nongovernmental trade associations, but many nongovernemtnal watchdog organizations that serve consumers have been skeptical of this voluntary compliance policy because they fear that violators will be given a free ride to flout the laws. The rationale for this approach, however, cannot be denied: given any agency's limited resources, companies must be encouraged to come forward to resolve environmental problems, using an effective compliance management program because of "Discovery of the Violation through an Environmental Audit or Due Diligence" [12].

Also in the United States, the Occupational Safety and Health Administration (OSHA) published in 1989 a set of voluntary management guidelines [9]. During OSHA's second decade, new administrative perspectives in OSHA greeted voluntary compliance and protection programs with energy and enthusiasm and as one of several vital examples of government in partnership with management and labor published in early 1989 [13]. OSHA reflected its progressively increasing recognition of the importance of effective in-house OSHA compliance program standards by recommending safety and health program improvements in conjunction with inspections and by crafting a regulatory agenda that made employer-initiated identification of hazards and self-inspection the centerpiece of its administrative efforts in its Voluntary Protection Programs (VPP). The guidelines were designed to assist employers to establish and maintain programs based on voluntary compliance.

From its inception, the VPP represented the evolution into a new generation of compliance programs. This path has been consistently followed, as stated by Jerry Catanzaro, acting chief of OSHA's Division of Voluntary Programs, in 1991:

> Employers . . . use OSHA's Safety and Health Program Management Guidelines to evaluate their program, to determine where they are strong and where they are weak, and to develop program(s) to maintain the strengths and strengthen the weaknesses. These guidelines developed from OSHA's experience with the VPP where it has been demonstrated that safety and health program management will provide for a safe and healthful workplace in a cost-effective manner [14].

OSHA emphasizes that commitment to occupational safety and health protection must be found in the words of the written policies and the deeds of top level managers [15]. In addition, commitment to these goals must be clearly visible so that employees have the freedom to implement workplace safety and health rules without stigma or reprisal, regardless of whether those rules were devised by in-house counsel as part of a team of industrial hygienists and medical professionals or were promulgated by the agency after a long rule-making process. This path has been consistently followed, as stated by Cathy Oliver, chief of OSHA's Division of Voluntary Programs, in 1997:

> OSHA's Safety and Health Program Management Guidelines are an excellent resource to focus employers and employees on improving workplace safety and health conditions. Performance-oriented and flexible, the guidelines have been successfully applied to a variety of industries and worksites: small and large, union and non-union. The Voluntary Protection Program (VPP) . . . recognizes worksites with excellent comprehensive safety and health programs, demonstrates the impact the guidelines can have educating . . . VPP works. Experience injury and lost workday case rates are 50% of the national average. Many VPP sites also boast production improvements, reduced absenteeism and lower workers' compensation costs—a real competitive advantage [15]!

Although controversy surrounds the program, agency staff asserts that VPP worksites experience injury and lost workday case rates at 50% of the national average production improvements, reduced absenteeism, and lower workers' compensation costs. But, like every OSHA program, the VPP has its critics. According to Franklin Mirer, former director of Occupational Safety and Health for the International Union of Automobile Workers, "VPP is a flawed program because it misdirects OSHA's limited resources, but it also has several very serious policy errors. The fundamental error is that VPP is based on a trade of employee rights to OSHA enforcement by employees (by taking the facility out of the inspection schedule), . . . entry depends upon a low employer-recorded injury rate (an incentive for falsification) and programs require an element of discipline for worker violation of safety rules" [16].

OSHA issued "Revisions to the Voluntary Protection Programs to Provide Safe and Healthful Working Conditions" [17] and requested comments from stakeholders [18] and the general public [19]. Decades later, in 2016, the guidelines remained in place and their possible update was the subject of public hearings in Washington, DC, but without substantive changes [20]. Adjusted for nanotechnologies, these solutions offer more than a "pretty face" [15c] for voluntary compliance and may prove to be an effective tool for the new approach to regulatory governance of risk under an international framework.

6.1.3 The Role of Noncorporate Stakeholders in Society

Communities around the world are flexing local muscles in order to find out more about the emerging risks of nanotechnology. One example is in affluent Cambridge, Massachusetts, home of major universities and consultants, discovering new facets of nanotechnologies. Its recent survey to local nanotechnology industries stated as its purpose, "Cambridge strives to be an exceptional setting for exploration into the frontiers and applications of science and technology. By establishing a transparent, responsible relationship with emerging technology firms to address reasonable public safety and occupational concerns we hope to build on good health and safety practices already in place and to encourage adoption of those practices universally" [21].

Understanding stakeholder roles is important for risk communication, but the state of the art remains poorly developed. Occupational health has a poor track record of cultivating stakeholder interest to muster political will for strong laws and stronger implementation of health protections at work. By contrast to the small public awareness of occupational safety and health programming, global efforts to eradicate diseases such as polio and malaria have captured the imagination of the general public and therefore have garnered major private funding. Political will that reflects stakeholder views is a key component for motivating corporate actors to engage in responsible development of nanotechnology (Fig. 6.1), and Shulte [22] demonstrates how stakeholder views can be obstructed by the complexity and inconsistent flow of data.

Stakeholder views involve having their say via meaningful opportunites to be heard. Stakeholder views also require clear pathways for the information in their views to be used within overarching policy efforts to develop responsible occupational health programs that apply nanotechnology. Stakeholder involvement is multifaceted, and the lines of communication are complex; thus Shulte's model suggests responsible development of nanotechnology is not a spectator sport [22]. Stakeholders, government partners, and specific entites who are the subject of regulation must proactively support the system and engage in exchanges of information in order for workable policies to emerge. Regarding dissemination of information, risk communication to stakeholders does not follow a straight-line path. Instead, the complex pathways for risk communication involve input from stakeholders, as well as bringing to the process of data flow their questions about exposure assessments, risk assessment, and risk management. Thus meaningful engagement of stakeholders goes beyond having their say and an opportunity to be heard, taking the next step to infuse their inputs into the complex loop of risk communication.

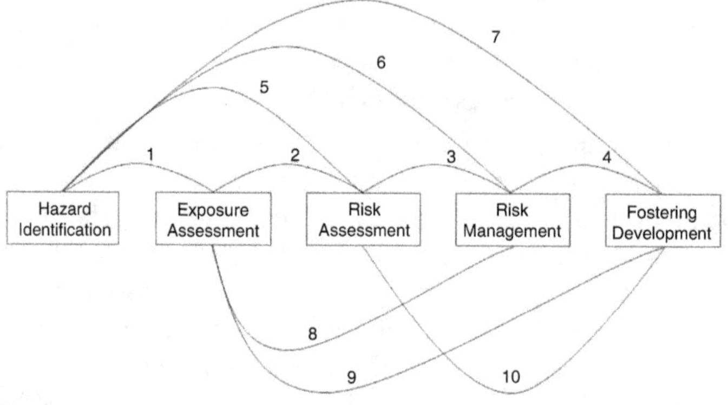

Figure 6.1 Occupational health and safety criteria for responsible development of nanotechnology. Source: US government.

Risk assessment has become a regulatory tool that attempts to quantify risks by relying on toxicology and epidemiological studies about chemicals, radiation, stress, and pathogens [23]. Relying upon toxicological and epidemiological studies that relate specific agents to adverse outcomes [24], risk assessment does not identify a connection between a hazardous agent and an adverse effect but tries to quantify adverse health impacts using measurable indices such as the air contaminant concentration, absorbed dose, or blood chemistry [24]. Inside a legislative cauldron or a policy arena, empirical factors concerning risk assessment based on sound epidemiology and robust toxicity assessments of various exposures will only play a role but will not determine the ultimate policy, which will also reflect an admixture of emotions and philosophical views of the stakeholders involved. Wise stakeholders will ask about publicly about the connection between the final codified legislative product and the role of stakeholder views and demand accountability for major disconnections between the goals and final legislative text.

Whomever the regulators, one must examine whether their authority is legitimate and whether the text makes sense. The abundance of regulation presents an issue that comes into focus slowly as a 21st-century problem, when any group can have a political voice on the web regardless of its underlying interest as a proponent of a philosophy or of the hidden agenda for another powerful group, such as a government or a major corporation. This raises an important political question about the role of the state, the legitimacy of rules and proclamations created by any single government, and which voices shall enjoy priority when designing or implementing rules about nanotechnology.

Figure 6.2 At the public forum in the Museum of the History of Sciences in Geneva, Switzerland, Nicola Furey, vice president of the Earth Focus Foundation, speaks about the role of the public in nanotechnology policy. Photo by Dominique Charoy for the Work Health and Survival Project, 2013.

At the public forum in the Museum of the History of Sciences in Geneva, Switzerland, Nicola Furey, vice president of the Earth Focus Foundation, asked [25], "Who speaks for the people, for workers, for consumers, who are exposed to any number of important toxic and hazardous substances without formal informed consent, in daily life? With the new applications for nanotechnologies, will we be able to rationally limit as well as cascade the dataflow? And how, if so, such that future generations can access knowledge, understand their rights and responsibilities . . . choosing from an array of data rather than simply scaling a mountain of unknown information pasted into an internet void?"

6.2 The Right to Know: A Conceptual Blueprint for Protecting Civilization

The blueprint for operationalizing precautionary principles under law is the so-called right to know about adverse health impacts of exposure in the workplace and in the environment. Access to data from the internet makes possible information dissemination about key, potential adverse health impacts and can give stakeholders access to Safety Data Sheets (SDS) that must travel with a product from manufacturers to suppliers. The right to know is actually a cluster of rights comprising a bundle of information, access to that information, and the ability to apply that data with immunity in order to change bad circumstances into good ones. Philosophers may debate and question whether such rights as a cluster are superior to other civil society rights. But, this philosophical belief is part of the core values that shape the regulatory framework for consumer product labels and the global system chemicals safety that offers information to consumers, customs authorities, downstream manufacturing users, and workers under international law. To support its legal justification for participating in the global system for chemical safety, the US government stated, "It is a longstanding agency position that employees have the right to know and understand the hazards" [26] and that "inadequate communication to employees regarding the hazards of chemicals constitutes a significant risk of harm" [27].

6.2.1 The Right to Know and Hamilton's Work Exploring the Dangerous Trades

The hero in any history about stakeholder empowerment to implement scientific precautionary principles under law is Dr. Alice Hamilton. Civil society owes a huge debt of gratitude to Dr. Hamilton (1869–1970; yes, she lived to be 101 years old!), who articulated these principles and their importance in the practical context of daily working life, in her pathbreaking text, *Exploring the Dangerous Trades* (1943) and in her work for the International Labour Organization (ILO) *Encyclopeadia of Occupational Health and Safety*. Dr. Hamilton was prescient in studying and writing about some of the most important industrial toxins of the twentieth century [28]. For example, benzene [29] was one subject of her concern [30] and the very problem she spotted early in the 20th century became the subject of major US Supreme Court opinions regarding the right of the government to regulate industrial toxins [31], in *IUD v. API* [32], which became the cornerstone of the law of significant risk for limiting exposures to workplace toxins [33]. She also addressed women workers' reproductive health in lead industries, an issue that remains controversial and fraught with policy concerns a century later [34]. In addition to writing the first authoritative work on industrial toxicology [35], through the conduit of the *ILO Encyclopaedia of Occupational Health and Safety* [36], Dr. Hamilton documented trades such as lead smelting [37], general discussions of the negative impacts from excessive exposure to lead in printing [38] and in society at large [39], lead refining, lead pigments [36] and paint [40], and manufacturing [34, 41]. She conducted studies regarding hazards associated with airplane manufacture [42], mercury [43], carbon monoxide [44], rubber [45], and chemicals in munitions industries [46] in the United States [47] and France [48]. She authored over 80 scientific reports on subjects such as lead [49], arsenic, brass, carbon monoxide, cyanides, turpentine [50], and toxins [51].

Often called the "Mother of Industrial Medicine in the United States," it is therefore no surprise that much of her life's work involved unrecognized thankless tasks. Dr. Hamilton lived half a century before she had the right to vote, and had no children. She was invited to hold the first university appointment in industrial medicine at Harvard University, even though she was a woman, and women were forbidden to attend Harvard Medical School. Women attained suffrage in 1919, when Dr. Hamilton was already 50 years old, but lacking the franchise did not stop her from lobbying the president of the United States and European heads of state in Geneva, Switzerland, when they created the Treaty of Versailles ending World War I. How Dr. Hamilton acquired the power and resources to garner their attention and have her say without the right to vote is an important mystery that inspires stakeholders who use information for empowerment [52]. Yet, her work empirically demonstrates a philosophical belief that she had a birthright to speak to world leaders about her opinions regarding life and death, during war and during peace, regardless of her disenfranchised status. Her work exemplifies the concept that she shared the right to have her say and to speak her mind, with people throughout the world, regardless of their level of education or their social status.

Exercising that right, however, required more than common knowledge; it required access to good data and large amounts of information in order to form one's own opinion. The first edition of the *ILO Encyclopaedia of Occupational Health and Safety* was an important part of her legacy to stakeholder empowerment. Published in the United States and Switzerland in 1930 [53], the encyclopedia codified the vision of a small cluster of doctors, lawyers, industrial hygienists, and governmental professionals, who collected the best thinking, reflecting the state of the art of their time, in two volumes—for anyone who could read!

The encyclopedia's first edition was created in a time when telephone communication across continents was difficult; most communications used letters and telegraphs. It was an era when penicillin, antibiotics, and big data based on an epidemiological understanding of disease had not yet been discovered, even though it is commonplace today. It was prepared in an era of widespread illiteracy, indentured servitude, and apartheid and very limited means of communication or now-commonplace medical care. For the first time since Rammazzini wrote his encyclopedia of occupational diseases, 200 years before, a vast gulf of ignorance was conquered by bringing technical information, in addition to writing the first authoritative work on industrial toxicology [35], through the conduit of the *ILO Encyclopaedia of Occupational Health and Safety* [36]. A small cluster of industrial hygienists, doctors, lawyers, international jurists, and policymakers brought together two volumes that were available to every public health practitioner in the world. By creating this resource for workers and their treating physicians, the encyclopedia writers employed a method for disseminating information and risk communication [54]. In so doing, they invented the right to know.

The right to know extends beyond worksites and also translates into meaningful risk communication for consumers, providing the legislative rationale for labels and public disclosure of information and public access to SDS regarding safe handling and storage of millions of toxic and hazardous products in daily use and under ILO C155, which sets forth the structure for national competent authorities to protect occupational health and safety by mandating disclosures about hazards [55]. Few commenters have accurately described the magnitude of Dr. Hamilton's impact on the philosophical underpinnings of the right to know. There were few organized risk prevention systems, and few laws required inspection or measurement of hazards that impacted work, health, and survival [56]. Rather than generalize by trade, Dr. Hamilton teased apart the components of job descriptions and distinguished between various assigned tasks [57]. For example, she noted the different exposures and physical stressors faced within printing trades (type *founders* were distinguished from type *setters* who had direct contact with the letters made of lead). This approach forecast the customized approach in the four points of the "hierarchy of controls" accepted in industrial hygiene practice [58] and modern precision medicine, discussed in nanotechnology literature [59] and in the recommended exposure limit guidelines by the National Institute for Occupational Safety and Health (NIOSH) for controlling workplace exposure involving nanoproducts [60], and also codified in ILO C187 concerning the accepted approaches for occupational health management and in its antecedent, the hazard communication regulations in occupational health [61]. Dr. Hamilton's footprints created the contours for these laws.

6.2.2 Consumer's Right to Know and the GHS

The WHO *Guidelines on Protecting Workers from Potential Risks of Manufactured Nanomaterials* underscore the importance of disclosure, not only as a tool for due diligence, but also as a mechanism for providing relevant clear and accurate information to consumers, because information about nanomaterials is transmitted along with products and they flow through the stream of commerce. By law in a host of countries that have agreed to implement the Globally

Harmonized System for the Classification and Labeling of Chemicals (GHS), transfer of hazardous materials including nanomaterials must be accompanied by SDS that can be read by any person who comes in contact with the materials, regardless of whether he or she is a consumer or a worker. WHO has relied extensively on the GHS as a mechanism for classification, training on safe handling, including training about key hazards, and a mechanism for control of risk and risk management. For this reason, linkages between manufacture, use in commerce, sales, and end users are tied closer together than ever before, blurring the distinction between producers and users. Linkages between workplace exposure and consumer exposure may seem conceptually remote, but the ubiquitous character of international commerce makes everyone a worker, a stakeholder, and a consumer at the same time. Consider, for example, the use of commercially packaged foods in a cafeteria in a university or government office. The university or the government is both an employer and a consumer in that context. It has both the right to know about problems in the storage, transport, packaging, and distribution of those commercial products and the obligation under law to protect the workers and end-user consumers from dangers to public health.

According to the World Health Organization (WHO), "More than 25% of the global burden of disease is linked to environmental factors, including chemicals exposures. For example, about 800 000 children each year are affected by lead exposure, leading to lower intelligence quotients . . . . Worldwide, lead exposure also accounts for 2% of the ischaemic heart disease burden and 3% of the cerebrovascular disease burden. Artisanal gold mining in developing countries remains a significant cause of mercury exposure, while mercury-containing medical instruments such as thermometers and sphygmanometers are a continuing source of exposure in both developed and developing countries. Some 9% of the global disease burden of lung cancer is attributed to occupation and 5% to outdoor air pollution" [62].

Consequently, the artificial distinction drawn in traditional regulatory frameworks that divides workplace exposure from consumer exposure in the community undermines environmental health because it is not practical to treat these groups as separate populations. Although in theory, these are distinct problems, actually any differences between these roles are blurred in real life. In a remarkable step forward to solve this policy dilemma of dual roles as a producer and consumer of data, under the GHS, over 25 UN agencies and regional governments such as the European Union (EU), national governments, and individual trade organizations such as the World Summit on Sustainable Development (WSSD) promote consistency among hazard and risk assessment, risk communication, and dissemination of information about chemical hazards [64].

Collaborating with public stakeholders, international organizations, and countries, the United Nations built one program to avoid the economic burden of confusion about labels and the right to know information in global commerce [65]. The GHS was formally adopted by the new United Nations Committee of Experts on the Transport of Dangerous Goods and the Globally Harmonized System of Classification and Labelling of Chemicals in December 2002. In 2003, the adoption was endorsed by the Economic and Social Council of the United Nations (ECOSOC). Since June 1992, the United Nations Conference on Environment and Development issued a mandate (Chapter 19 of Agenda 21) [66]. A wide array of stakeholders have been brought together to develop feasible international standards for the handling and disclosures surrounding recognized dangerous or hazardous chemicals in use throughout global trade. The rationale behind the GHS provides consistent and predictable data from manufacturers and suppliers to downstream users. Each link in the supply chain must operationalize its right to know by asking producers upstream for high-quality data using SDS and then support the next layer of the right to know by providing training about materials and attaching that information to its products [66].

Although the notion of streamlining regulations has been popular in many nations since the 1980s, the GHS represents a first cut at tackling intenational regulatory confusion, untying the knot across many commercial endeavors worldwide. The GHS brings together a hodgepodge of rules from rival agencies within the United Nations, the EU, the United States, Asia, China, and Canada—each with its own powers and symbols for chemical hazard warning and analysis—in order to avoid the confusion that inevitably follows when jurisdiction overlaps. The combined referent power of these global players means that few actors in commerce can afford to ignore the GHS. For nations that have not ratified the terms of the GHS protocols, the GHS can nonetheless become international customary law and practice [66].

Hazard classification: Provides specific criteria for classification of health hazards and physical hazards, as well as classification of mixtures. The GHS offers pictogram symbols for each classification, and those symbols are employed in commerce worldwide.

Labels: Chemical manufacturers and importers are required to provide a label that includes a harmonized signal word, a pictogram, and a hazard statement for each hazard class and category. Precautionary statements must also be provided.

SDS: These have a specific 16-section format, which lists key components, methods for storage and handling, and warnings about possible acute exposures and long-term adverse health effects and require the use of internationally accepted symbols for hazard classification. This format is used throughout the world. When a material travels in commerce, it keeps this same SDS.

Transmitting SDS to users along the supply chain is required, thereby creating a fail-safe mechanism for achieving disclosure without local government inspections. Because any part of the supply chain that does not receive an SDS can inquire upstream why there is none, many SDS are available on the web free of charge. Among those stakeholders who might be expected to wring their hands and complain about any regulation, the GHS has been surprisingly viewed by many companies as a collective sigh of relief. Enabling producers and end users to exchange information without confusion, the GHS's approach to internationally harmonized hazard communication represents an important step forward in relieving the global burden of disease caused by preventable but harmful exposures to toxic and hazardous substances. The GHS therefore represents an elegant model for a flexible framework protecting world health and promoting the unification of best practices under international law. Nanotechnology and applications of nano-enabled products for consumers might also fit neatly into this pre-existing matrix.

6.3 Hamilton's Legacy: Government Obligations Operationalizing the Right to Know

The notion that information about toxic and hazardous exposures was required for disclosure as a matter of right was novel, perhaps revolutionary, in Hamilton's time. This philosophical belief is the foundation for the right to know. This value is key to shaping the new international governance that followed and remains important in 21st-century laws in every nation [67]. Consistent with the implementation of precautionary principles that operationalize the right to know, OSHA's Hazard Communication Standard (HCS) [68] was first issued in 1983 and covered the manufacturing sector of industry [69]. In 1987, the agency expanded the scope of coverage to all industries where employees are potentially exposed to hazardous chemicals after three years of litigation about the scope and effect of such disclosures. In 1994, OSHA made several minor amendments to the HCS. Modern laws such as the GHS translate that right into useful dissemination of written information, thanks to Hamilton's influence, so that the right to know reaches far beyond workplaces, protecting every stakeholder.

6.3.1 Seoul Declaration on Occupational Safety and Health

"Recalling that the right to a safe and healthy working environment should be recognized as a fundamental human right and that globalization must go hand in hand with preventative measures to ensure the safety and health of all at work." The Seoul Declaration on Occupational Safety and Health recognizes that the positive ramifications of sound safe and healthful work for productivity, economic development, and social development, are a "societal responsibility," benefiting society as a whole. Adopted during the Summit at the XVIII World Congress on Safety and Health at Work in June 2008, high-level representatives from governments, employers and workers, and senior stakeholders from around the world committed unanimously to protect this fundamental human right through the implementation of the declaration. This declaration brought together a variety of experts from the sciences and academia to the World Congress, but outreach to captains of industry was the declaration's primary focus. It was signed by the chief executive officers (CEOs) of major multinational corporations as well as leaders from governments and therefore represents an example of transnational stakeholder involvement to promote voluntary compliance with precautionary principles protecting health.

6.3.2 Right to Know Embracing Nanotechnology

In the case of nanotechnology, internationally codified principles of workplace right to know may prove to be crucial to saving the life of marginal employers and also protecting public health. The argument against application of standards is equally clear: nanotechnologies and synthetic or engineered nanoparticles were never contemplated by the drafters of these instruments, and therefore holding either the states' parties or their social partners accountable to the standard when applying nanotechnology might exceed the scope of the permission granted by the existing law. Although problematic, this is where the "rubber meets the road" for promoting responsible development, fostering innovation, and protecting the human right to health by applying nanotechnology to amazing new products. Given that nanotechnology is fraught with uncertainty, despite its obvious international benefits to humanity, the call from workers and consumers, and the array of stakeholders to be given more information, one question is therefore likely to arise in the near future: whether the infrastructure contemplated by the chemical safety international agreements also applies to nanotechnology. The case favoring the GHS application to nanotechnology remains the compelling need in global commerce for a unified system of identifying, classifying, and transporting materials safely, thereby avoid the economic costs and human costs from accidents when people using the materials are unaware of correct methods for handling or for avoiding hazards.

A reasonably comprehensive list of basic audit tools:

- Customized manuals and checklists
- Practical education and regular communication by various media
- identifying, controlling, and placing warnings on major danger areas

- Legal audits (i.e., identifying and prioritizing applicable laws)
- Controls over contractors and distributors (to ensure they don't violate laws)
- An effective consumer complaint system
- Effective reporting systems by two streams:
  - o Hotlines for prompt rectification of all failures of the system
  - o In-house enforcement, including disciplining of those responsible for breaches and rewarding whistleblowers
- Records, statistics, and informatics (internal reviews)

WARNING! Checklists are a useful servant but a dangerous master. They should be used with discretion, after thinking, updating, and making them custom-tailored to your needs!

References

1. National Science and Technology Council Committee on Technology, Subcommittee on Nanoscale Science, Engineering and Technology, *National Nanotechnology Initiative: The Initiative and Its Implementation Plan*, July 2000, Washington, D. C., Report to the President of the United States.
2. Vladimir Murashov and John Howard, National nanotechnology partnership to protect workers, *Journal of Nanoparticle Research*, **11**:1673–1683, DOI:10.1007/s11051-009-9682-2, 2009.
3. *What Civil Society Can Do to Develop Democracy Presentation to NGO Leaders*, Feb. 10, 2004, stanford.edu/~ldiamond/iraq/Develop_Democracy021002.htm.
4. *Borel v. Fibreboard Paper Products Corporation et al.*, National Surety Corporation, intervenor-appellee, 493 F.2d 1076 (5th Cir. 1973) U.S. court of appeals for the Fifth Circuit, Sept. 10, 1973.
5. Jay A. Sigler and Joseph E. Murphy, *Interactive Corporate Compliance: An Alternative to Regulatory Compulsion*, Quorum Books, Greenwood Press, CT, 1988.
6. Ilise Feitshans, Brenton Saunders, and Joseph E. Murphy, *Corporate Compliance Programs: An Effective Shield Against Civil Penalties*, Bureau of National Affairs (BNA) Healthcare Fraud Reporter, March 1997.
7. Julia Smith, Kent Buse, and Case Gordon, Civil society: the catalyst for ensuring health in the age of sustainable development, *Globalization and Health*, **12**:40, 2016.

8. Ilise Feitshans, Positive incentives for compliance: balancing new tools and their limits, *Preventive Law Report*, **3**(14):10, 1995.
9. Ilise Feitshans, *Designing an Effective OSHA Compliance Program*, Thomson Reuters (available on Westlaw.com), 1990 and 2013.
10. Ilise Feitshans, The corporate compliance extravaganza guidelines, *Corporate Conduct Quarterly*, Rutgers University, Camden, NJ, 1st quarter, 1996.
11. Environmental Protection Agency, Incentives for self-policing: discovery, disclosure, correction and prevention of violations, Part Iii 60 Fr 66706, *Federal Register*, **60**(246), December 22, 1995, final policy statement.
12. Under Section D(1), the violation must have been discovered through either (a) an environmental audit that is systematic, objective, and periodic, as defined in the 1986 audit policy, or (b) a documented, systematic procedure or practice that reflects the regulated entity's due diligence in preventing, detecting, and correcting violations.
13. OSHA Safety and Health Program Management Guidelines, 54 *Federal Register* 3904 to 3916 (1989).
14. Jerry Catanzaro, cited in a letter reprinted in the first edition of Ilise Feitshans, *Designing an Effective OSHA Compliance Program*, Clark Boardman Callahan, 1991.
15. (a) Jerry Catanzaro and Judith Weinberg, Answers to some frequently asked questions on VPP, *Job Safety & Health Q*, **22**(Summer), 1994; (b) Ilise Feitshans and Victoria Bor, eds., *Occupational Safety and Health Law 1995 Supplement*, Bureau of National Affairs (BNA) Washington, D. C., 1995; (c) Ilise Feitshans and Cathy Oliver, More than just a pretty program: OSHA Voluntary Protection Programs improve compliance by facing problems head-on, *Corporate Conduct Quarterly*, Rutgers University, 1997; (d) OSHA Safety and Health Program Management Guidelines, Issuance of Voluntary Guidelines [Docket No. C-02] 54 *Federal Register* 3904 (Jan. 26, 1989).
16. Ilise Feitshans, *Designing an Effective OSHA Compliance Program*, Thomson Reuters (available on Westlaw.com), 2013, citing interviews about the VPP. See also Vladimir Murashov and John Howard, Essential features for proactive risk management, *Nature Nanotechnology*, **L4**, 2009.
17. OSHA, Revisions to the voluntary protection programs to provide safe and healthful working conditions, 65 *Federal Register* 45,649 to 45,663 (July 24, 2000).

18. Public comments submitted to OSHA in response to its October 12, 1999, notice: OSHA received comments from 15 respondents. These included eight VPP participating companies, two professional associations, two trade associations, two private consultants, and the Voluntary Protection Programs Participants' Association. OSHA, Announcement of Voluntary Protection Programs (VPP), 47 *Federal Register* 29,025.
19. OSHA, Voluntary Protection Programs (VPP), draft revisions and requested comments from stakeholders and the general public, Notice 64 *Federal Register*. 55,390, October 12, 1999.
20. (a) Ilise Feitshans stating, "Sound occupational health policies are the grease for the machinery of commerce," comments updating the OSHA Voluntary Protection Plan Guidelines Testimony on behalf of the Work Health and Survival Project, before the US Department of Labor, Washington, DC, March 10, 2016; (b) Ilise Feitshans, *OSHA Management Guidelines: Proposed Revision Response to Request for Public Comments*, testimony before the US Department of Labor, Washington, D. C., March 10, 2016, Work Health and Survival Project, 2016.
21. Cambridge Survey for Nanomaterials is reproduced in Ilise Feitshans, Chapter 9, *Global Health Impacts of Nanotechnology Law*, Pan Stanford, 2018.
22. P. A. Shulte, C. L. Geraci, V. Murashov, E. D. Kuempel, R. D. Zumwalde, V. Castranova, M. D. Hoover, L. Hodson, and K. F. Martinez, Occupational safety and health criteria for responsible development of nanotechnology, *Journal of Nanoparticle Research*, **16**:2153, 2014 (published online Dec. 7, 2013).
23. W. W. Lowrance, *Of Acceptable Risk: Science and the Determination of Safety*, William Kaufman, Los Altos, CA, 1976.
24. US Congress, Office of Technology Assessment, Chapter 3 of *Preventing Illness and Injury in the Workplace*, Washington, D.C., 1985.
25. Nicola Furey, *The Role of Science in Our Society: Can't Our Governments Do MORE?* At the conference "Law and Science of Nanotechnology: Perfect Together? A Public Discourse about the Emerging Issues of Our Times," Saturday, August 17, 2013, available on YouTube and in DVD format. At the public forum in the Museum of the History of Sciences in Geneva, Switzerland, as discussed in Ilise Feitshans, Nanotechnology governance debated along Geneva's lake, *The Monitor*, **13**(2), 2014, American Society of Safety Engineers (ASSE), reprinted in additional ASSE publications.

26. US Department of Labor, Occupational Safety and Health Administration (OSHA), 29 CFR 1910, 1915, and 1926, Hazard Communication Final Rule 77FR17574 at 17600, quoting 58FR29826 (Aug. 8, 1988).
27. Ibid., at 17575.
28. Ilise Feitshans, Alice Hamilton, in *American Women Writers: A Critical Reference Guide from Colonial Times to the Present*, Vol. 1, Lina Mainiero, ed., Langdon Lynne Faust, assoc. ed., Frederik Ungar, New York, 1978.
29. Alice Hamilton, The lessening menace of benzol poisoning in American industry, *Journal of Industrial Hygiene*, **X**:227–233, 1928.
30. Alice Hamilton. The growing menace of benzene (benzol) poisoning in American industry, *JAMA*, **78**(9):627–630, 1922.
31. Ilise Feitshans, Law and regulation of benzene, *Environmental Health Perspectives*, **83**, Aug. 1989, National Institute of Environmental Health Sciences, Research Triangle Park, North Carolina, publishers commenting upon *Industrial Union Department v. American Petroleum Institute (The Benzene Case)*, 448 U.S. 607 (1980), United States Supreme Court.
32. *Industrial Union Department v. American Petroleum Institute (The Benzene Case)*, 448 U.S. 607 (1980), was a case heard before the United States Supreme Court. This case represented a challenge to the OSHA practice of regulating carcinogens by setting the exposure limit "at the lowest technologically feasible level that will not impair the viability of the industries regulated." A plurality on the Court wrote that the authorizing statute did indeed require OSHA to demonstrate a significant risk of harm (albeit not with mathematical certainty) in order to justify setting a particular exposure level.
33. US Department of Labor, Occupational Safety and Health Administration (OSHA), 29 CFR 1910, 1915, and 1926, Hazard Communication Final Rule 77FR17574 at 17574.
34. Alice Hamilton, *Women in the Lead Industries, Bulletin of the US Bureau of Labor Statistics*, Government Print Office, Washington, D.C., *Industry Acid and Hygiene Series*, **253**, 1919.
35. Alice Hamilton, *Industrial Toxicology, Harper Medical Monographs*, Harper and Brothers, New York, 1934.
36. Alice Hamilton, Some new and unfamiliar industrial poisons, *New England Journal of Medicine*, **CCXV**(10):425–426, 1936.
37. Alice Hamilton, Lead poisoning in the smelting and refining of lead, *Bulletin of the US Bureau of Labor Statistics*, Government Print Office, Washington, D.C., *Industry Acid and Hygiene Series*, **141**(4), 1914.

38. (a) Alice Hamilton and C. H. Verrill, Hygiene of the printing trades, *Bulletin of the US Bureau of Labor Statistics*, Government Print Office, Washington, D.C., *Industry Acid and Hygiene Series*, **209**(12), 1917; (b) R. V. Luce and Alice Hamilton, Industrial anilin poisoning in the United States, *Monthly Review of the US Bureau of Labor Statistics*, **2**(6):1–12, 1916; (c) R. V. Luce and Alice Hamilton, Anilin poisoning in the rubber industry, *JAMA*, **66**:1441, 1916; (d) R. W. Holmes, Alice Hamilton, C. Hedger, C. Bacon, and H. M. Stowe, The midwives of Chicago, *JAMA*, **L**(7):1346–1350, 1908; (e) H. M. Thomas and Alice Hamilton, The clinical course and pathological histology of a case of neuroglioma of the brain, *Journal of Experimental Medicine*, **II**(6), 1897.

39. (a) Alice Hamilton, The economic importance of lead poisoning, *Bulletin of the American Academy of Medicine*, **XV**:299–304, 1914; (b) Lead poisoning in the United States, *American Journal of Public Health*, **iv**:477–480, 1914.

40. (a) Alice Hamilton, Lead poisoning in Illinois, *JAMA*, **lvi**:1240–1244, 1911; (b) White lead industry in the United States, with an appendix on the lead oxide industry, *Bulletin of the US Bureau of Labor Statistics*, Washington, D.C., Government Printing Office, *Industry Acid and Hygiene Series*, **95**:189–259, 1911; (c) Industrial lead-poisoning in the light of recent studies, *JAMA*, **lix**:777–782, 1912; (d) Lead poisoning in potteries, tile works, and porcelain enameled sanitary ware factories, *Bulletin of the US Bureau of Labor Statistics*, Washington, D.C., Government Printing Office, *Industry Acid and Hygiene Series*, **104**(1), 1912; (e) Hygiene of the painters' trade, *Bulletin of the US Bureau of Labor Statistics*, Washington, D.C., Government Printing Office, *Industry Acid and Hygiene Series*, **120**(2), 1913; (f) Lead poisoning, *JAMA*, **772**, 1913.

41. Lead exposure and its harmful impact upon pigment workers and battery workers remained controversial throughout the twentieth century. See Jack Levy, Ilise Feitshans, and John Kasdan, on behalf of the Industrial Hygiene Law Project, Brief Amicus Curiae, July 1990, in the Supreme Court of the United States *IUAW v. Johnson Controls* regarding compulsory sterilization of women workers in lead smelting and pigments; in the century following the prescient work of Hamilton, sadly little had changed: Alice Hamilton, The storage battery industry, *Journal of Industrial Hygiene*, **IX**:346–369, 1927.

42. (a) Alice Hamilton, Dope poisoning in the making of airplane wings, *Monthly Review of the Bureau of Labor Statistics*, **V**(4):18–25, 1917; (b) **6**(2):37–64, 1918; (c) Industrial poisoning in aircraft manufacture, *JAMA*, **lxix**:2035–2039, 1917.

43. Alice Hamilton, Mercurialism in quicksilver production in California, *Journal of Industrial Hygiene*, **V**:399–407, 1924.

44. Alice Hamilton, Carbon-monoxide poisoning, *Bulletin of the US Bureau of Labor Statistics*, Government Print Office, Washington, D.C., *Industry Acid and Hygiene Series*, **291**, 1922.

45. Alice Hamilton, Industrial poisons used in the rubber industry, *Bulletin of the US Bureau of Labor Statistics*, Government Print Office, Washington, D.C., *Industry Acid and Hygiene Series*, **179**(7), 1915.

46. Alice Hamilton, Industrial poisons used in the making of explosives, *Monthly Review of the US Bureau of Labor*, Bureau of Labor Statistics, **IV**(2):177–199, 1917.

47. (a) Alice Hamilton, Industrial poisons used or produced in the manufacture of explosives, *Bulletin of the US Bureau of Labor Statistics*, Government Print Office, Washington, D.C., *Industry Acid and Hygiene Series*, **219**(14), 1917; (b) Toxic jaundice in munition workers: a review, *Monthly Review of the US Department of Labor*, Bureau of Labor Statistics, **V**(2):63–75, 1917; (c) Industrial poisons encountered in the manufacture of explosives, *JAMA*, **lxvii**:1445–1451, 1917.

48. (a) Alice Hamilton, Dinitrophenol poisoning in munition works in France, *Monthly Labor Review*, Bureau of Labor Statistics, **7**(3):242–250, 1918; (b) Practical points in prevention of TNT poisoning, *Monthly Labor Review*, Bureau of Labor Statistics, **8**(1):248–277, 1919.

49. Alice Hamilton. Lead poisoning in the manufacture of storage batteries, *Bulletin of the US Bureau of Labor Statistics*, Government Print Office, Washington, D.C., *Industry Acid and Hygiene Series*, **165**(6), 1915.

50. Alice Hamilton, Recent changes in the painters' trade, *Bulletin of the US Department of Labor*, Division of Labor Standards, Government Print Office, Washington, D.C., **7**, 1936.

51. Alice Hamilton, P. Reznikoff, and G. H. Burnham, Tetra-ethyl lead, *JAMA*, **84**:1481-1486, 1925; Treasury Department, US Public Health Service, proceedings of a conference to determine whether or not there is a public health question in the manufacture, distribution, or use of tetraethyl lead gasoline, Government Print Office, Washington, D.C., *Public Health Bulletin*, **158**, 1925.

52. Deborah Salerno and Ilise Feitshans, Alice Hamilton, MD: gaining visibility for industrial medicine, *Journal of Epidemiology and Community Health*, **57**(10):791, 2003, http://www.ncbi.nlm.nih.gov/pmc/articles/PMC1732313/pdf/v057p00791.pdf.

53. *Health at Work: A Basic Human Right Brought to Daily Life by the ILO Encyclopaedia*, March 18, 2009, keynote speech at the "Moving to

Sky" ASSE MEC Conference, Ilise Feitshans, JD and ScM, Coordinatrice Encyclopaedia SAFEWORK, ILO, Geneva, Switzerland.

54. Ilise Feitshans, *One to Five: The Remarkable Moment in Intellectual History That Created the First Edition of the ILO Encyclopaedia of Occupational Health and Safety*, prepared for and presented at the International Commission on Occupational Health (ICOH) 4th International Conference on the History of Occupational and Environmental Health, University of California, San Francisco, California, June 2010. Abstract no. 205 in *At Work in the World, International Conference on the History of Occupational and Environmental Health*, Paul Blanc and Brian Dolan, eds., University of California Medical Humanities Press, San Francisco, 2012.

55. ILO Convention on Occupational Safety and Health No. 155 and ILO Convention on Occupational Health Services No. 161, the Protocol to Convention (2002), No. 155, framework requires recording of accidents and diseases, an inspection system, and training about safe handling of hazardous substances.

56. (a) Alice Hamilton, Prophylaxis of industrial poisoning in the munition industries, *American Journal of Public Health*, **VIII**:125–130, 1918; (b) Causation and prevention of trinitrotoluene (TNT) poisoning, *Monthly Review of the US Department of Labor*, Bureau of Labor Statistics, **VI**(5):237–250, 1918; (c) Industrial poisoning in American anilin dye manufacture, *Monthly Labor Review*, Bureau of Labor Statistics, **III**(2):199–215, 1919; (d) Inorganic poisons, other than lead, in American industries, *Journal of Industrial Hygiene*, 89–102, 1919; (e) Industrial poisoning by compounds of the aromatic series, Journal of Industrial Hygiene, 200–212, 1919; (f) Hygienic control of aniline dye industry in Europe, *Monthly Labor Review*, Bureau of Labor Statistics, **IX**:1–21, 1919; (g) Industrial poisoning in making coal-tar dyes and dye intermediates, *Bulletin of the US Department of Labor*, Bureau of Labor Statistics, Government Print Office, Washington, D.C., *Industry Acid and Hygiene Series*, **280**, 1921; (h) A discussion of the etiology of so-called aniline tumors of the bladder, *Journal of Industrial Hygiene*, 16–28, 1921; (i) Trinitrotoluene as an industrial poison, *Journal of Industrial Hygiene*, **2**:102–116, 1921; (j) The industrial hygiene of fur cutting and felt hat manufacture, *Journal of Industrial Hygiene*, 137–153, 1922; (k) Industrial diseases of fur cutters and hatters, *Journal of Industrial Hygiene*, 219–234, 1922; (l) Scope of problem of industrial hygiene, *Public Health Reports*, **XXXVII**(42):2604–2608, 1922.

57. (a) Alice Hamilton, The prevalence of industrial lead poisoning in the United States, *Medicine*, **IV**(1 and 2, Section XX):218–226, 1925; (b)

Enameled sanitary ware manufacture, *Journal of Industrial Hygiene*, **XI**(5):139–153, 1929; (c) Industrial hygiene, *American Journal of Public Health*, **XXIII**:332–334, 1933; (d) Industrial poisons, *New England Journal of Medicine*; **CCIX**(6):279, 1933; (e) Occupational poisoning in viscose rayon industry, *Bulletin of the Department of Labor*, Division of Labor Standards, Government Print Office, Washington, D.C., **34**, 1940.

58. International Labour Organization (ILO), *My Life, My Work, My Safe Work: Managing Risk in the Work Environment*, ILO.org/Safework/Safeday, supported by the International Social Security Association, World Day for Safety and Health at Work, April 28, 2008. The hierarchy of engineering controls, recognized internationally as the cornerstone of best practices in industrial hygiene, has four basic components According to the ILO, four key steps to reduce risks include but are not limited to (1) eliminate or minimize risks at their source, (2) reduce risks through engineering controls or other physical safeguards, (3) provide safe working procedures to reduce risks further, and (4) provide, wear, and maintain personal protective equipment. The hierarchy of controls was also discussed in Jack Levy, Ilise Feitshans, and John Kasdan, Brief Amicus Curiae, July 1990, in the Supreme Court of the United States *IUAW v. Johnson Controls* on brief for behalf of the Industrial Hygiene Law Project regarding compulsory sterilization of women workers in lead smelting and pigments.

59. Cassandra D. Engeman, Lynn Baumgartnert, Benjamin M. Carr, Allison M. Fish, John D. Meyerhofer, Terre A. Satterfield, Patricia A. Holden, and Barbara Herr Harthonr, The hierarchy of environmental health and safety practices in the US nanotechnology workplace, *Journal of Occupational and Environmental Hygiene*, 487–495, 2013.

60. Department of Health and Human Services (NIOSH), *Current Strategies for Engineering Controls in Nanomaterial Production and Downstream Handling Processes*, Publication No. 2014-102, cdc.gov/niosh/docs/2014-102/.

61. Ilise Feitshans, Hazardous substances in the workplace: how much does the employee have the "right to know"?, *Detroit College of Law Review*, **1985**(3).

62. Sixty-Second World Health Assembly A62/19 provisional agenda item 12.14, April 23, 2009, Strategic Approach to International Chemicals Management.

63. Report by the Secretariat Importance of Sound Management of Chemicals for the Protection of Human Health.

64. Plan of Implementation of the World Summit on Sustainable Development. Document A/Conf.199/20, Annex. 3. International

Labour Conference, Ninety-Fifth Session, Geneva 2006. Provisional Record 20A.; WHA60.26; Rio Declaration of Principles.

65. WHO, International Programme on Chemical Safety. WHO/IPCS meeting on strengthening global collaboration in chemical risk assessment in conjunction with the 9th meeting of the Harmonization Steering Committee, www.who.int/ipcs.

66. GHS, adopted June 27, 2007, unece.org/trans/danger/publi/ghs/ghs_rev00/00files_e.html. For updated information, see https://unitar.org/cwm/portfolio-projects/globally-harmonized-system-classification-and-labelling-chemicals. The GHS has been developed to enhance and standardize this protection across the globe. Also see Training for Capacity Building on the Globally Harmonized System of Classification and Labelling of Chemicals (GHS) Training Materials Package at https://www.unitar.org/cwm/ghs/AseanProject/GHS2011Training/TrainingMaterialsPackage.

67. Ilise Feitshans, *Designing an Effective OSHA Compliance Program*, commenting on 29 USC 651 OSH Act, regulations in 29 CFR 1910.1200 and the US Supreme Court case, *Whirlpool v. Marshall, Whirlpool Corp. v. Marshall* No. 78-1870, argued January 9, 1980, decided February 26, 1980, 445 U.S. 1 Section 11(c)(1) of the Occupational Safety and Health Act of 1970 prohibits an employer from discharging or discriminating against any employee who exercises "any right afforded by" the act, including the right to refuse hazardous work in cases of imminent danger.

68. 29 CFR 1910.1200, 1915.1200, 1917.28, 1918.90, and 1926.59 but was subject to litigation for the first three years. The rule has been fully enforced in all industries covered by OSHA since March 17, 1989 (54FR6886, Feb. 15, 1989).

69. Hazard communication standard, notice proposed rulemaking, 48FR53280, Nov. 25, 1983, revised http://www.dol.gov/compliance/laws/comp-osha.htm, 2010.

Chapter 7

Risk Governance for Regulatory Management of Nanotechnology

7.1 International Efforts to Harness Nanomaterials under Law

The notion that no one will regulate nanotechnology is a concept from the past. Popular demand for some type of nanotechnology regulation has caused a rethinking of the role in society played by regulatory governance of risk and risk management programs. New issues range from the questions of standardization using robust science to the policy opportunity for consciously removing embedded sexism and racism from regulatory frameworks. Despite a wide variety of opinions about the type of risk and methods for prevention, experts agree that precautionary approaches are necessary. The new paradigm for labor law in the 21st century must reflect that reality. Bringing together public health principles and existing workplace standards will require education awareness and outreach to nontraditional groups, such as professional associations, nongovernmental organizations (NGOs), and research institutes.

Nanotechnology's revolution for commerce has resulted in an international call for partnerships between governments and the producers of nanotechnology. In response, there have been rapid developments toward the codification of nanotechnology regulations

Global Health Impacts of Nanotechnology Law
Ilise L. Feitshans

ISBN 978-981-4774-84-0 (Hardcover), 978-1-351-13447-7 (eBook)
www.panstanford.com

on several levels throughout the world: governments within their own nations or working with several governments, under the auspices of international organizations, and trade associations and NGOs also preparing text for use as guidelines, model acts, and laws preventing not only liability but also a variety of additional legal concerns. Every nation and several regional organizations offer a legislated nanotechnology program whose underlying enabling legislation authorizes appropriations, following a strategic agenda. These developments will call into question concepts of statehood and governance, asking whether these concepts are antiquated or, instead, are still so vibrant that they can apply to the rules that will govern nanotechnology. It remains unclear whether national governments or the superstructure of international governmental organizations in the United Nations (UN) will be the correct place to address emerging laws that will govern nanotechnology.

The 21st century's emerging style of governance has already demonstrated that a unique admixture of public power, combined with a strong infusion of procedures, data, and policies from corporate and private resources, has profound implications for future models of governance. Throughout the world, people are questioning whether democracy remains a vital form of governance or merely an antiquated path for those whose societies are rooted in the heritage for ancient Greece [1] antecedent to the Roman Empire. Meanwhile, new forms of governance, including royal-based governance with democratic components, are taking hold across the world. The question which if any of these styles will be the mode of fashion or the enduring model? creates jobs for political theorists [2]. This is surprisingly relevant for the creation of a regulatory regime that will foster the development of nanotechnology, while protecting public health, because the selection of the forum to create that regime may be as important as text of the law itself. In the 20th century, dynamic forces accomplished sweeping social changes that brought governmental funding of programs involving colossal unquantified risks, such as the risk of civil war that followed the creation of nationalized civil rights under law; mass social protection programs offering the middle class pensions, health care, and aid to people living in poverty; and the risks associated with funding "big science," including extravagant weapons. In the realm of science, dynamic social forces precipitated the development of programs designed to create, detonate, manufacture, store, decommission, and dismantle nuclear weapons—an expensive endeavor that consumed three lifetimes of research, from the 1930s to the present, including the Cold War. The thriving status of both the military–industrial complex and the world's civilian population testifies to the ability to balance profits and risks in a manner that promotes commerce and protects public health in civil society.

In the 21st century, several innovative approaches have been taken to introduce the information and concerns of end users, who are corporations, workers, multinational enterprises, multinational governments, national governments and federations, and consumers, into the discourse about nanotechnology's potential risks and how to address those risks under law. NanoImpactNet of the European Union (EU), for example, aspired to create "responsible development of manufactured nanoparticles" and to support the definition of regulatory measures and implementation of appropriate legislation in Europe by developing "strong two-way communication" to stakeholders and the EU [3]. These developments and the growing trend toward multinational treaties and agreements governing applied nanotechnology represent an end to traditional control by governments in policymaking and implementation. This new era in governance started small, with attention to developments for terminology and regulations but foretells the possibility of a transnational approach to regulation of business and global commerce.

A unique feature of emerging nanotechnology regulatory frameworks is their transnational collaboration, both within the EU and inside privately funded NGOs such as the International Organization for Standardization (ISO). Achieving the goal of including stakeholder views in nanotechnology risk governance, various legislative initiatives target key problems and then evaluate the inadequacies and gaps in compliance with a goal of demonstrating how performance can be improved through positive incentives for compliance. Model legislation features awards for excellence, preferential treatment favoring program participants, reduced sentencing, special risk pools in insurance, and protection of information generated by internal audits through the use of the self-evaluative privilege. Legislation enabling corporations and end users of nanomaterials to enjoy self-evaluative privilege gives discretion to protect special information concerning intellectual property and trade secrets, while making good-faith efforts to obey the law. As such, self-evaluative privilege may be useful, especially as it might apply to the work of third-party auditors such as awards commissioners or accredited bodies under international customary agreements. More traditional tools for encouraging compliance also exist, with mixed results depending upon the prestige available in its proper context, for example, activity regarding recognition of compliance activities through the use of an award for excellence also empowers corporate compliance practitioners, such as technical experts in auditing, safety engineering, design, and industrial hygiene, and provides a place to discuss conditions on-site with impunity for people who might otherwise be unlikely to provide information [4]. Well-planned legislation that addresses risks, while balancing the development of commerce, is the hallmark of 20th-century scientific precedents; emerging laws for nanotechnology can ensure that every stakeholder (manufacturer, researcher, consumer, or worker) will have the very best information at their fingertips when making decisions about nanotechnology use and its implications for the future.

7.1.1 The Council of Europe

With 47 member nations, embracing 800 million people, the Council of Europe (CoE) (Fig. 7.1) views its role as the health and human rights vanguard for law governing the right to health and consumer protection throughout Europe. Its human rights court has remained a leading model for jurisprudence throughout the world. In 2012, the CoE Parliamentary Assembly began the first steps toward nanotechnology regulation that would require its member nations to apply safeguards to consumer health and the environment, consistent with its mandate to follow scientific precautionary principles. Pan-European and international reciprocal agreements were offered to the Parliamentary Assembly in the report *Nanotechnology: Balancing Benefits and Risks to Public Health* [5]. The report outlines potential avenues for CoE legislative activity around nanotechnology in commerce such as a study commission, standards, or a binding legal instrument. Rejecting previous calls for moratoria, the CoE would create a multinational regulatory structure that incorporates scientific precautionary principles into real-world technological processes without sacrificing research and innovation, while "protecting the human rights to health and to a healthy environment of everyone."

Figure 7.1 Inside (top) and outside (bottom) the headquarters of the Council of Europe (CoE). Source: Council of Europe website (www.coe.int).

The CoE will use the report as a resource for determining which path it will follow regarding informed consent for consumers who use products of nanotechnology applications and workers exposed to nanomaterials from production to downstream end users and for harmonizing regulatory frameworks, including risk assessment methods, risk management protecting researchers and workers, and labeling. According to CoE procedures, the final product must be negotiated in an open and transparent process, involving multiple stakeholders (national governments, international organizations, the Parliamentary Assembly, civil society, including registered NGOs, various experts, and scientists). The scope of its proposed text would apply across borders, across the origins of nanomaterials (synthetic, natural, accidental, manufactured, engineered), and across the functional uses and biological fate of the nanomaterials, and although focused on pan-European issues, stakeholders could be engaged in a worldwide dialogue.

Debated and accepted by the Parliamentary Assembly in Strasbourg, France, in April 2013, the report outlines essential European legal concepts for laws concerning nanotechnology safety and the regulation of nanotechnology in commerce, while implementing precautionary principles within the rubric of international law. The mandate from the report also includes bioethical questions surrounding informed consent of consumers who use products of nanotechnology applications, especially nanomedicine. Jurisdiction for the report is rooted in the *Final Declaration of the Third Summit* of the CoE "to ensure security for our citizens in the full respect of human rights and fundamental freedoms" and to meet, in this context, "the challenges attendant on scientific and technical progress" in 2005. Projected deliverables from a study commission could first take the form of a Committee of Ministers Recommendation and later be transformed into a binding legal instrument if the majority of member states so wished.

The CoE's Oviedo Convention regarding bioethical standards to apply new scientific research mandates "a culture of precaution incorporating the precautionary principle into scientific research processes, with due regard for freedom of research and innovation" [6]. Calling for harmonization in the area of nanotechnology regulation and its implementation, CoE staff said, "Unfortunately there seems to be far too little communication between the different stakeholders and even less effort at harmonisation. The result is a cacophony, and a current legislative environment which either tends to stifle innovation or is short on risk control. . . . The resultant lack of application of the precautionary principle may yet prove to be dangerously short-sighted, as it has the potential to cause enormous harm both to human health and to the environment. It is thus clear that there is a need to create a common standard to properly balance potential risks and benefits of nanotechnology by harmonising the legislative base with the precautionary principle in mind. What is also needed is not just communication and debate between the different stakeholders in achieving this goal, but also an informed public debate, including consumer information and education (including labelling requirements where appropriate). The Council of Europe, with its unique human rights mandate and Convention-based standard-setting, may be able to achieve such transparent harmonisation where other stakeholders with a narrower mandate have, so far, failed" [7].

7.1.2 Convention on the Prevention of Counterfeit Medicines and Medical Devices ("Medicrime") and Convention on Biomedicine and Human Rights

Nanotechnology is involved both in the cause and the prevention of medicrime because counterfeit medical products falsely represent the identity, source, active agent, labeling or packaging, or accompanying documents of medical products. The ability to copy documents and active ingredients threatens the integrity of nanomedicines and traditional medicine alike. For example, 3D printing and other nano-enabled strategies make it easy to copy not only the active ingredients but also the actual labels and certification embedded into patented medical products and devices. At the same time, nanomedicine will create new devices and medical discoveries that must be protected against counterfeit versions that lack product liability assurances for the general public and regulatory oversight for quality and distribution. "Even more profitable than drug trafficking, this new form of crime has an undeniable advantage for criminals: they go largely unpunished or receive only mild sanctions," wrote Gabriella Battaini-Dragoni, deputy secretary general of the CoE [8] in the handbook shown in Fig. 7.2.

The CoE's medicrime convention is the first international agreement to define acts of counterfeiting medicine as a crime under international law. The convention prohibits creating, assembling, or distributing counterfeit medicines as criminal acts [9]. In this context, the term "counterfeit" corresponds to "falsified" or "fraudulent" but does not determine intellectual property rights, which is beyond the scope of the medicrime convention. Even if illegal products contain the same ingredients as the original product, some batches may have an excessive or insufficient dose of active ingredients; quality is not guaranteed. Inappropriate storage conditions may lead to deterioration of the product. As a model for international cooperation to promote consistent standards that protect public health, the convention fosters pharmavigilance partnerships by setting forth principles for harmonizing laws of different nations in order to stop the manufacture, transport, and distribution of counterfeit medical products marketed globally. Since no single nation can attack medicrime alone, large industrialized

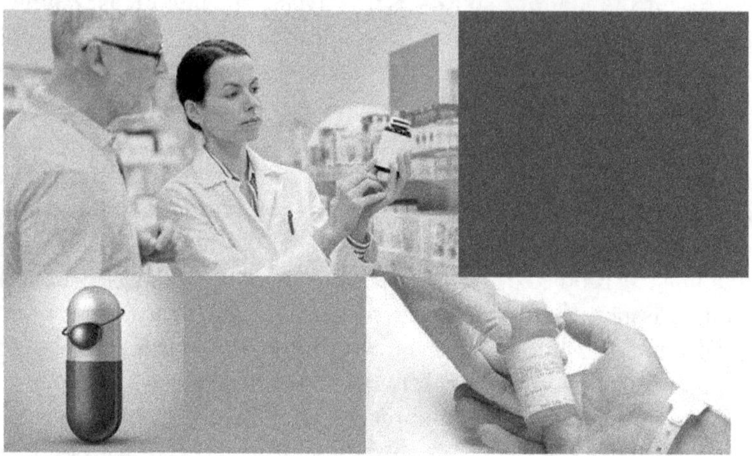

Figure 7.2 Cover of *Handbook for Parliamentarians: The Council of Europe Convention on the Counterfeiting of Medical Products and Similar Crimes Involving Threats to Public Health*. The handbook, published by the Secretariat of the Committee on Social Affairs, Health and Sustainable Development of the Parliamentary Assembly, the European Directorate for the Quality of Medicines and Health Care, and the Directorate General of Legal Affairs and Human Rights of the Council of Europe, is available in English, Spanish, Russian, and French free of charge (www coe.int). Source: Council of Europe.

nations such as France and small developing nations agree on the importance of ratification of this convention. Burkina Faso has deposited its instrument of ratification to the medicrime convention, which entered into force in January 2016. Signatory states may use expertise coordinated by the European Directorate for the Quality of

Medicines and Healthcare (EDQM) for follow-up, targeting a problem that threatens Africa as well as North America, Asia, and Europe. "The Convention against Medicrime presents the rare opportunity for corporate pharmaceuticals, scientific researchers international law enforcement and human rights activists to work together to defend civil society against organized crime," stated Mme. Claude Chirac of the Fondation Chirac, Paris, France. "Fake medicines not only hurt the unsuspecting patients who are victims of fake medicine and undermines public confidence in the integrity of public health delivery systems, but medicrimes also make profits that fuel efforts against governments and civil society by funding terrorism" [10].

The treaty envisions multilateral collaboration across nations and with experts from international governments such as Interpol, Europol, the World Customs Organization (WCO), and the World Health Organization (WHO) in order to effectively detect fake medicines before they enter regular streams of commerce, as seen in Fig. 7.3. The convention provides criminal law penalties and victim protection measures based on cooperation at national and international levels and establishes a monitoring body to oversee international cross-sectoral collaboration between public health professionals, law enforcement, and the judiciary.

Figure 7.3 Experts from around the world who met to attack medicrime. Source: Council of Europe.

Similarly, only the CoE has attempted to codify biomedical principles into international law. The CoE's Oviedo Convention on Human Rights and Biomedicine, created in 1997, does not specifically mention nanotechnology in its text but may be both implemented and undermined using nano-enabled products, depending on the context. Nanotechnology plays a major role in this sweeping social change, bringing new approaches to physical properties of matter, new uses of human stem cells from oneself, and ultimately new definitions of key legal terms. When the convention was drafted and debated, scientists were still mapping the human genome and bioethicists were worried about what society would do with information revealed at the chromosomal level. Article 13 attempted to address this problem and arrest development of eugenic changes to the human species or the creation of a new subhuman species by stating, "An intervention seeking to modify the human genome may only be undertaken for preventive, diagnostic or therapeutic purposes and only if its aim is not to introduce any modification in the genome of any descendants." Since it can be argued that any changes are ultimately designed to influence destiny, civil society faces the multibillion dollar question, "Are these changes okay or in violation of the convention or not covered by it at all?"

Scientific realities, not envisioned at the time of its writing, therefore turn the convention's vision of bioethics upside down: frozen eggs and remarkably accurate in vitro harvesting impact numerous individual choices and decisions about human reproduction that viewed en masse will alter human populations as a whole. Nanotechnology offers new tools for genetic testing and gene therapies. Manipulating particles at the nanoscale is smaller than DNA. Therefore nanotechnology and nano-enabled devices can be applied to genetic engineering, gene therapies, and a variety of new approaches to reproduction in humans and other species. Nanotechnology can be used to alter nanoparticles within DNA and RNA, which is a subset of a subset at the chromosomal level, thereby enabling geneticists to alter or create new proteins by manipulating DNA or the protein corona that surrounds a cell. Regarding ongoing genetic testing, gene editing, and the development of replacement organs from an individual's own stem cells, or other aspects of personalized medicine, the natural question is whether application of nanotechnologies is juridically viewed as "an introduction of modifications in the genome." In the alternative, it could be argued that such developments are excluded from the convention, because there is no mention of nanotechnologies in the text itself. This raises the possibility that thanks to scientific progress, researchers, commercial entities, and prospective customers who wish to be parents can circumvent the constraints of the convention.

A classic question of legislative drafting appears at the edge of the convention's jurisdiction: whether plain meaning of the text limits the implementation of purposes and principles, unless a reasonable interpretation can be applied to embrace nanotechnologies, or whether amendment is needed. When discussing genetic manipulation of the human genome or related species, some people might have the mistaken impression that nanotechnology and genetics are distinct disciplines, but nanotechnology actually offers a new tool for addressing the convention's concerns. The impact of nanotechnologies on the scope and vibrant application of the convention will depend very much upon *how* nanotechnologies are applied in the genetics context. And if they are okay under the convention's plain meaning but violate its principles, is it the text or the principles that must change? Or if they are not covered by plain meaning and violate its principles, what shall we do? Amend? Ignore? Write a different convention?

Cancer drug delivery presents a deceptively easy example. There is likely to be wide popular consensus that society agrees to alter the genetics of cancer in order to kill cancer cells. Some people, including bioethicists, may not even care if exterminating cancer cells also has a long-term genetic impact on a terminal patient who enjoys an extended survival curve because that comparatively small damage means he or she still has years more to live. A person who is not expected to reproduce cannot, by definition, face an impact on the human genome, even if there would be alteration or damage to his or her individual genetic material. Thus, altering the genome of cancer cells is a desirable outcome, even though it might, at first reading, run afoul of the terms of the Oviedo Convention. Another context in which embracing nanotechnology would advance the convention's purposes involves those instances when nanomedicine and nano-enabled medical devices are used for developing synthetic organs for use in humans, grown from the patients' own stem cells, despite the convention's prohibition on tinkering with individual genetics. Such applications of nanotechnology can be very effective in implementing some of the biomedical concerns addressed by the CoE's convention. For example, synthetic organs may reduce trafficking in human organs once the price of grow-your-own organs becomes commercially marketable. It is also not difficult to imagine that people will be able to use their own stem cells to create eggs or sperm, and thus have only one parent for a newborn child who will be completely of their choosing if not design, with their genetic imprint plus a few alterations via gene therapies or fetal surgery in an artificial womb. Thus, interpreting nanotechnology as acceptable under the terms of the convention has many facets to explore. Society may like this interpretation to embrace nanotechnology when it offers targeted drug delivery assaulting a cancer cell or ends trafficking of human organs.

But perhaps some other types of manipulation run afoul of the convention? Since no legislature has authorized a list of forbidden genes, and no popular vote has been taken to determine which diseases and genetic conditions are by majority of political will forbidden or undesirable, avoiding discussion of whether the biomedicine convention applies to nanotechnologies that did not exist at the time of its writing runs the risk of defeating the purposes of the convention. Thanks to scientific progress, individual researchers, potential parents, or government authorities imposing the conditions for an acceptable fetus may be able to circumvent the constraints of the provision. At present, new nano-enabled technologies such as "three parent" manipulation of genetic material in the maternal mitochondria fly in the face of its provisions that ban "and only if its aim is not to introduce any modification in the genome of any descendants." This unexpected outcome undermines the purposes and intention of the convention, leaving it equally vulnerable to new questions not clearly covered by its text and without criteria to differentiate between acceptable genetic alteration in contrast to genetic impacts that are unacceptable. Or impacts that constitute a major trade-off for someone who is very sick but not considered terminal and therefore might live struggling with problems such as Parkinson's or Alzheimer's or broken bones without nanomedicine?

Another blank spot in the blanket interpretation of the convention is equally disconcerting: What about human volunteers in nanotechnology experiments (which exist in some European countries)? Are such experiments exempted because the convention does not discuss nanotechnology? These open questions are so numerous and important that it is reasonable to ask whether the CoE should amend or modify the CoE's Convention on Biomedicine and Human rights, but in reality integrating expert information about state-of-the-art nanotechnology would avert such an expensive and destructive path. A working group could be empowered to examine these issues with a view to creating interpretive guidance documents. There is a credible argument to support an opinion that application of nanotechnologies can be considered "an introduction of modifications in the genome," but there is no text referring to nanotechnologies that offers jurisdiction over nanotechnologies in order to scrutinize them. These developments are, however, the first of many more in a sea change of nano-enabled alterations of evolution, suggesting additional future challenges to sustaining the impact of those achievements. If experts anticipate these issues then civil society can effectively address them without writing new law.

7.1.3 International Collaboration for Nanosafety Research: A Global Issue

The Nanosafety Cluster Report 2013, Strategic Research Agenda, considers International collaboration as a fruitful platform for having a larger impact and obtaining benefits in research as well as in aspects related to governance and safety issues of nanotechnology. The globalization of research is proceeding rapidly, and this is having significant implications for the European nanosafety research landscape. Large projects involving a set of demanding multidisciplinary, hypothesis-driven research endeavors require international collaboration because in most cases the required expertise or resources may not be available in any one single country. Furthermore, groundbreaking innovations often take place in the interface or cross-roads of different scientific disciplines and research environments, which extend beyond national borders. Instead, production of new scientific knowledge is shifting from national to international arenas. International partnerships create unique opportunities for enhancing scientific excellence, physical and intellectual research environments, and innovative training.

The European Commission's "Nanosciences and Nanotechnologies Action Plan for Europe 2005–2009" called attention to issues such as nomenclature, metrology, common approaches to risk assessment, and the establishment of a dedicated database to share toxicological and ecotoxicological as well as epidemiological data by setting forth the priorities illustrated in a chart in Fig. 7.4. Progress has been achieved in many respects to identify the areas requiring joint efforts, but EU or global-level coordination is far from achieving the goals of adopting international standards, despite important steps in that direction. To encourage accomplishing these goals, the EU 7th Framework established several research initiatives across borders. On the basis of the working assumption that toxicity data is continuously becoming available, the relevance to regulators is often unclear or unproven, but "the shrinking time to market . . . drives the need for urgent action by regulators" [11]. Interdisciplinary programs are committed to:

- Provide answers and solutions from existing data, complemented with new knowledge
- Provide a tool box of relevant instruments for risk assessment, characterization, toxicity testing, and exposure measurements
- Develop, for the long term, new testing strategies adapted to innovation requirements
- Establish a close collaboration among authorities, industry, and science, leading to efficient and practically applicable risk management approaches for nanomaterials and products

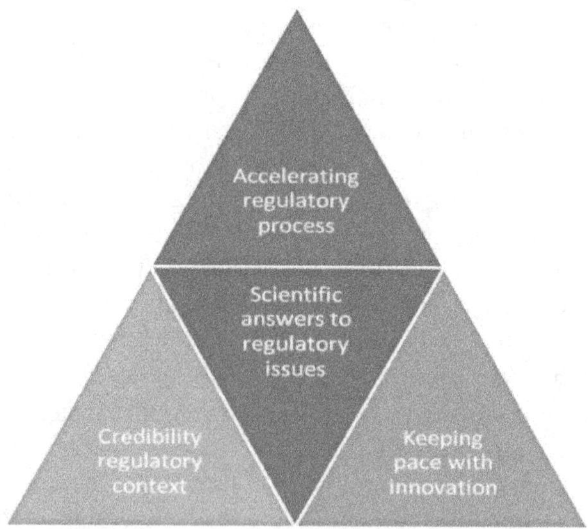

Figure 7.4 NanoReg official priority chart from the European Commission. *Source*: NanoReg.

The European Commission, equally keen that nanotechnology applies responsible development and mindful of the safety needs for innovation and development, stated, "Ensuring the safe development of nanotechnologies, through a sound understanding of their potential impact on health or the environment, and through the development of tools for risk assessment and risk management is a key factor to fully harvest the benefits of their development" [12] and therefore created the Nanosafety Cluster [13]. The main objectives of the NanoSafety Cluster are:

- To facilitate the formation of a consensus on nanotoxicology in Europe
- To provide a single voice for discussions with external bodies
- To avoid duplicating work and improve efficiency
- To improve the coherence of nanotoxicology studies and harmonize methods
- To provide a forum for discussion, problem solving, and R&D activity planning in Europe
- To provide industrial stakeholders and the general public with appropriate knowledge on the risks of nanoparticles and nanomaterials for human health and the environment

For example, MARINA developed specific reference methods for all the main steps in managing the potential risk of engineered nanomaterials (ENMs). It addressed four central themes in the risk management paradigm for ENMs: *materials, exposure, hazard*, and *risk*. The methods developed by MARINA are (i) based on beyond-state-of-the-art understanding of the properties, interaction, and fate of ENMs in relation to human health and the quality of the environment and are either (ii) newly developed or adapted from existing ones as reference methods for managing the risk of ENMs [14]. MARINA developed a strategy for risk management, including monitoring systems and measures for minimizing massive exposure via explosion or environmental spillage. Networking and sharing of information have become a fundamental task for the Nanosafety Cluster. According to the editors of the first issue of their newsletter, "Research outcomes must actively be communicated. Scientists are well aware of this and regularly publish in scientific journals: scientists talking to scientists. While this is essential, it is not enough. Research must also be communicated to the outside world: to innovators, to the public, and to decision makers" [15].

7.1.4 European Union–U.S. Partnership

The Community of Researchers (CoR) for the United States–EU working group has tackled many preliminary obstacles concerning methodology within the sciences and read across among science disciplines. To achieve this important goal, the Management of Risk CoR has begun key dialogue across disciplines. As discussed in the presentation "The Need for International Harmonization of Nantechnology Law" at the National Science Foundation USA EU CoR 2013, "Bridging NanoEHS Research Efforts," in Washington, DC, prepared for the CoR working group on control and management of risks, there is an unprecedented need for global harmonization of law [16]. Each partner in this collaboration has a mature infrastructure that can generate new rules and enforce compliance with law and make use of the program for the Globally Harmonized System of Classification and Labeling of Chemicals (GHS), as is strongly recommended by the 2017 WHO guidelines regarding workplace exposure to manufactured nanomaterials (MNMs) [17].

Amid the confusion between existing laws, draft projects, and future narrow texts to address specific developments in medicine, food, or intellectual property, there is no shortage of law but no clear analytical path. Recognizing that many nations and international entities are working toward creating the same thing, a regime for nanotechnology regulation, the first steps have been made to bring together rival activities, "Bridging NanoEHS Research Efforts" in the United States and the EU, with a view toward multilateral standardization. The CoR also met at the US National Science Foundation in 2015 to brainstorm in a "scrimmage" session, where researchers from different disciplines evaluated selected problems in the same roundtable session. Realizing a coherent regulatory governance framework that will be at once realistic and flexible to nanotechnology will require the admixture of multidisciplinary parameters for examining key policy issues. Using new tools for big dataset analysis, computer modeling of representative scenarios, and gap analysis, adaptable rules will enable new information and consensus standards to be folded together and placed into one decisional cauldron.

The goal is to extend these bridges in order to translate good science, as understood and accepted by respected scientists, into actual law that will influence policy decisions among decision makers in academia and commerce. The strategic roadmap planned by the working group addresses the new disciplines of nanoinformatics, which brings together, organizes, and analyzes disparate big datasets. The nanoinformatics framework is intended to be a resource where researchers who generate data can use a repository for their raw datasets and where secondary researchers can enjoy cost savings by refining the data and cleaning it, rather than collecting new data, by combining information gathered from researchers in nanoscience, biomedical science, environmental engineering, and environmental science.

Since data has already been collected, the project plans to avoid the expense of new data collection by sharing information. The ISA-TAB-Nano file-sharing format was developed and used by the EU NanoReg project, and incorporated into the eNanoMapper logic, to create ISA-TAB-Nano-Expanded Data Submission Templates. "The templates are being designed to encompass the needs for research data nanoinformatics organization of as many scientists as possible. Thus, stakeholders' comments about template functionality and completeness as well as suggestions for their improvement is crucial," said Dr. Christine Hendren, Center for the Environmental Implications of NanoTechnology (CEINT), Duke University [18].

The EU has NanoReg, NanoReg2, Nanocluster, and Nanosafety programs. Each with a responsibility for different aspects of the magic of nanotechnology applications has the responsibility for engaging stakeholders, especially academic and private sector partners, by focusing on specific points of contact in different work packages. Representatives from these programs also contribute to the transatlantic exchanges for CoR. The CoR efforts underscore the crowded legal landscape where new laws are emerging with increasing need for the benefits of predictability and consistency that are derived from harmonization of laws and the reality that eventually harmonization will be inevitable to protect commerce from expensive conflicts of laws and inconsistent standards, which can pose obstacles to the free flow of nano-enabled products, even among trading partners.

7.1.5 ERA-NET SIINN Harmonizing European Standards Regarding Nanotechnology Precautionary Principles and NanoReg2

The European ERA-NET "Safe Implementation of Innovative Nanoscience and Nanotechnology" (SIINN) [19] tries to facilitate rapid transfer of nanoscience and nanotechnology research into industrial applications. SIINN brings together a broad network of ministries, funding agencies, and academic and industrial institutions to create a sustainable transnational program for safe use of nanotechnology without creating barriers to innovation. The first priority of SIINN is to develop a consolidated framework for addressing nanorelated risk management for humans and the environment by investigating nanotoxicity. The consolidated framework for environmental and health safety is intended to be used by European and national policymakers, stakeholders, and decision makers before precautionary actions and regulations, including industry, researchers, European governance, international organizations, insurance companies, risk assessment, and certification organizations. Their text presents a condensed and accessible gateway to the identification of:

- Best practices
- Synergy potentials and the elaboration of recommendations for future collaborations on the strategic and operational levels, addressing precautionary measures
- Prenormative work
- Steps toward regulations
- Common actions and an overview of best practices for workers' safety and environmental protection
- Identification of knowledge gaps and recommendations for improvements regarding safe handling of nano-objects in process and product R&D
- Safe processes, products, and transport and safe end-of-pipe processes

Policy analysis within the program has examined working definitions relevant for "nanomaterials" by different organizations or countries and correctly framed the drafting puzzle: various definitions of nanomaterials give a general size frame for nanomaterials in both external and internal dimensions or refer to unique physicochemical characteristics of a specific material under discussion, but none of the existing definitions offer an "unambiguous" description of nanomaterials [20]. SIINN's forthcoming deliverable grapples with this unresolved problem.

7.2 The New Model for Risk Governance: Risk Management and Risk Mitigation

Choices to Be Made When Regulating Nanosafety

Key public health policy questions to explore nanotechnology and nanomedicine risks that require regulation and labeling include:

1. **Context**: What are the expected uses that will lead to exposure?
2. **Characterization**: What are the different impacts that may be realized due to form, shape, and size?
3. **Functionalization**: Are there times when the material is dormant or when its interaction with other materials triggers a specific risk response?
4. **Cumulative Impact**: What is the likelihood that the nanomaterials will remain and add to the impact of subsequent exposure?
5. **Synergy**: What is the likelihood that a specific nanomaterial will interact with other materials that can cause harm, even if each substance is harmless independent of the other?
6. **Life Cycle Assessment**: Do the materials degrade or remain stable, and does society want them to remain unchanged or to break down? What is the half-life of the material after it is no longer useful, and what special steps for disposal and recycling might be necessary?
7. **Liability**: Are there acceptable risks or harms that are not considered important enough to generate liability? Or do nanomaterial manufacturers protect end users with strict liability?

One remarkable facet of nanoregulation involves the strong preventive tone in all of these efforts. Each group involved in the emerging regulations has a strong concern for anticipating problems and developing robust methods for measurement, data collection, and big data analysis of multiple datasets, even though no catastrophic problem has been identified and no crisis has occurred. Another startling facet of these concurrent developments for nanoregulation concerns their legitimate political authority.

Traditional models for regulation follow a very different path compared to the emerging nanoregulations. Laws are traditionally created by legislative entities within different jurisdictions. Various laws cover different subjects, may give rights such as a license or permit to different segments of the population, and may embrace more than one issue by governing a very wide array of activities. Comparative law is rare and usually driven by the needs of a client in a particular well-proscribed context. For example, to fight a parking ticket, it is not necessary to know the parking laws of all nations. But to ship goods from one country to another, it is important to know the prohibited items for sale in the destination country; any requirements regarding packaging, shipping, or handling the goods; and taxes or customs. Globalization has made such simple operations as shipping goods abroad exponentially more complex. Few producers rely entirely on domestic goods; almost everyone in commerce sends or receives goods from other countries as components of the final products. Nanotechnology is no exception to this rule; if anything, it will take the process of globalization a step further.

Globalization of commerce requires that scientific consensus be built in at the outset in order to provide clear guideposts about which methods, levels of exposure, outcomes, and types of toxicity are either permitted or illegal. Many entities, such as NanoReg, are designed to create regulations, but the organizations themselves are not formally part of any one government. Such regulators derive their power from funding and delegated authority in the regional government, for example, in the case of NanoReg it is the EU. But participation by researchers is not elected; the participants are self-selected or appointed by colleagues. Few have any legal training at all. In theory, this new model for regulatory power in the hands of scientific constituency will generate robust, high-caliber science as a driver for new laws, but it is not bottomed upon democratic principles such as representation of specific groups or geographically defined populations. So in Europe, for example, there are several regulatory entities who have delegated authority from the overarching government, and these groups do not necessarily connect with one another. This new approach to creating laws for regulatory governance of risk is fascinating and may produce better laws from the standpoint of scientific reasoning for future legal requirements. But it is an open question how these science regulators will use their power from the perspective of established legal principles that govern procedures and constitutional law. There are many systems in the development phase but few efforts toward harmonization, which will be the necessary next step if such regulations are going to foster, rather than serve as a roadblock to, medical progress and commerce. Regardless of how well any of these laws have been crafted, it is impossible to conduct commerce if all of these laws apply at the same time. The resulting roadblock to commerce can inadvertently choke the very industries that should thrive.

Most of the applications for the new technology cannot be imagined when writing the laws, but as in the WHO Constitution, a good framework can embrace new problems anyway. Decisions regarding the potential areas of high risk and the potential long-term risks of small exposures to nanoparticles are no exception in this regard. Policymakers must decide without good data to fund and supervise these projects before risks are known; taking into account current political will, while leaving room for the benefits and mistakes in new discoveries that will reshape the policy response to these issues, requires a flexible regulatory framework. The framework cannot simply be a one-shot firecracker approach that looks at a situation, arguably finds problems, but then ceases to monitor the situation for long-term effects. Regulatory efforts must be reviewed periodically to refresh the program, daring to ask, "What is the question that we are not asking ourselves, and when we finally ask it, are we confronting it properly?"

One salient trait that functions like a compass to reveal a workable path across this swamp in the legal landscape, riddled with broken laws, false leadership by unauthorized policies and codes that offer little real guidance, and lofty proclamations by the higher branches of civil society, is that the best laws, codes, and guidance for practical implementation have a flexible framework at the heart of their features, regardless of their scope or subject. The emerging nanoregulations are aware of this facet of legislative drafting and are struggling to develop juridical tools to offer flexible drafting.

Laws and best practices from occupational health offer a model that has been successfully applied in the past and that can be folded into the emerging nanotechnology framework by having an outcome-oriented approach to compliance. Instead of offering numbers as cutoffs for determining whether the regulations have been obeyed, a compliance system provides alternative measures of effectiveness so that priorities under law can be shifted as new products, new information about risk, new methods about risk assessment and risk management, and commercialization of new products develop. How to include these flexible approaches into scientifically based regulations that involve nanotechnologies are discussed next.

7.2.1 Design Components of Risk Management Systems

Risk management systems were invented to control risks that cannot be quantified but are nonetheless real. When risk management systems work properly, embedded values also function as a safety valve, preventing the flow of inappropriate concepts from gaining currency throughout the system [21]. Significantly, if the correct values have been embedded into the infrastructure, illegal or irresponsible activities will be prevented by the system itself. This applies not only to manufacturers, chemical plants, and steel mills—those businesses that have a greater degree of physical risk for workers—but also to all other businesses that have an established workplace with key staff. For this reason, too, there is a golden opportunity to create positive social change favoring gender equity and equal opportunity for older workers and people with disabilities: harmful embedded values in the workforce can be excluded from the new system, and new values can replace existing working assumption that are impacted by race, gender, age, or ethnicity in the new system created to regulate nanotechnology. The concept of embedding values into the employer's infrastructure is not novel or new: due diligence requires developing committee structures and then the regular use of a smooth-flowing, ongoing infrastructure for improved communications for training and for reporting problems.

Each in-house compliance program must decide for itself the area of emphasis that will best foster compliance specific to the workplaces and the demands of the tasks at hand. This requires examining the relative advantages and weaknesses of different components of effective occupational health programs. A clear organizational structure allocates in-house compliance tasks and makes in-house staff accountable for the achievement of compliance program goals and objectives [22]. These goals and objectives should be stated in written policies, followed by action and hotlines, committees, and other compliance tools to prevent systemic failures. One key tool for maintaining due diligence is the internal audit. Systematic voluntary evaluation, review, or assessment of compliance with relevant laws include discussion of risk assessment and risk management choices about the data that was reviewed. The in-house compliance program for safety and health cannot be treated as if it were a trade secret; it must be open to everyone to participate, and everyone's complaints as well as praises must be heard.

7.2.2 Selling the Program: Communication Is a Two-Way Street

The starting point for creating or auditing any compliance system is understanding that compliance is a results-driven task. Ultimately, the only thing that really matters is whether the system is actually effective in minimizing breaches of the law in practice; proving due diligence requires ongoing and overt commitment. This notion of commitment involves three interrelated spheres of activity that traverse the corporate structure "from the boardroom to the mailroom." In the Occupational Safety and Health Administration's (OSHA) words, these prongs are active employee involvement, visible management commitment to compliance efforts for worker protection, and sustaining of compliance [23]. How does anyone know whether these efforts are meaningful in the eyes of regulators or, more importantly, actually achieving anything?

Communication is the "blood" that circulates to turn *management commitment* into *active involvement.* Communication means frequent, reader-friendly reminders that are to the point, attractive, and brief. Regular reporting requires two streams of communication. First, operating managers need to report up the management line on compliance issues in a frank way that will ensure that problems are addressed. As part of the review process, reports from industrial hygienists and safety personnel to the corporate counsel and to the compliance team, as well as follow-up reports, should be amplified by the work of outside consultants. The major features of these larger reports should be summarized into a one- or two-page memo for general distribution. The openness of this general approach underscores commitment to occupational safety and health programs and reassures staff. Without full-time, up-and-down-the-line support for the implementation of these three elementary components, success is extremely unlikely.

Commitment is easy to discuss, but it can be difficult to achieve. Due diligence requires more than just a written compliance philosophy and saying that everyone must adhere to it. Therefore, the role of the compliance staff as gatekeeper of information and a clearinghouse for new ideas, solutions to problems, and a host of compliance activities is increasingly becoming the linchpin of a successful program [24].

Employee involvement in communication is the second key component of these two streams for communication. Employee involvement includes employee activities in training regarding the proper handling of toxic or hazardous substances and outreach to employees who actually confront hazards to ferret out potential problems. Employees are often the ones who really understand problems with machinery or premises. Well-informed employees in the workplace in direct contact with hazardous emissions or other potential hazards are a crucial resource for such information; their insights about actual working conditions in the facilities at the enterprise are invaluable. As discussed in industrial hygiene literature, employee participation and cooperation is pivotal in shaping successful industrial hygiene programs because without the workers' trust and support for the tasks set forth by the industrial hygienist, there is no chance for launching a successful program.

7.3 The Nuts and Bolts of Risk Management: Corporate Compliance Programs

Occupational health and safety management laws are implemented through several channels for intervention to prevent workplace harms and thereby reduce corporate liability, such as risk management programs. Risk managers, industrial hygienists, and many additional health professionals use well-honed prevention strategies to stop problems from becoming catastrophically large. Such in-house programs and the supporting infrastructure of regulatory agencies work together to do much more for the economy than merely reduce the costs of accidents and the overall burden of disease in society. Avoiding downtime, replacement costs, and displacement of families who lose a major or primary wage earner, with the attendant social problems for family survival and the reduction of misery from preventable death, thereby saving money, is among the key goals of compliance programs. In turn, these efforts benefit all of civil society. But these savings are merely the tip of the iceberg. Applying the best practices and well-understood methods of reducing risks prevents waste and saves money.

Sound in-house corporate compliance programs therefore provide a lifeline, keeping marginal employers afloat in turbulent economic times. Rightly prioritized, at the top of the chief executive officer or any employer's agenda, occupational health management systems can save the life of a company from the brink of bankruptcy by avoiding loss as well as preventing liability. Achieving compliance objectives or goals to improve working conditions or protect occupational health compliance requires an internal risk management system controlled by the company, regardless of whether those goals are simple or ambitious. Having a structure that responds to occupational health questions is essential for many reasons, including the added value for satisfying downstream customers, thereby preventing product liability. These features of the justification for occupational safety and health in-house programs have therefore slowly emerged as invaluable added value for internal corporate governance infrastructures, instead of mere add-ons for a well-documented program.

Accepted models for occupational safety and health corporate compliance infrastructure are especially useful for documenting and managing the data for compliance with standard features of regulatory requirements worldwide: (i) targeting problem areas before or after formal complaints, (ii) troubleshooting before potential violations of law, and (iii) analyzing job hazards. The internal management system is also important for logging the incidence and duration of injuries and illness, performing internal audits, reporting improvements and violations to in-house staff, and informing all staff of the hazards and their rights (as required to be disclosed under law). Although some experts may argue that this is a function of in-house politics within an enterprise (because enforcement officials and regulators may be underfunded and understaffed), pragmatists note that no agency should have adequate appropriations to verify every work process of every employer every day.

Thus, effective compliance programs seek to motivate participants on the basis of philosophies and incentives that come from within the corporate structure itself, as exemplified by old Environmental Protection Agency (EPA) policies for voluntary compliance and self-policing in the United States [4]. Implementing a management system to formalize the compliance process may therefore often require going "beyond" compliance from the standpoint of the letter of the law in order to ensure that precautionary principles embedded in the law are met and that proof of so-called due diligence toward compliance can be well documented [25].

A long list of compliance tools can be used, as appropriate, to help achieve good compliance. The cornerstone for any effective compliance system is real, ongoing, and overt commitment. Without a clear and readily discernible focus on these three elementary components, success is extremely unlikely. Commitment is easy to say but difficult to achieve. Beyond simply issuing a written compliance philosophy and saying that everyone must adhere to it, commitment is evinced by actions. A written policy is necessary, of course, and it must come from the top levels of management. Legalistic (especially concerning the emerging information technologies) must be practical, intelligible, and clear. A team including safety engineers and industrial hygienists among the vice presidents of major employers [26] should choose those aspects of all the available tools that will be the most suitable to a given set of operations for a given business, management systems, active employee involvement to prevent or control safety and health hazards at the worksite, and ongoing feedback into systems.

7.3.1 How?

The vital question of *how* sounds simple but is not easily answered. Strategies are a long-term core value of a company, a value reflected in decisions relating to production, maintenance, and performance evaluations. Emerging consensus about how to manage risk in order to protect public health despite the immature state of the art of nanotechnology involves accepted engineering controls, as exemplified by the 2013 National Institute for Occupational Safety and Health (NIOSH) document that specifically outlines the basics for control of workplace exposures to nanomaterials [27]. Management's commitment to the program can be shown by money expended for compliance; the quality of the materials used in training; the high caliber of the professionals who are brought in as consultants to the program; accessibility of program requirements such as a code of conduct for part-time workers, independent contractors, and the general public; and a willingness to participate in program activities, inspire the organization to obey the law that is the driver behind the program, and prevent "paper programs" [11, 23, 28].

Figure 7.5 Framework for risk management decision making. Source: US government.

Since there is no shortage of directives and statutory authority pointing to the employer's primary responsibility for any liabilities, there are few excuses for inattention to a compliance program. Without the power and support of senior management authority, it is impossible for compliance staff to acquire a detailed knowledge of operations to ensure a high standard of compliance with the law [23]. Compliance staff must be well educated and have clout to change internal processes by shutting down dangerous conditions or by changing corporate culture in favor of occupational safety and health awareness. Responsibility for specific areas, embedded via custom-tailored training about the key hazards and best practices for avoiding unwanted consequences, written into each job description and followed up by internal enforcement is key proof of managerial commitment to the compliance program.

Current Challenges

- **The traditional risk assessment and management strategy requires an occupational exposure limit (OEL).**
- **OELs for radioactive materials are based on a unified concept of dose.**
- **Such a unifying concept is not available for all hazards that can affect total worker health.**
- **How can an effective hygiene program for all stressors be developed and implemented in the absence of comprehensive OELs?**
- **What control approaches are feasible and effective?**

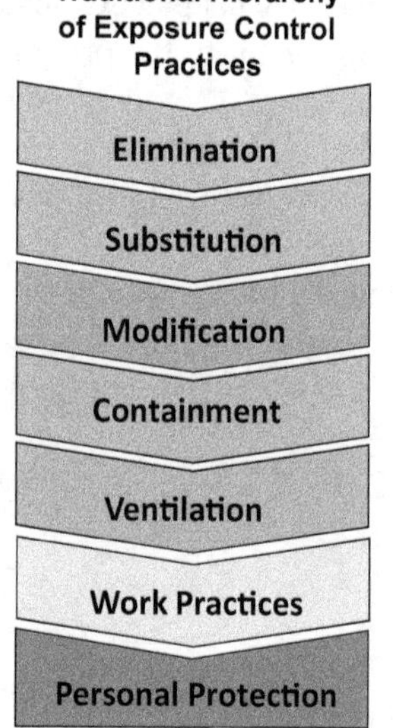

www.cdc.gov/niosh/topics/ctrlbanding/ 28

Figure 7.6 Description of current challenges to the traditional hierarchy of exposure control, as a justification for control banding. Source: US NIOSH.

Effective in-house risk management programs also require the best-possible lines of communication in order to be successful. To work, lines of communication must be established long in advance of any emergency. Such efforts can also facilitate compliance with new changes or modifications in the governing law because of their inherent flexibility. The strongest systems are cyclical, as noted in the chart by Mark Hoover [29] in Fig. 7.5. The cycle makes reporting normal as well as required, with built-in follow-up as well that ensures long-term accountability. With the help of corporate anthropologists and outside consultants, staff can be briefed about key issues, when designing right-to-know training, tailored to the corporate culture of a specific enterprise and thereby creating a specialized system with effective results [30]. Written policies may also be required to withstand scrutiny in a court of law in the event of a violation of national or local health codes, tort litigation, or environmental litigation, and therefore their text should reflect the presence of embedded compliance values. Simply giving staff lectures, without backup resources or problem-solving workshops in anticipation of hazards or without considering the new occupational health effects that the sector of the economy may face, is shortsighted. Normal operating procedures double-check against system failures. And documentation offering proof of such safeguards against system failure is the key to a solid defense during litigation.

7.3.2 The Hierarchy of Engineering Controls

The hierarchy of engineering controls that has been traditionally recognized by safety engineers, industrial hygienists, and a variety of allied health professionals internationally remains the cornerstone of best practices. This handy matrix that is commonly used around the world has four basic components. These well-honed principles have already been codified in many national laws and international standards, reflecting the commitment of many countries to improve safety and health, and it is likely that a compliance program would be asked to explain why the system was not used in the unlikely event that its approach is not useful for managing risks associated with nanotechnology applications or nano-enabled products that impact global health for workers, consumers, or inadvertently exposed populations.

According to the International Labour Organization (ILO) publication *My Life, My Work My Safe Work* (2008) and echoed by WHO in the organization's *Guidelines on Protecting Workers from Potential Risks of Manufactured Nanomaterials* (2017), this traditional tool remains vibrant for planning strategies to minimize the risks from exposure to potential dangerous substances, including nanomaterials. According to the blog posted by NIOSH director Dr. John Howard, "Increasing production of manufactured nanomaterials (MNMs) and their use in consumer and industrial products means that workers in all countries will be at the front line of exposure to these materials, placing them at increased risk for potential adverse health effects" [31].

Experience gained during the development of these guidelines contributed to the NIOSH effort to advance a systematic review process tailored for the occupational safety and health topics (for more information, see NIOSH Science Blog on the *systematic review for occupational safety and health questions*. The guidelines were very rapidly posted on the US CDC NIOSH website. In the introduction to the US posting, Dr. Howard pointed out that NIOSH has been at the forefront of research on working safely with nanomaterials. NIOSH plans to continue supporting this WHO effort at the guideline implementation phase, which will focus on turning these guidelines into practice. WHO's International Agency for Research into Cancer (IARC) has classified certain nanomaterials as "carcinogenic to humans" or "possibly carcinogenic to humans" in the NIOSH documents for protecting workers who are exposed to carbon nanotubes (CNTs) and MNMs, the components of the four key steps to reduce risks include but are not limited to:

1. Eliminate or minimize risks at their source.
2. Reduce risks through engineering controls or other physical safeguards.
3. Provide safe working procedures to reduce risks further.
4. Provide, wear, and maintain personal protective equipment [32].

As in the case of any list, this approach cannot be definitive. But this approach to prioritizing hazards and responding to their features as a way to manage or mitigate risk is sufficiently well respected that using this approach can be viewed as reasonable in light of the state of the art.

7.3.2.1 Eliminate or minimize risks at their source

This important first step aims to remove or minimize risks before they enter the workplace. Manufacturers and suppliers of work equipment and substances may be able to provide innovative approaches to solving these problems, but history teaches that some substances are difficult to replace. For example, it may be possible to substitute a hazardous chemical with a less hazardous one that achieves the same purpose. Asbestos, for example, is a very hazardous substance, whose use has been banned in some countries, but there is a great debate raging over whether asbestos should be used or whether it should be replaced by certain substitutes, even though those substitute products are often inadequate. In this instance, the lesson of asbestos is very difficult to accept, and in the eyes of some people, it remains unacceptable. That lesson is that sometimes no one can find a substitute that is suitable, and in such cases, special protections must be in place to protect workers and consumers. A regulatory framework geared to special protections in instances where there is no scientific or regulatory consensus about the substitute can contain special provisions about such ultrahazardous activities, including a system to detect, report, and discuss such circumstances.

In that case, the specific job hazard analysis plays a very important role in determining which substitutes, if any, can be prescribed. The quantity of the dose and the duration of exposure also play a key role in this first step. Separate and apart from the old technological issues, the paradigm for understanding the impact of engineered nanoparticles upon human health must take into account several traditional hazards that are likely to be present in the workplace, and then attempt to confront those hazards to reduce their impact in combination with new nanomaterials and new sources of exposure. For example, in the labs where nanoparticles are used, there may be additional mundane issues of noise and vibration emissions from work equipment, which must be controlled whether or not they interact with nanoparticles to impact human health. One could not forsake common hearing protection simply because glitzy nanomaterials are in use.

Figure 7.7 (Left) Workers on top of a building under construction. Photographer: Harris & Ewing, Inc. (Right) Three construction workers, suspended by ropes, with pneumatic hammers, cutting into a wall or mountain during construction of the Grand Coulee Dam, Columbia River, Washington (http://hdl.loc.gov/loc.pnp/ppmsca.17402). Source: US Library of Congress, Prints & Photographs Division, Washington, DC.

7.3.2.2 Reduce risks through engineering controls or other physical safeguards

As illustrated in Figs. 7.8 and 7.9, the face of the workforce has always included all races and women as well as men. Many risks may be unique to subgroups of the working population, but eventually the need for sound engineering controls to remove key hazards is the same. Simple hazards from heavy lifting can be removed through automation whether by creating electronic books or hydraulic machines inside coal mines.The forethought about job hazard analysis and the desire to reduce potential harm is equally essential in either set of tasks.

Whether or not risks can be eliminated or minimized at their source, they may often be further reduced through effective physical safeguards. These can be relatively simple, such as guardrails protecting against falls from scaffolding or protective covers for electrical equipment. Good ventilation also provides protection against risks from harmful substances, for example, as in a hospital operating room, where nurses and physicians should be protected from waste anesthetic gases. Such engineering controls require ongoing attention in order to be effective. They must be properly maintained if they are to remain effective, and therefore it is internationally accepted by scientific experts that setting up a program is not enough: maintenance procedures must become integrated into the overall management system [33]. The current working assumption is that traditional products for safety equipment that provide fall protection, machine guards, and guardrails may be made from nanomaterials but that these materials will not have a negative health impact by migrating or exhibiting deterioration. This question will eventually become important as nanomaterials remain in commerce and are used across the working life of safety products. This question about migration and degradation is equally important for disposable safety products, such as gloves, respirators, and protective clothing that may benefit from the advantages of nanomaterials.

Figure 7.8 Young African miners, c. 1941. Source: US Library of Congress, Prints & Photographs Division, Washington, DC.

Figure 7.9 Librarians Lena Notson and Alyce Thomas had to sort and file periodicals by hand using tools like the binders and broadside racks shown in the image. Today, periodicals are managed with electronic databases, allowing the Wirtz staff to search thousands of publications with the click of a button. Source: Archives of the Library of the US Department of Labor, Washington, DC.

7.3.2.3 Provide safe working procedures to reduce risks further

Good planning and organization are always important but particularly so for some activities. For example, maintenance work or blockage clearing in machines require safe isolation procedures to prevent accidental start-up; many workers have been injured during such operations. Work with ionizing radiation also needs careful planning and organization, with radiation badges or monitors worn by those who have to carry out such work and the use of appropriate physical barriers.

Nanosensors for many chemical hazards may become a very useful part of the in-house safety and health compliance system for the use of materials that pose a known risk of an adverse health impact. Warning signs and signals can be effective preventive measures, but they need to be supported by other precautions and are only useful when they are visible, clearly written, or audible and in a language that everyone understands. This reality explains the significance also of the emerging global approach to labels, warnings, and chemical safety training in the GHS. Although no system of nanosensors has yet been developed for engineered or manufactured nanoparticle exposures, and it is not yet clear that such nanosensors would be necessary, it is nonetheless true that nanomaterials pose novel health hazards and therefore should be transported and handled following some special training. The established symbols and procedures that are accepted worldwide under the GHS may prove very useful for nanotechnologies and engineered or MNMs inside nano-enabled products.

7.3.2.4 Provide, wear, and maintain personal protective equipment

Providing personal protective equipment (PPE), such as dust masks and hearing protectors, is the least reliable form of protection since its effectiveness relies on proper selection, training, use, and maintenance. Therefore, traditional occupational health and safety discussions underscore that PPE should be used only as a last resort. This fundamental principle avoids the inefficiency of relying on individuals to implement compliance and recognizes that a system cannot stop simply because of human error. This emphasis on avoiding systemic failure rather than blaming individuals does not change the fact, however, that such equipment is required for some operations. For example, no ventilation system can protect against all the exposures firefighters confront in an emergency, but the system tries to do its very best. Likewise, hearing protection may be required for working in noisy environments, even though all other means have been successfully used to reduce noise exposure as far as practicable. Additionally, people are different, and therefore, PPE must be suitable for the workers concerned and properly maintained so that it remains effective. There was much discussion in the 1990s, for example, regarding whether respirators and other protective clothing were available on the open market in women's sizes. The failure of the market to provide correctly fitting PPE kept women out of high-paying dirty jobs.

Nanotechnology efforts at prevention add a new wrinkle to this complex custom-tailored system. In particular, nanoparticles are so small that they cross otherwise impermeable borders to enter the space between skin cells or to traverse the weave of protective clothing. The unknown effect of these exposures, and the question of how to measure such tiny doses of a dangerous substance or hazardous chemical, are very important. Therefore, applying the hierarchy of engineering controls when nanotechnologies are concerned requires rethinking long-accepted methods of filtration and respiratory protection. How to craft appropriate protective clothing remains an open question when preventing unwanted exposure to nanoparticles. It is also unknown at this time which substances are more likely to cross otherwise impermeable borders in the weave of fabric or the seal of respirators or the small spaces between cells in the liver, spleen, or lung. Applying precautionary principles and examining the question of whether such hazards, as have been described in the scientific literature, and whether potential risks are sufficiently recognized or understood to trigger statutory protections is therefore a vital first step toward planning the research-to-practice phase of nanotechnology applications, even when those risks are not well understood.

Figure 7.10 Destitute pea pickers in California. Mother of seven children, age 32, Nipomo, California, c. 1934 (http://hdl.loc.gov/loc.pnp/pp.print). In the 21st century, this pea picker would be given personal protective equipment to protect against pesticides, bugs, and ultraviolet rays. Photographer: Dorethea Lange. Source: US Library of Congress, Prints & Photographs Division, Washington, DC.

In conclusion, the best tools required for risk mitigation and protection of worker health exist already, until new research reveals specific hazards that are presently unknown. If applied with forethought when rethinking the two vital social values of consolidating health programming to reduce administrative waste and charting a new strategy to protect global public health, two sets of benefits can be realized by civilization at the same time, not as competing interests but as one invaluable social change: implementing improved health care delivery systems and medical surveillance at work, while applying new nanotechnology.

References

1. Jorge Nef and Bernd Refter, *The Democratic Challenge: Rethinking Democracy and Democraticization*, Palgrave Macmillan, 2009, p. 12: "Direct participation by the citizens of Athens 2,500 years ago did not extended [sic] to the majority of the population constituted mostly by slaves and foreigners. Women were excluded too."
2. Richard Heffner, *A Documentary History of the United States*, Mentor Books, New American Library, New York, p. 286, 1952, in contrast to Nef and Refter, *The Democratic Challenge*. See, in particular, their discussion of Madison's writings in *Federalist Number 10*, where Madison warns of the tyranny of the majority (p. 17). "Madison claimed that any group unrestrained by external checks would tyrannize others, depriving them significantly of their natural rights."
3. NanoImpactNet, *The European Network on the Health and Environmental Impact of Nanomaterials*, www.nanoimpactnet.eu, Coordinator, Michael Riediker Institut universitaire romand de santé au travail, Lausanne Switzerland, in *Compendium of Projects in the European NanoSafety Cluster*, pp. 102–114 at 103, 2011.
4. Jay A. Sigler and Joseph E. Murphy, *Interactive Corporate Compliance: An Alternative to Regulatory Compulsion*, Quorum Books, Greenwood Press, CT, 1988.
5. Committee on the Environment, Agriculture and Local and Regional Affairs Commission de l'environnement, de l'agriculture et des questions territoriales, *Nanotechnology: Balancing Benefits and Risks to Public Health*: A Preliminary Survey for the Council of Europe, revised draft report, rapporteur Ilise Feitshans, May 29, 2012, final approval April 23, 2013, based on AS/ENA (2011) 35, September 22, 2011,

Aena11_35. *Nanotechnologies: A New Danger to the Environment?* Preliminary draft report, rapporteur Valeriy Sudarenkov, Russia, SOC.

6. Council of Europe, Strasbourg, France, report to Committee on the Environment, Agriculture and Local and Regional Affairs Commission de l'environnement, de l'agriculture et des questions territoriales, *Nanotechnology: Balancing Benefits and Risks to Public Health: A Preliminary Survey of Possible Legislative Approaches to Nanotechnology for the Council of Europe*, preliminary draft report, rapporteur Ilise Feitshans, May 29, 2012.
7. Comments by Tanja Kleinsorge, Head of the Secretariat, Committee on Social Affairs, Health and Sustainable Development, Parliamentary Assembly of the Council of Europe, at the public forum "Law and Science of Nanotechnology: Perfect Together?" Museum of the History of Science, Geneva, Switzerland, August 17, 2013, youtu.be/vEmorvYOJz8.
8. Council of Europe, *Handbook for Parliamentarians: The Council of Europe Convention on the Counterfeiting of Medical Products and Similar Crimes Involving Threats to Public Health* (CETS No. 211), Introduction, Council of Europe, Strasbourg, France, 2015.
9. Directorate General Human Rights and Rule of Law (Action against Crime Department, Criminal Law Division). Committee of Ministers: Council of Europe Convention on the Counterfeiting of Medical Products and Similar Crimes Involving Threats to Public Health (CETS No. 211). See also EAMI Permanent Forum on International Pharmaceutical Crime, http://pfipc.org/.
10. Ilise Feitshans, *An International Public Health Threat Requires a Global Solution: Counterfeit Medicines and the Council of Europe Convention against Medicrime*, MS JD writers in residence blog, "So You Want to Be an International Lawyer," April 2016.
11. http://www.nanoreg.eu/index.php/media-and-downloads.
12. Nicolas Segebarth and Georgios Kagarianakis, Foreword, in Michael Riediker and Georgios Kagarianakis, eds., *NanoSafety Cluster: Compendium of Projects in the European Nanosafety Cluster*, 2011 edition, Brussels, Belgium.
13. Michael Riediker and Georgios Kagarianakis, eds., *NanoSafety Cluster: Compendium of Projects in the European Nanosafety Cluster*, 2011 edition, Brussels, Belgium.
14. http://www.marina-fp7.eu/.
15. Georgios Katalagarianakis and NanoSafety Cluster newsletter, Nicolas Segebarth, *NanoSafety Cluster Newsletter*, Issue 1, Oct. 2013 at 1.

16. BE Etats-Unis 348 15/11/2013 Politique scientifique: Gestion des risques liés aux nanotechnologies. Une coopération Europe-Etats-Unis, http://www.bulletins-electroniques.com/actualites/74320.html.

17. World Health Organization. WHO guidelines on protecting workers from potential risks of manufactured nanomaterials, World Health Organization, Geneva, 2017. *Public Health, Environmental and Social Determinants of Health*, cluster of climate and other determinants of health, license: CC BY-NC-SA 3.0 IGO, www.who.int/phe.

18. Personal correspondence, Ilise Feitshans from Dr. Christine Hendren, CEINT, Duke University. Commentary on the U.S.–EU CoR's following webinar, March 22, 2017: "Risk Assessment, Legal Governance and Regulatory Connections: Achievements of the USA-EU Community of Researchers 2013 to 2017" and forthcoming article to be prepared by CoR.

19. siinn.eu/en/.

20. Karl Hoehener and Juergen Hoech, *Deliverable D2.6 Draft (M30) Consolidated Framework for EHS of Manufactured Nanomaterials*, ERA-NET SIINN, July 7, 2013, p. 9/123.

21. Ilise Feitshans, Bridging the gap between occupational health and general compliance: international trends from four nations (United Kingdom, USA, Australia and Canada), *Preventive Law Reporter*, Denver, Colorado, 1999.

22. Brian Sharpe and Ilise Feitshans, The nuts and bolts of occupational health management systems for compliance in Australia and the USA, *International Conference Paper, 33rd Conference of the Ergonomics Society of Australia*, November 1997.

23. Ilise Feitshans and Cathy Oliver, More than just a pretty program: OSHA Voluntary Protection Programs improve compliance by facing problems head-on, *Corporate Conduct Quarterly*, **5**:69–73, 79, 1997.

24. Ilise Feitshans, Bridging the gap between occupational health and general compliance: a portrait of compliance in Canada, *Safety and Health Practitioner*, London, England, 1997.

25. Ilise Feitshans and Joseph E. Murphy, Positive incentives for compliance in occupational health: addressing workplace violence, based on presentations at the 25th International Congress on Occupational Health (ICOH), Stockholm, Sweden, 1996, and at the request of the Assistant Secretary of Labor for OSHA, Washington, D.C., University of Denver, *Preventive Law Reporter*, March 1997.

26. Ilise Feitshans, *Designing an Effective OSHA Compliance Program*, Thomson Reuters (available on Westlaw.com), 1990 and revised in 2013.

27. Department of Health and Human Services (NIOSH), *Current Strategies for Engineering Controls in Nanomaterial Production and Downstream Handling Processes*, Publication No. 2014-102, cdc.gov/niosh/docs/2014-102/.

28. Jerry Catanzaro and Judith Weinberg, Answers to some frequently asked questions on VPP, *Job Safety & Health Q*, **22**(Summer), 1994; Ilise Feitshans and Victoria Bor, eds., *Occupational Safety and Health Law 1995 Supplement*, Bureau of National Affairs (BNA) Washington, D. C., 1995; OSHA Safety and Health Program Management Guidelines, Issuance of Voluntary Guidelines [Docket No. C-02] 54 *Federal Register* 3904 (Jan. 26, 1989); Judith Weinberg, OSHA consultation: a voluntary approach to workplace safety and health compliance, *Corporate Conduct Quarterly*, **5**(2).

29. Mark D. Hoover, *Some Key Elements for Assessing and Managing Ideas for Discussion: Exposures to Occupational Hazards*, NIOSH, Morgantown, West Virginia, draft date October 27, 2010.

30. *Cargill, Inc. v. U.S.*, 173 F.3d 323, 18 O.S.H. Cas. (BNA) 1685, 1999 O.S.H. Dec. (CCH) P 31814 (5th Cir. 1999).

31. https://blogs.cdc.gov/niosh-science-blog/2017/12/15/who-nano/.

32. International Labour Organization (ILO), *My Life, My Work, My Safe Work: Managing Risk in the Work Environment*, ILO.org/Safework/Safeday, supported by the International Social Security Association, World Day for Safety and Health at Work, April 28, 2008. Also available in French: Ma vie, mon travail, mon travail en sécurité -Gestion du risque en milieu de travail, Geneva, 2008; Spanish: Mi vida, mi trabajo, mi trabajo en seguridad - Gestión del riesgo en el medio laboral, Geneva, 2008.

33. Ilise Feitshans and Brian Sharpe, The nuts and bolts of compliance programs, in *Bringing Health to Work*, Emalyn Press, 1997; Ilise Feitshans, Necessity of the program, in *Designing an Effective OSHA Compliance Program*, Thomson Reuters (available on Westlaw.com), updated annually.

Chapter 8

Bioethical Impacts: Nanotechnology Transforming Disability into Health

8.1 Rethinking Traditional Policies: Nanotechnology Impacting Health and Work as Part of the Human Condition

Society's need for health at work is as ubiquitous and perennial as civilization and the DNA of life itself. Amid modern complexity, there remain constant underlying basic human needs: work health and survival have been inextricably linked throughout the history of human civilizations. Without work, society cannot survive, and no work can perpetuate society without health. No society has survived without producing things, without work. We enjoy the fruits of many past civilizations today as we draw upon their architecture such as the Pyramids, the Parthenon, and the Great Wall.

Civilizations can be brought to a halt in times of plague and pestilence; and even the most impressive of collective efforts can be stopped when injuries overtake any individual's ability to work. Society therefore needs both, working people and healthy people, for civilization to survive. But these classifications are not dichotomous or mutually exclusive. The fluid categories of sickness and health, which fluctuate within individual abilities and deficits and within the life of any given individual across time, hold implications for

Global Health Impacts of Nanotechnology Law
Ilise L. Feitshans

ISBN 978-981-4774-84-0 (Hardcover), 978-1-351-13447-7 (eBook)
www.panstanford.com

every worker in every job description, ranging from dignitaries in the highest offices of leadership, celebrities, and heads of state in North America or Europe to the laborers tearing apart old ships in the shipyards of Asia and from the boardroom to the mailroom. Philosophies and values embedded in ancient cultures can touch, even today, our daily lives. Indeed, the remnants that survive from ancient cultures are found in architecture, statues, and pottery artifacts of the skilled crafts and creative labors of lost societies. There has always been work as long as there has been human society. Without the work of architects, builders, and the slaves, who were driven by underpaid overseers and who once in a great while died in riots or revolutions, we would not have the wonderful artifacts of history and past cultures that form the foundations of our society today.

Figure 8.1 Work and health are inextricably linked to survival. Cover of *Occupational Health from an International Perspective*, senior scholars honors thesis, Barnard College, Columbia University. Thesis and photo copyright: Dr. Ilise L. Feitshans.

Without the hard work of steel workers and construction workers and oil and chemical workers, no one would have the marvelous modern buildings and simple structures that serve our society today. Since the time when people chipped stone to fashion tools, occupational diseases, as in this example, the respiratory ailment derived from crystalline free-silica dust, (silicosis), threatened public health, although silica is still the subject of debate regarding its regulation. Every civilization has left records of occupational deaths in hunting, healing, construction, agriculture, and industry. Although still in use in the 20th century without much thought given to its major hazards, lead was among the first metals known to the early Egyptians, Hebrews, and Phoenicians: lead colic and paralysis were mentioned by Dioscorides in the first century AD. People may debate whether one type of work is more important than another; some people will argue that such values are socially determined by the economic worth of a particular job; a job that is worth the minimum wage is valued less in society than a highly paid job, and therefore society must have a greater need for the highly paid job.

None of these types of work, not the great monuments, not the writings or the arts, could exist without a modicum of human health. Work is both the key to sustaining daily life and civil society's legacy for posterity, as seen in Fig. 8.2, a scene from the video "Lessons Learned from Three Centuries of Occupational Health Laws." In this scene, actors recreate the work of indentured servants in colonial life, cooking and serving in the pre-revolutionary United States. The video was made thanks to the Indian King Tavern Museum in Haddonfield, New Jersey, and is archived by Digital 2000 Productions. The museum and its historic local environs provide the backdrop for discussing the importance of early health and sanitary regulations regarding trade and commerce in food in 18th-century United States. These visions of the role of work may change across time due to social changes in society, and the types of jobs that are available may change because of nanotechnology, but the link will remain between work, health, and the survival of society.

Nanomedicine will require society to rethink ancient notions that are the building blocks of social constructs regarding the nature of disease, treatment, and the prejudices encountered by people who suffer from illness, regardless of whether those sources of disability are inherited, crated by happenstance, or come from work. The most startling aspect of this revolutionary approach, called "personalized medicine" or "precision medicine," is the ability to prognosticate: nanomedicine and nano-enabled tools will make it possible to predict with remarkable accuracy the presence of illness in large populations and enjoy early diagnosis and prophylaxis of diseases.

This, in turn, will change the social notions of health that determine which types of illness constitute "disease" that cause "disability."

Figure 8.2 Indentured servants cooking in 18th-century United States, from the video "Lessons Learned from Three Centuries of Occupational Health Laws." Copyright: Dr. Ilise L. Feitshans.

Figure 8.3 Cover of the WHO *World Report on Disability*. Source: WHO, 2009, Geneva, Switzerland.

Disability, although universal in its likely incidence in the life span of any given human being, also challenges the operationalization of a fundamental tenet of equality: that every person is the same and, consequently, all people—men , women, and children (as protected in separate international conventions)—have equal rights and should be treated the same.

In 2012 the World Health Organization (WHO) *World Report on Disability* (Fig. 8.4) asserted:

- "Health" and "disability are social constructs

  but *NOT*
- medical determinations based on empiricism or fact.

The notion that social conventions and public health policy should seek equity and fairness instead of equality is an important contribution of disability law to international health jurisprudence. For example, it may sound as if everyone is treated equally if they must walk up a flight of stairs to get a free vaccine, but for the people who cannot walk this is unfair; they require assistance or reasonable accommodation in order to have the opportunity to be treated fairly. And if they must hire someone to help them, the vaccine for them is no longer free. In this context, the absence of an equal opportunity to have a free vaccine is a consequence of their impairment. When impairment becomes substantial, it may be considered a disability under law.

Figure 8.4 Roadway construction, Bahrain, 2009. Photo by Dr. Ilise L. Feitshans.

The logical extension of this theory that treating everyone the same is not always fair is important for public health programming. Under this theory, what constitutes "healthy" or "disabled" is actually a construct created by society on the basis of social conventions, not created by medical notions of health or disability. This social construct is based on the needs of society as well as the fitness of individuals and therefore may change when large populations become able to recover and perform tasks that they could not do when they were disabled and also as science using nano-enabled devices uncovers new forms of illness and potential disability. The ancient question of whether everyone must accept treatment prescribed by social conventions will surface again as new diseases may bring a demand for the right to refuse treatment when the potential illness is predicted and preventive measures are recommended or required by new laws.

Several examples illustrate this point:

- Smoking was encouraged by society years ago (it was viewed as sexy in movies!). Now smoking is treated by society as an addiction and a public health threat for nonsmokers who are exposed to passive smoke; smoking has become illegal in many public places and on airplanes.
- Food is also a matter of social constructs about health: when food is sparse, people fear the public health threat of malnutrition; when it is overabundant, people worry about obesity. But whether the attributes of being fat or thin are attractive is culturally determined without a direct correlation between disability, illness, or healthy living.
- In genetics, the role of social constructs discussed by WHO becomes even clearer. For example, occupational diseases such as coal miner's black lung were considered the product of genetic traits among the families of coal miners in the 19th century. Since the late 20th century, coal mining is empirically measured and controlled with strong monitoring of exposures and duration of exposure, with a diverse workforce of many races, and subject to penalties when dangerous conditions or toxic exposures exceed the limits under law.
- The regulatory apparatus in place for black lung and other occupational diseases would not work to prevent disease, however, if the 19th-century societal view that black lung is a product of genetic traits were true.
- Another striking example is the sickle cell trait, which was used in the 1950s as a reason to exclude black air force pilots

from flying at high altitudes. Thus, the genetic trait functioned to separate pilots on the basis of race in an era in parallel with a legal system where segregation separating black and white races was acceptable under law. By contrast, recent research praises the presence of the sickle cell trait as a natural defense against malaria. Therefore the same genetic trait that was misused as a tool for destructive discrimination in one context is viewed as an advantage that may prove to be lifesaving under a different social view of the same disease [1].

8.1.1 What Is Health?

Answering the ancient question what is health? is the greatest challenge for the new generation of epidemiology and the occupational, environmental, and public health sciences in light of the developments in nanomedicine.

The simple textbook answer to this ancient question is, "Health is a state of complete physical, mental and social well-being and not merely the absence of disease and infirmity" [2].

More than a generation after the writing and ratification of these paradigm-defining words in the WHO Constitution, this phrase will have renewed meaning in the case of nanotechnology because nanomedicine will redefine old concepts such as disability and health [3]. This definition will change dramatically because of nanotechnology, because early detection and presymptomatic treatments will change the meaning of "absence of disease or infirmity." New treatments will cause a paradigm shift in the stage of illness that confirms diagnoses, and consequently, people who are viewed as "healthy" in the greater society will be "treated" as if they are already ill [4]. This may produce excellent health outcomes, but there also looms large the specter of uninsured populations increasing in size and new forms of discrimination against a class of people who were not previously viewed as disabled.

Convergence of new genetic technologies and the nanomedicines that will transport the products of our understanding of genetic processes across cell walls will reshape our societal concepts of sick or healthy populations. Nanomedicine may signal the demise of "one size fits all" regulatory standards, replacing them with personalized assessments based on flexible regulatory criteria that will redefine populations on both sides of the border: people who will soon be easily "cured" and people whose illness is so small it seems to be invisible.

These concerns must be addressed not only without bankrupting health care systems or saddling unsuspecting third parties with liability but also without creating an underclass of people who lose their employability due to stigma, discrimination, potential future injury, or lack of access to good medical care. For these reasons, convergence of new genetic technologies and nanomedicines may redefine our collective understanding of "safety," "health," or "disability" and may challenge both fundamental legal principles of equity and fairness and also the scientific underpinning of existing standards. The 20th-century heritage of one-size-fits-all standards will give way to different standards for special needs and vulnerable populations as nanotechnologies enable precision medicine.

Nanotechnology will therefore force a redefinition of health and disability, in daily life and under law. Using the promising techniques of nanomedicine, it will be possible to treat diseases sooner, possibly even while the people who are expected to become ill are still apparently healthy. If so, nanotechnology provides people around the world with the exciting opportunity to bring together health sciences, policymaking, and law to address prophylactic treatment on a mass scale as never before. It is not certain, however, whether nanomedicine will be useful for everybody; if an individual's health status reveals a problem that is not clearly covered by existing law, then society must decide whether insurance law and societal protections against discrimination will bend or stretch to accommodate the patient.

8.1.2 What Is Work?

There is a simple answer to the question what is work? "Work is anything you don't want to do, but that you do it anyway." Oscar-winning movie star and popular singer Madonna scandalized her peers by describing her workplace on the movie set during the filming of *Evita* [5]. Madonna's diaries document her discomfort with makeup contact lenses and lapses of security that made her both fear and confront threatening crowds of adoring fans. "On the drive from the airport I twice saw graffiti painted on the walls that said, 'Evita Lives, Get Out Madonna.' How's that for a welcome? . . . Today was the day from hell, sort of. First I slept like shit. Shakespeare this was not. And why should they sleep? Everyone is unemployed; no one has to get up and go to work in the morning. The only people making any money are the press, and they will go to any extreme to get a picture or any information about me" [5].

This concept that work is a hardship can be tested in the daily lives of people who are not movies stars, enduring the highly paid hardships of working on location: just ask a teenager to clean someone else's room or his or her own bedroom. Getting that teenager motivated to actually perform the task is work; the teenager is likely to storm out of the household angrily rather than completing the assigned tasks. Yet, work is also a liberating, invigorating source of achievement, pride, and rewards, which boost self-esteem. Although work can be isolating from family and friends, it is also and potentially the greatest source of intimate contact with the rest of humanity. Work takes us away from personal life, and yet, it is the embodiment of all our hopes and aspirations, which are, in turn, transferred to our children in their work in the next generations. And thus the paradox of "working for a living" despite the hardships, the risk, the sacrifice of leisure time, even if working with close family and friends. The chance that one might die or become injured while toiling to make life better for one's family, friends, or oneself is at the same time a reality that there is no such thing as zero risk; the goal therefore is to minimize risk and to mitigate its impact through new approaches to risk management , some as simple as roadway construction warnings in Fig. 8.f. This well-established notion that healthy working conditions are linked to the survival of civil society will return to the policy arena for nanotechnology.

> Sound occupational health risk management programs are the grease for the machinery of powerful economic engines. Information provided through occupational health programs helps employers survive, because accidents and disease are not simply expensive but wasteful. No one can afford waste in our economy. The fat to be trimmed, however, is not the same as the grease for the wheels and machinery that makes smooth commerce.

There is another component that is also overlooked about work: not all work is paid. Work may require payment under some labor laws. That definition is incomplete from the standpoint of legislative drafting to prevent or reduce risk for two reasons: First a definition that is based on payment leaves out unpaid labor, especially slaves. Second, the definition excludes volunteers who are unpaid. Yet, even when domestic servants are hired, the need to perform unpaid household tasks remains [6]. Long ago, people recorded causal relations between exposures to dangers and injury in the workplace, even though they did not have the investigative and regulatory tools that exist now to correct these problems. Some of these notions survive from ancient times, along with other artifacts of the work of the ancient civilizations. For example, Pliny the Elder noted that workers in dusty environments should wear fish skins to protect them, an early form of respirators. The impact of sleep deprivation was also understood centuries ago. Adam Smith wrote in the 18th century, in a manner that remains valid, "Workmen [sic] . . . when paid by the piece are very apt to overwork themselves and to ruin their health and constitution in a few years. A carpenter in London is not supposed to last in his vigor above eight years, and something of the same kind appears in many other trades" [7].

Smith's comment was part of an economic argument against slavery; he argued that slavery cost far too much for maintenance and the risk of loss of property compared to hiring overworked and underpaid free men. By the 19th century, legislators understood that factories and coal mines ought to be inspected to prevent dangerous conditions because even if one or two people were successful in avoiding the risks of such work by choice, in the aggregate, society needed hundreds of thousands of people to work in the coal mines and factories. The struggle for replacement of people who died working in these dangerous but essential professions was the hallmark of the Industrial Revolution; the overall disease burden also took its toll, costing money to nonworkers, such as the owners and operators of coal companies, as well as the greater society. In response, therefore, laws were written in which common law theories were modified, not only to protect the harmed workers and their dependent families, but also to ensure the smooth flow of commerce throughout society.

Occupational health literature, which so minutely details the peaks and valleys of exposures to so many toxins in the workplace, surprisingly fails to discuss society's willingness to reconcile the presence of potential harm, injury, or risk while working for a living. Nor does the literature discuss the meaning of these exposures in life with family or life in the home from the standpoint of their impact on other family stressors and the overall well-being that shapes individual health.

Figure 8.5 International Labour Organization (ILO) *Effective Protection for Domestic Workers: A Guide to Designing Labour Laws* book cover. Source: ILO, 2012, Geneva, Switzerland.

Some of society's most important work, such as child rearing and other forms of intrafamilial nurturing, is unpaid work. There is no salary for performing such tasks as cooking a special meal for one's lover or oneself, but it is nonetheless a necessary job. Family caretaking is a necessary job, whether paid or unpaid, fraught with risks. For example, much of the labor performed by housewives, who use harsh chemicals to clean their house, are exposed to the risk of disease as caretakers for the sick or who are sleep deprived are also underpaid or undervalued by not being paid at all [8], with attendant risks: Caring for a sick child exposes adults to communicable disease; driving a friend to the mall for clothes or groceries introduces the hazards of driving and perhaps other hazards in the transport of commodities and contact with the general public. So, too, new technologies must be taken into account [2, 9]. As nanomedicine opens the door of employment opportunities for women as surrogate mothers with implanted eggs from different countries and at various stages of fetal development, the questions must be asked whether the terms of the new convention protecting domestic workers will apply to these trades, too. Even if such workers are sometimes outside of their own household, or are required by contract to stay in baby farms where their activities, medical status, and fetal growth can be carefully monitored, there remains a concern that this form of household servitude also has an impact on the life and health of the unborn child because of transplacental transfer of nanoparticles. The exposures to nanotechnology will embrace all parts of society. To be effective, the concept of "work" must be redefined under law as broadly as possible in order to capture the range of risks. Nanotechnology is redefining work by creating new approaches to ancient tasks and new tasks that never existed before, and therefore will force society to revisit these perennial issues under existing law and by drafting new laws.

8.1.3 What Is Nonhealth or Disability?

Everyone has a disability. Everyone has a gift. Your job is to find the gift and remove the obstacles of disability

—Sylvia Feelus Levy, 1974 [10]

An emerging issue for policymakers that will gain increasing significance with the successes of nanomedicine concerns the life of so-called "healthy disabled" people. In every society there have always been people who are considered disabled: Individuals who qualify for national insurance or social assistance due to a specific severe handicap or cluster of impairments with attendant comorbidities but who are not expected to share in the productivity of the general population at large. Due to illness, these populations are often excluded from clinical trials and research efforts because it is presumed that they will not be able to withstand experimental treatments and that their experience is not representative of the larger, seemingly healthy population. Within this group, however, many people may conform to traditional parameters for the social construct of being considered healthy [11]. For example, someone with palsy who cannot walk may have normal-range blood pressure and not be at risk for negative health impacts of diabetes, obesity, or coronary failure. Nanotechnology applied to medical care and nanomedicine will enhance the importance of the healthy disabled population for a great spectrum of research, including the influence of social expectations following diagnosis of illness on psychological status, social well-being, and longevity in relation to an expected survival curve.

If applied with forethought when rethinking these vital social values, two sets of benefits can be realized by civilization at the same time, not as competing interests, but as one invaluable set of societal change: The miraculous developments that sound like science fiction to those people who eagerly anticipate these medical products, combined with the new social dimension of protecting rights of people with disabilities, will reshape all of civil society. Thus an unprecedented opportunity exists to benefit from the simultaneous nanotechnology revolution and the revolutionary social change that recognizes individual human potential by promoting the equal opportunity for people with disabilities under law.

8.2 Voluntariness or Choice in Major Life Decisions

Fundamental issues of justice, fairness, and equality that exist in science law and policy are often rooted in a belief system about "choice." As Smith remarked centuries ago, poor working conditions and low wages for long hours may be viewed as a matter of free choice because workers are not obligated to accept those terms or conditions. Modern studies about hidden costs of disease, revealed in the calculations of the disease burden (GDB) by WHO, offer evidence that the harm from unchecked freedoms in contract hurts all of society by wasting money and increasing costly health care burdens upon society as a whole. It is in this context, too, that the fabric is woven to produce a safety net of *corporate social responsibility* designed to foster sustainability and innovation, both simultaneously needed by society. Efforts toward responsible development of nanotechnology reflect both the learning from modern tools of risk management and the free-choice heritage.

Perhaps the most overused word in the law and ethics vocabulary is the word "choice." Fundamental issues of justice, fairness, and equality that exist in eugenics, just beneath the glitzy research and "big science" used to map the human genome, have outcomes that seem to depend upon choice or the quest to attain a wider array of choices. Choice may be relevant when standing in a department store, facing three different dresses that all fit perfectly at a reasonable price, but is an inapposite term when characterizing medical decisions. People may select a medicine or decide to continue a pregnancy to term, but the random underpinning of having such an opportunity is undervalued by use of the word "choice." Genetic studies revealed that societal decisions about what constitutes good or desirable traits suggest that humanity is moving along a spectrum from chance to choice [12], but little attention has been given to what choice really means in terms of baseline information that humans can neither control nor change. It is like fighting gravity—sometimes there is no point in pretending that the outcomes of decisions are the product of anything more than a very narrow spectrum of choice.

Although some people may advocate an oversimplistic view that justice is merely a distribution problem, whether fairness is a question of distribution, a fundamental value, or an elusive social goal, "equal opportunity" has a benefit to the greater civil society that ancient paradigms rooted in racism, sexism, classes, and stigma for disability have been deprived. The line between health and disease will be redrawn by society, but the question must be asked whether people who refuse treatment will have had a genuine choice if, for example, their decision is based on economic constraints and whether failure to comply with the new accepted social construct for treatment will be punished, either indirectly by social stigma or directly by a subsequent refusal of the health care system to provide obviously needed treatment when the illness becomes manifest. And, people who decide to do something ambitious even though it is difficult should not be punished or chastised for having made a "choice," if in the end all posterity will benefit because they have pursued their difficult goal. Nanotechnology cannot answer these ancient riddles, but the transformations new technology brings to society enable people to consider rethinking the old problems and redefining the correct approaches to choice when measuring outcomes and shaping societal goals.

8.2.1 Principles of Exposure to Risk: Employer Responsibility for Recognized Hazards

Across the world, hundreds of federal, state, and international bodies of law are bottomed upon the philosophy that work-related illnesses are an avoidable aspect of industrialization and that consumer protection against risky products is a cornerstone of public health. Employer acceptance of responsibility to provide safe and healthful working conditions is part of the paradigm in which an employer makes the choice to run an enterprise and employ people. But it is also clear that not every event that follows from the decision to grant employment is also a product of clear and discernible choice. It makes sense, however, that employers should be accountable and be society's repository for liability. Therefore, tools are needed to assist in the achievement of goals made by choice and reducing the risks that follow from those choices.

Occupational health and safety management laws are implemented through several channels for intervention to prevent workplace harms and thereby reduce corporate liability, such as risk management programs. Risk managers, industrial hygienists, and many additional health professionals use well-honed prevention strategies to stop problems from becoming catastrophically large. Such in-house programs, and the supporting infrastructure of regulatory agencies, work together to do much more for the economy than merely reduce the costs of accidents and the overall burden of disease in society. Nanotechnology laws, whether licensing new products or setting forth criteria for risk governance, can provide an opportunity to redirect the resources within these systems and thereby provide a more flexible but comprehensive system for protection of health than civil society has experienced in prior generations.

8.2.2 Redefining Choice in Maternal and Child Health Laws

Protecting reproductive health for all requires a new paradigm. Jurisprudence surrounding reproductive health has been carved into small portions by courts and legislatures, with only piecemeal protections for pregnant workers and their offspring. International human rights law does not take into account the special needs of women during pregnancy nor the randomness of events following the decision to carry a pregnancy to term, erroneously labeled as a "choice." Conversely, abortion jurisprudence relies at its root upon a woman's right not to choose to have children and therefore analytically stops at the termination decision. Consequently, important opportunities to protect health and ensure economic well-being of mothers and their children have been lost to humanity [13]. On the microscale, these issues become intensely evidence specific and problematic [14]. Yet, when viewed from the broad perspective of occupational accidents, death from fire, or other types of disasters that can be avoided through adequate planning [15], and from greater appreciation of the basic societal need to have a new generation populated by the healthy offspring of the previous working generation, the notion of choice becomes an artificial construct for refusing to grapple with social issues that are the product of larger problems such as sexism and embedded discrimination [16]. Since not all mothers have consistently adequate protection of health or job security, many consequences follow that cannot be blamed upon individuals or mothers as a matter of "choice."

These issues cry out for a new approach to solve these ancient conundrums [17]. Piecemeal approaches have been taken for this problem in the past [18], but nanotechnology offers new avenues for legal resolutions to the important policy questions surrounding the impact of working conditions on the health of the next generation as it potentially supports or undermines human life. Responding to these issues properly requires rethinking about these problems from a perspective that is maternalistic, rooted in the needs of working women who cannot control the destiny of their pregnancy [19].

A maternally driven model, starting with the working assumption that whether or not they are paid, all mothers work, can better address the uniquely individual but paradoxically universal need for health protection. The appropriate model for asking these questions, and for assessing these answers, employs a new paradigm: taking into account the special needs of women during pregnancy without the unfortunate tendency of society to blame mothers for social ills that may impact their children. From a perspective that is maternalistic, advocates, researchers, and policymakers can use risk governance in emerging nanotechnology frameworks such as EC4safenano to address infrastructural determinants health and illness for childbearers and their offspring. A new maternalistic model can reflect a deeper understanding of the inextricable link between health at work and the health of posterity. Nanotechnology and the breakthroughs in nanomedicine that rely on transplacental drug delivery offer the opportunity to create a maternally driven model for addressing the need for reproductive health protections. The new model for reproductive health can replace myths and prejudices that blame mothers or pretend that high-level professionals are not workers, replacing these errors from the past with a universal approach to occupational health that can become the vehicle for more efficient protection for health, life, and posterity of civil society.

8.2.3 ART in Laws about Parenting, Including LGBTI Communities

Nanotechnologies that produce nano-enabled assisted reproductive technologies (ART) and innovations in nanomedicine that help infertile people will cause a redefinition of parenting. As pointed out at the Brocher Foundation, Hermance, Switzerland [20], there is a need to engage academics, professionals, governments, and all stakeholders in a discussion about the global market of cross-border human embryo and stem cell transfers, egg cell and sperm donations, and surrogate mother arrangements. Nanotechnologies as applied to ART enable many more people to have biologically related children than in previous generations and also involve adding third or fourth people to the reproductive equation of parenting. Nanoscience discovering information about, for example, transplacental transfer of materials increasingly points to a biological role of people in the chain of custody for embryonic materials during the ART process, even though laws may not recognize their role. This aspect of nanomedicine opens the door for parenting to new constituencies who may not have participated in human reproduction in the past. Nanotechnology applications and regulations about ART therefore offer society an opportunity to give a distinct parental role to people throughout the ART process, including allowing same-sex families to participate in new social paradigms around parenting.

8.2.4 Autonomous Choices about Prolonging or Ending Life

Nanomedicine in combination with improved genetic testing will alter the boundaries of presymptomatic discovery of disability, thereby holding important implications for legal and cultural mechanisms to offer or require treatment for people who are expected to soon become ill. New opportunités for treating terminal patients will raise ancient questions about the *value of one day more of life.*

Bioethical studies involve asking these perennial human questions that nanotechnology will once again bring to the fore. Informed consent questions about the scope and possible long-term harms from studies already occurring with human volunteers are already in play in Switzerland, even as molecular biologists work with nanotechnology applications to replace animal experimentation with synthetic biology paradigms for studying toxicity and predicting the likelihood and progress of disease. As noted by the Work Health and Survival Project (WHS) in its proposal for the MOM Project grand challenge, "Since nanotechnology is a revolution by every measure, we possess a once in a millennium opportunity to uproot embedded errors in our methods of creating and administering health care that concretizes or exacerbates health disparities. Or, we can use the new technology to eliminate or reduce those disparities and provide gender equity in health outcomes at last."

The scope of experimentation and understanding of the entire process of translating scientific findings and data into legislative policy also raises important bioethical concerns about fairness to producers and consumers when (i) limiting or controlling exposures to toxic substances, (ii) regulating the shipping and application of nanotechnology in global commerce and workplace exposures, (iii) measuring the health impact of environmental bioaccumulation, and

(iv) determining the content for consumer labels for products using nanotechnology. Sylvia Feelus Levy was a patient with terminal cancer, who died three days before her only son's ornate wedding party. There had been a simple wedding ceremony in the hospice where she stayed during the last two weeks of her life. But it was the big party three days after her death that she had lived all her life wanting to see: live music and dancing and flowers, a party, well-wishers, and fancy food that are common among wedding traditions that she had known for all of her life. The legendary moment when she would dance with her son at his wedding was just three days away. She had waited for that moment all of her life, but instead her life was gone. She missed that lifelong cap to her maternal achievements by a few hours. Surely everything she owned as material wealth paled in value compared to the ability to see the big party and dance at the wedding of her only son. Three days.

Three days meant the difference between being there or being a ghost of a memory for few guests at the dancing with live music and catered food by the man-made lake and being someone whom no one asked about because they assumed she was very ill. Because conversations about terminally ill people are often awkward, there is little one can say of genuine comfort, and everyone is powerless to change the outcome, no one talked about death at the big party, as if she never existed at all. She had the dress purchased and hanging in the closet, but she had often said she knew she would never wear it. People who attended the funeral just days before were mute. A mom who worked all her life to see the wedding of her children, and when at last it happened, no one noticed she was not there. Surely, she would have given anything she owned to be there alive and well and dancing the first dance with her son, just as his father-in-law took the opportunity to dance with the bride. A priceless use of time, if science could have prolonged the survival curve 70, 80, or 100 hours in the life of one cancer victim.

Disability among humans confounds traditional notions of equality because the random nature of genetic traits, propensities, and contact with disease do not treat everybody equally. Every individual in society may be ill, recuperate, and regain health or lose health again many times in his or her lifetime. So the population to be considered "disabled" is fluid, unlike populations defined by gender or race. Under present social constructs that reflect the limits of existing medicine, benefits of prolonged life are weighed against the cash outlay. It is often said that the last year of life is the most expensive and that the quality of life is very poor. So there are few incentives to prolong life in terminal disease. In instances where mere hours can dramatically alter the individual's life story, despite disease, there will be incentives to provide such services and unclear limits for inappropriate practices unless there are governing regulations.

Disability poses profound challenges to international human rights laws that are predicated on equality [21]. Core populations that are considered "disabled" will change across time, even for people with long-term conditions that are disabling; no person lives an entire lifetime devoid of illness, infirmity, or physical disability or impediments to the quality of life from genetic conditions or the accidents of nature, daily modern life, or war. Equality, and the notion that everyone should be treated the same as part of the concept of "fairness," is a fundamental belief [20] and a working assumption in most international human rights mechanisms. Underlying social values can be viewed in this sense as a distributor of equal opportunity and justice. Paradoxically, disability presents the inherent challenge of understanding, accepting, and allowing society to benefit from the most individualized of all individual rights without applying the concept of equality [23]. Ill-health, disease, injuries, and genetic information therefore compel the international system to squarely and candidly confront the unique nature of individual differences that cannot be replicated, that make each individual a human with his or her own set of memories, gifts, limits, and experiences. But no set of experiences with disability are the same; neither their severity nor their impact can be considered equal.

Nanomedicines and nanotechnology applications for health care services offer the promise of changing this health status calculus dramatically. The bioethical issue will soon arise, concerning some opportunities offered by nanomedicines: whether patients can store and then harvest replacement parts in the case of disease in an attempt to buy priceless once-in-a-lifetime moments, and if so at what price? Health econometrics will need predictive models that can measure what cost to society will be involved when accepting this new facet of the GDB and how much patients should be expected to fairly pay to have such opportunities. Existing disability conditions will play a role in social governance of these decisions to some extent, and the race, class, age, and gender of individuals will also be important factors to consider. Questions concerning whether health status can be guaranteed to some purchasers of replacement services or whether there should be no liability for rendering services that prolong life for a specific deadline regardless of a duty of care will then become prominent. Stakeholder input will be needed in order to enlighten the policy discussions about these major life issues, which will use innovation to promote health and commerce at the same time.

8.3 Who Needs Law?

There is widespread consensus that nanotechnology will change the world by creating products that are smaller, faster, stronger, safer, and more reliable, even for small and midsize enterprises, the so-called SMEs [24]: What does this mean for the rule of law? And, more importantly, who needs law?

Technology can spread information—fast—but is it a thoughtful, reasoned analysis of the law that our technology spreads or just some propaganda that turns "law" into a bad word [25]? The law is beautiful, one of the most precious gifts in our society. But the ability of contemporary culture to pervert the rule of law by depriving citizens of correct information about cases or statutes is sad. In the 1920s, every US student was required to study civics in high school. This provided everyone with a basic, rudimentary understanding of the law regarding social order, civil liberties, and basic societal concepts of right and wrong. Citizens could not graduate high school without demonstrating some understanding of the law. In that time, young citizens were required to learn about their government and laws before they first exercised their franchise. That requirement vanished sometime between the two world wars and the generation that followed. By the next generation, the greater society's desire to have citizens who are aware of the law somehow became perverted by political forces that understood that a body politic is more easily manipulated when citizens are unaware of the law. By the 1970s, studying law was not even standard fare in college (although such courses may be offered at the law school on campus, credited toward a college degree). Now information about law has fallen aside in the information age—rarified knowledge possessed by few. Yet, citizens who are unaware of the law bump into it in the oddest moments when they are trying to do something. It may feel strange to be required to give other religions or people with disabilities or unusual ethnicities equal rights, equal employment opportunity, or equal media coverage for seemingly inane political views.

Furthermore, when people bump into the law for the first time, without legal education, typically the law is saying no!

No, you can't walk across the street at the red light when you want to; no, you can't tell those people not to go to your school; no, you can't run your business the way you want to; no, you don't have the right to control your land and neighborhood the way you thought you could. No, you can't drink as much alcohol as you want before driving; no, you can't drive along open roads as fast as you want. Inevitably, someone else is standing there yelling about their rights under law, wanting money or wanting to curb personal liberty to make up for some invisible harm. To a citizen who has lived years in a nation without studying the law, it must be a shocking introduction—easy to dislike the law. By contrast, citizens who are aware of the law rarely wake up in the morning with goals of breaking the law. Few citizens start the day deliberately running red lights, followed by deliberately making racist comments to colleagues and then deliberately hurting people. Instead, people who are aware of the law gladly comply with the social contract that ultimately also helps them. So, too, if there is a clear regime to protect public health, people will consistently look to the law and apply sound principles of risk management when earning their daily bread by applying nanotechnology, the newest slice of global daily life.

8.3.1 The New Paradigm for Rights of People with Disabilities under International Law

Disability is a universal, puzzling, and pervasive facet of the human condition. Everyone is different, yet everyone must get the opportunity to be treated the same. Universality is a fundamental cornerstone of all human rights norms, so disability protections, including the freedom from prejudice that harms the implementation of civil rights for persons with disabilities, would seem natural, if not positively codified, under human rights norms. National and local laws prohibiting discrimination based on disability and the United Nations Convention of 2006 that complements them typically also protect people who are otherwise considered "healthy" from discrimination if another person or institution has harmed them because of a mistaken belief that they are disabled or if the illness or disease is "not manifest."

Such terms will take on a whole new meaning in a generation when treatments may be required or commonplace using medicines that depend upon nanotechnology. "There is a fine line of difference between sanity and insanity," said the late stand-up comedian, prescient social commenter, and author George Carlin in his classic routine "Class Clown," "*and I have erased it*" [26].

Two decades later, the US Congress followed his lead. There was a time in world history [27] when it was legal to draw a line and segregate people with disabilities into institutions, away from the rest of the population. But the US Congress, the United Nations Convention on the Prevention of Discrimination against People with Disabilities (UN PWD, 2006), and hundreds of national legislative bodies have erased that line of separation. In July 1990, the first President Bush signed into law the Americans with Disabilities Act (ADA), following the Individuals with Disabilities Education Act (IDEA) and underscoring the role of state and local human relations laws prohibiting discrimination based on disability. Those statutes also became the model for the international legislation by the United Nations, which has been ratified by the majority of nations in the world.

Against this backdrop of preexisting recent social change regarding the rights, nature, and social behaviors surrounding health and disability, nanomedicine will change the rules of the game of disability treatment, the definition of disability for insurance, and the nature of long-term prognosis for rehabilitation. Cures and manageable treatments, often for presymptomatic patients who can anticipate problems on the basis of their genetic profile, life experience, and age also means an increased role for rehabilitation. Nanotechnology applications will therefore challenge the existing social constructs about health and disability, because these social constructs shape the demand and use of health care delivery systems, purchases of devices and medicines for treatments, and the needs assessment as new demands will impact use of services.

Nanomedicine's major breakthroughs in health care will transform the lives of so many patients that it is worthwhile to reflect with forethought about policy implications on a population level. More patients who have a given illness will be sicker longer or in the secondary phase of long-term illness, when they will be able to function at a much higher level compared to people who had the same illness and might have functioned decades or centuries before. But they will need treatment. When thousands and eventually millions of people will be able to overcome the obstacles of long-term illness and return to their former life, will the society outside the hospitals, nursing homes, and rehabilitation centers be ready and willing to absorb them into the workforce and their homes?

If applied with forethought when rethinking these vital social values, two sets of benefits can be realized by civilization at the same time, not as competing interests, but as one invaluable set of societal change.

8.3.2 Bringing Health to Work for People with Disabilities

Society has been radically changed, hopefully irretrievably for the better, by the creation of laws that promote the rights of people with disabilities and prohibit discrimination against them. Disability poses profound challenges to international human rights laws [21]. Due to the revolutionary change in discrimination laws that require hiring, promoting, and protecting people with disabilities as a vital part of the workforce [28], several fundamental aspects of workplace design, implementation of industrial hygiene protections, training for occupational health programs, and "the way we do business" will fundamentally change. The arrival of nanotechnology in the workplace provides an outstanding opportunity to implement such change, because the changing demands of work that will come about through the application of new technology also require redesigning the workplace. It is possible, therefore, with forethought, to create opportunities that maximize the benefits of both the social change in disability laws, combined with demographic changes, by rethinking old values and reshaping social constructs, in light of nanotechnologies that will transform disability into health.

At last, an unprecedented opportunity exists to benefit from both the nanotechnology revolution and the revolutionary social change that recognizes individual human potential under international laws preventing discrimination against people with disabilities. The arrival of nanotechnology, praised and heralded as a welcome revolution reshaping industry, provides the perfect opportunity for rethinking workplace design and blending into the weave of the workplace fabric antidiscrimination goals, thereby folding into the fabric of the typical workforce people whose special needs may have previously placed them at the margins. The maturity into the workforce and reproductive age of a generation of people with disabilities who have been raised with an understanding of their rights is an important social change that cannot be ignored. Despite medical disadvantages, such students become professionals, parents, and workers, but they do not lose their consciousness of their rights as people with disabilities.

For the first time under international health laws, there exists a cohort comprising an entire generation of people with identified disabilities who would have been living in institutions a generation or two ago but, instead, are armed with rights as well as a renewed sense of self-worth. Demographic change based on the new social constructs for the role of people with disabilities under law, combined with new avenues for treatment using nanomedicine, will require a different methodology to tease apart the cause and effect between workplace or environmental exposure and health outcomes. The new workforce that is implicitly different compared to the totally "able bodied" workforce of the previous generation holds implications for social theories of aging that will change the target population for treatments using nanomedicine. The presence of disabled workers using nano-enabled assistive technologies and nanomedicine treatments before they experience commonly visible symptoms will change the nature of many job descriptions, as only the "essential functions" of the job will be necessary under law. These combined developments have attendant implications for job design, specialized methods of risk communication, and job hazard analysis. It should not be surprising therefore that the abundance of new methods for accommodations for the young generation will also have intergenerational impacts for the working health of older people, who may wish to remain in the workforce even while undergoing treatment using nanomedicines.

Around the world, law now requires equal opportunity for people with disabilities. Implementing laws promoting the rights of people with disabilities has expanded the definition of disability and the obligations of everyone in society to create opportunities for them, regardless of cause of injury. New methods to measure the cause and effect between workplace or environmental exposure and health outcomes, including the use of big data combining several datasets, will be developed using nanotechnologies in order to understand the role of new exposures in combination with existing disability. The new disability paradigm under law combined with nanomedicine will change the daily lives of chronically ill populations by redefining societal notions of health. This is not an abstract problem, because there are very stiff fines and penalties for failure to comply with disability laws preventing discrimination, and nano-enabled methods will allow analysis of individual cases much more precisely. The rightful presence of an identified disabled population within the workforce will change the nature of many job descriptions because only essential functions of the job will be necessary, as mentioned earlier. Jobs will then be custom-tailored to accommodate deficits and to maximize individual productivity.

These new members of the vibrant workforce are empowered by refined tools of self-advocacy under law, who possess the assistive technology as well as the regulatory mechanism to operationalize their rights.

Simultaneously, job analysis and job descriptions will be required to take into account inexplicable but nonetheless a common coupling of some diseases together, called comorbidity. Commonly, when a parent or caretaker is asked to assume some responsibility for a person with an identified disability, the caretaker is briefed about the panoply of expected problems that come with the main disability. For example, many people with invisible disabilities may also have an obsessive compulsive disorder. Although nanosciences may not provide clear explanations of why some diseases or illnesses are often found together, the ability to use nano-enabled technologies to bring disabled people to the workforce means that comorbidities will also be taken into account when fashioning job descriptions, even if this health impact may be diminished due to applications of nanomedicine..

Consequently, nanomedicine will accelerate society's increasing interest in providing custom-tailored job analysis and attendant health care supporting work, in a manner that is similar to the controversial techniques that are proposed by pharmaceutical complies to apply to "personalized medicine" using nanotechnology. For these reasons, convergence of new genetic technologies and nanomedicines may redefine our collective understanding of "safety," "health," or "disability" and may challenge both fundamental legal principles of equity and fairness and also scientific underpinnings of existing standards. Antiquated prejudices embedded in existing paradigms for addressing these issues will be discarded in order to construct valid working assumptions. This population shift requires deciding who will pay for treatments, when will they become considered standard without a consent required, when will individual patients be allowed to exercise their right to refusal, and when undergoing treatment for presymptomatic illness will there be a major wave of hard policy choices to be made regarding which people be considered disabled with all the legal protections available to people with disabilities. Additionally, long-standing but incorrect working assumptions regarding safety and health and the format of preventive programs will be challenged by the presence of an entire new cohort of people with disabilities. The juridical heritage of one-size-fits-all standards will give way to an individualized approach, applying standards that attempt to achieve performance-based outcomes. And the sea change represented by modern disability laws, operationalized by applying nanotechnologies, will acquire the force of a tsunami.

8.3.3 Redefining Life and Work

During the 20th century, increased regulation of occupational health is credited as one of several variables that improved life expectancy at birth among US residents (increased by 62%, from 47.3 years in 1900 to 76.8 in 2000) [29]. In April 2015, the US government announced, "Safety and health at work is a basic human right and a necessity for . . . development" [30].

As described by Jean-Claude Javillier, the jurisprudence of international labor standards has come to a crossroads [31]. Predicted changes in working life that were forecast in testimony in 1998 are brought to the fore by nanotechnology's revolution for commerce [32].

Risk governance programs must have conduits for input and feedback from all stakeholder constituencies. To be effective, these portals for stakeholder input should also include the views of people who work with nanomaterials. This requires listening to the voices of individuals, professional societies, and their affiliated nongovernmental organizations (NGOs), often in the informal sector or in the leadership and management of small- and medium-size enterprises. Nanotechnology provides fertile terrain for new start-ups in small business enterprises. These nanomaterial workers cannot be represented by the needs of the antiquated model for labor representation, which offers individual stakeholders a narrow voice in hierarchical trade unions. Instead, defining and operationalizing rights for stakeholders in the arena of nanotechnology policy must offer new opportunities for the voices of workers to be heard, even in the managerial staff that is excluded from classical models for workers' right to know and related risk communication. Labor movements and their opinion leaders remain ill-equipped to respond to the social changes that have dragged them along the road. Javillier comments upon Alston's correct view of the "shrinking" role of labor law and the legal community's quest for new incentives to apply juridical norms in daily life.

New methods for workplace exposure measurement, and controls [33], possibly requiring new precautionary approaches handling nanomaterials [34] will reshape induistrial processes and once again reorganize the structure that governs work. To meet this demand, existing systems for occupational health protection must radically transform the definition of the terms "worker", "workplace," "occupational hazard," "occupational illness," occupational injury," and "workplace death" when rethinking the notion of who was in harm's way as a worker and the remarkably limited control that the employer had to prevent occupational death. The system that is revolutionized by nanotechnology must confront an inevitable issue concerning implementation of best practices and sound regulatory frameworks: Given all these laws, no one has the time, money, or expertise to monitor and enforce their implementation, how to incentivize the reluctant, making benefits of compliance available to all.

The state of the art of understanding about emerging risks of nanotechnology in general, and the potential adverse health effects on the skin, lungs, and reproductive health of workers from occupational exposure to carbon nanotubes (CNTs) and nanofibers in particular, requires closer scrutiny. Implementing a management system to formalize the compliance process may therefore often require going «beyond» compliance from the standpoint of the letter of the law in order to ensure that precautionary principles embedded in the law are met and that proof of so-called due diligence toward compliance can be well documented.Additionally, long-standing but incorrect working assumptions regarding safety and health and the format of preventive programs will be challenged by the presence of an entire new cohort of people with disabilities. These new members of the vibrant workforce are empowered by refined tools of self-advocacy under law, who possess the assistive technology as well as the regulatory mechanism to operationalize their rights. The most pressing challenges involve the integration of a new generation of disabled workers, protected by law, but without work experience, into the cultural matrix of occupational health and safety—a field that rarely had the opportunity to work with people after the onset of disability until recent changes in national statutory protections for people with disabilities and also the development of the UN Draft Protocol on Rights for People with Disabilities. The presence of a large population of people with disabilities will therefore require the development of different datasets compared to existing data, with new working assumptions to protect their special needs.

Another area of transformative attention due to nanotechnologies is the relationship between work and reproductive health. There is only sparse literature concerning the risks faced by women workers or potential parents in the workplace compared to the greater body of occupational epidemiology that scrutinizes occupational health. Significantly, even fewer of the studies in this field and its subbranch of reproductive epidemiology examines the role of all work, whether paid or unpaid, at home or in an office, in relation to reproductive health. Yet, throughout the history of humanity, women have worked during pregnancy. Anyone who has driven a car while experiencing morning sickness with a small child in the backseat who was screaming and crying loudly with futile protest against being strapped into a child safety seat knows that pregnant women face distinct risks that impact future life. Whether using strong household chemicals to clean the future nursery and paint it to make it a fresh home for a newborn or simply lifting siblings while engaging in a variety of child-rearing chores, characterizing work at home as safer for women or the unborn would not be a fair comparison to some types of sedentary work that are highly paid. Yet, pregnant women do not enjoy the privileges such as infamous special parking places or access to support from society that are extended to disabled persons under law.

Nanotechnology may offer the opportunity to alter these old paradigms. Occupational health policies in many modern societies have previously reflected a misplaced notion that occupational health and thus the vital issues of reproductive health at work only touch a small group of workers, pregnant women, young women, or, at best, when speaking in gender-neutral terms, workers who expect to have children. The arcane analysis has been costly to civil society, a maternally driven new model to address the need for reproductive health protections to challenge the dearth of effective programs. The basic human need for health and work that preserves civilization [35] requires rethinking existing problems that have seemed intractable in the past but may be solved by applying new science if the policymakers are willing to question and possibly replace unsuccessful outdated methods from the past. Thus, the well-established notion that healthy working conditions are linked to the survival of civil society will return to the policy arena for nanotechnology. Rights to health protections are ageless, and therefore changing paradigms for patient choices and informed consent in light of personalized medicine, which applies nanotechnology techniques to preexisting genetic and proteomic information about the individual patient, will also require a redefinition of access to care and the right to refuse treatment culturally and under international laws. Traditional models for health will be revised in order to respond to the needs of a transcultural workforce comprising older workers, women who were not previously represented in the workforce in great numbers, and workers who work for several employers, if not at the same time then certainly during their lifetime.

Nanotechnology confronts these millennial questions and offers new approaches for protecting public health . . .

References

1. Commonwealth of Pennsylvania, comments at the request of State Senator Jane Earll regarding informed consent for genetic testing, SB684, June 18, 2003; *Work Health and Survival: The Future of Genetic Testing at Work*, OEM report, May 2002; Spider silk jeans or spider silk genes? The future of genetic testing in the workplace, *New York Law School Journal of HR*, **18**(1), 2001, based on invited lectures at Yale University; *Genetic Destiny: Todays Laws, Tomorrows Technology*, MCLE for the Moseley Institute, 1999.
2. World Health Organization (WHO) Constitution.
3. Ilise Feitshans, The bright path ahead for workplace safety and nanomedicine, special presentation for the American Society of Safety Engineers ASSE 2012 meeting, Denver, Colorado, published in conference proceedings by ASSE, Des Plaines, Illinois, 2012.
4. Dennis Thomas, Fred Klaessig, Stacey L. Harper, Martin Fritts, Mark D. Hoover, Sharon Gaheen, Todd H. Stokes, Rebecca Reznik, Elaine T. Freund, Juli D. Klemm, David S Palk and Nathan A. Baker, Impact on medicine and biomedical research, in *Informatics and Standards for Nanomedicine Technology*, www.wiley.com/wires/nanomed 2011.
5. Madonna, The Madonna diaries, *Vanity Fair*, Nov. 1996.
6. Ilise Feitshans *Doing That Job That Moms Do for Free*, based on her work for the ILO white paper on domestic workers (2008), course materials for gender and globalization, Geneva School of Diplomacy, 2010.
7. Adam Smith, *An Inquiry into the Nature and Causes of the Wealth of Nations*, electronic classic, Pennsylvania State University, 2005.

8. International Labour Organization Convention on Domestic Workers, adopted June 2011.
9. Navi Pillay, Opening remarks, *Human Rights: The Next 20 Years, UN High Commissioner for Human Rights*, Geneva, Switzerland, Dec. 5, 2013.
10. Sylvia Feelus Levy, 1974, cited in Ilise Feitshans and Jay Feitshans, *Walking Backwards to Undo Prejudice: Report of the US Capitol Conference*, including disabled students, what works, what doesn't, Emalyn Press, 2001; Ilise Feitshans, *Gifts and Disabilities: Two Sides of the Same Coin*, essay reprinted for presentation at Barnard College Department of Education, November 2004.
11. A. Nordström, *Changing the Perspective: From Disease Control to Healthy People*, 2013, http://globalhealth.thelancet.com/2013/06/25/changing-perspective-disease-control-healthy-people.
12. Ilise Feitshans, From chance to choice: genetics and justice, invited review, *New England Journal of Medicine*, September 14, 2000.
13. Ilise Feitshans, Protecting posterity: the occupational physician's ethical and legal obligations to pregnant workers (state of the art reviews STAR), in *Ethics in the Workplace*, September 2002.
14. See generally Tee Guidotti and Susan G. Rose, *Science on the Witness Stand, Evaluating Scientific Evidence in Law, Adjudication and Policy*, OEM Press, 2001, and specifically Ilise Feitshans, Evidentary needs in occupational health law, in *Science on the Witness Stand*, OEM Press, 2001.
15. Ilise Feitshans, *Protection Means Exclusion? The Law against Fetal Protection Policies*, presentation at the Alice Hamilton Conference Center, National Institute for Occupational Safety and Health (NIOSH), Cinncinnatti, Ohio, November 1997.
16. Ilise Feitshans, Job security: prohibiting wrongful discharge of pregnant employees under the Model Employment Termination Act, *Annals of the American Academy of Political and Social Sciences*, **536**:119, 1994.
17. Ilise Feitshans, Is there a human right to reproductive health?, *Texas Journal of Women & the Law*, **Fall 1998**.
18. Ilise Feitshans, Legislating to preserve women's autonomy during pregnancy, *International Journal of Medicine and Law*, **14**(5/6), 1995.
19. Ilise Feitshans, Lecture about womens' health, Global Alliance for Womens' Health, United Nations Fourth World Conference on Women, China, 1995.

20. Brocher Foundation Summer Institute, ART, May 30–June 5, 2015, Hermance, Switzerland.

21. Ilise Feitshans, JECH invited submission: Speakers Corner Embracing the Universality of Disability in Order to Accept Difference: Can We Achieve Equal Treatment in Workplace-Based Prevention Programs under Law? November 1, 2006, paper presented at the Columbia University Human Rights Seminar, invited paper speaker series *Protections for Neurodiversity and Physical Disabilities Under International Human Rights Law*.

22. Declaration of the Rights of Man and Citizen, France, 1789, US Constitution, 1793. Consistent with this heritage, the notion of rights of men has been expanded to embrace the rest of the world. The United Nations' founding documents, the United Nations Charter, and the United Nations Declaration on Human Rights call for "respect for human rights and for fundamental freedoms for all without distinction as to race, sex, language or religion."

23. Ilise Feitshans, *Neurodiversity and Human Rights*, invited paper, Columbia University Seminars at the Center for the Study of Human Rights, 2006, http://www.columbia.edu/cu/seminars/seminars/society/seminar-folder/human-rights.html.

24. An SME is defined as a manufacturing business that employs fewer than 100 persons or a nonmanufacturing business that employs fewer than 50 persons in the Hong Kong Special Administrative Region (HKSAR). The number of persons employed includes individual proprietors, partners, and shareholders actively engaged in the work of the company and salaried employees of the company, including full-time and/or part-time salaried personnel directly paid by the company, both permanent and temporary (TID online information, SME funding schemes, http://www.smefund.tid.gov.hk/eng/eng_main.html, April 7, 2006).

25. Ilise Feitshans, Who Needs Law? Letter to the editor, *Haddonfield Sun*, Haddonfield, New Jersey, September 5, 2007.

26. George Carlin, "Class Clown," Columbia Records, 1969.

27. See United Nations, Standard Rules on the Equalization of Opportunities for Persons with Disabilities, Gen. Assembly Res. A/RES/48/96 (Dec. 20, 1993), http://www.un.org/esa/socdev/enable/dissre00.htm>; United Nations, Ad Hoc Committee on a Comprehensive and Integral International Convention on the Rights and Dignity of Persons with Disabilities, Gen. Assembly Res. 56/168 [P 1] (Dec. 2002), http://www.ohchr.org/english/issues/disability/convention.htm; UN GAOR,

56th Sess. Agenda Item 119(b), UN Doc A/RES/56/168 (2002), revised August 2006, Ad Hoc Committee on a Comprehensive and Integral International Convention on the Protection and Promotion of the Rights and Dignity of Persons with Disabilities Eighth Session, New York, 1425 August 2006, Draft Convention on the Rights of Persons with Disabilities and the Draft Optional Protocol to the International Convention on the Rights of Persons with Disabilities to be adopted simultaneously with the convention.

28. The rights of children with disabilities are protected in Article 23 of the Convention on the Rights of the Child, Gen. Assembly Res. 44/25, Annex, U.N. GAOR, 44th Sess., Supp. No. 49, at 167, U.N. Doc. A/44/49 (1989), http://www.unhchr.ch/html/menu3/b/k2crc.htm, hereinafter CRC. Disability is a prohibited ground of discrimination in Article 2(1). Id. art. 2(1) See, for example, Americans with Disabilities Act (ADA), as discussed in Ilise Feitshans, *Designing an Effective OSHA Compliance Program*, Thomson Reuters (available on Westlaw.com).

29. Ilise Feitshans, Law in Public Health Practice, Second Edition, Richard A. Goodman and Richard E. Hoffman, Wilfredo Lopez, Gene W Matthews, Mark A Rothstein, and Karen L Foster, eds., 570 pp., Oxford University Press, New York, 2007, invited review *New England Journal of Medicine*, May 2007.

30. ILAB news release, 15-0803-NAT, US Department of Labor, April 2015.

31. Jean-Claude Javillier, Responsabilité sociétale des entreprises et Droit : des synergies indispensables pour un développement durable, in *Gouvernance, droit international et responsabilité sociétale des entreprises Governance, international law and corporate societal responsibility*, L'Institut international d'études sociales (IIES), Geneva, Switzerland.

32. Ilise Feitshans, NACOSH testimony voiced in Bringing Health to Work and based on comments in Ilise Feitshans, *Rethinking Our Values in Occupational Health*, Eighth International Conference on Thinking, Alberta, Canada, July 1999.

33. Craig A Poland , Rodger Duffin I Kintoch, Andrew Maynard, WA Wallace A Seaton et al, "Carbon Nanotubes introduced into the abdominal cavity of mice show asbestos-like pathogenicity in a pilot study, Nature Nanotechnology 2008, 3:423-428

34. Ilise Feitshans, review and comment on *NIOSH Current Intelligence Bulletin: Occupational Exposure to Carbon Nanotubes and Nanofibers*, Docket No. NIOSH-161, "Legal Basis and Justification: Recommendations Preventing Risk from Carbon Nanotubes and

Nanofibers," prepared in response to the question presented by NIOSH: whether the hazard identification, risk estimation, and discussion of health effects for carbon nanotubes and nanofibers are a reasonable reflection of the current understanding of the evidence in the scientific literature." Stakeholder response prepared on behalf of the International Safety Resources Association (ISRA), U.S., 2010.

35. Ilise Feitshans, Who Needs Occupational Health? International Laws Protecting Occupational Health and Safety for Everyone, 1st International Conference on Occupational and Environmental Health, National Institute of Occupational and Environmental Health 1B Yersin stR., Hanoi, Vietnam, November 2003, and presented at WHO, Geneva, Switzerland, January 2012.

Chapter 9

Forethought Beats Afterthought

9.1 "There Oughtta Be a Law."

There is a magic moment in time when an idea becomes a law . . . the signing of the law by the executive, the witnesses to a wedding before civil authorities, the vote taken by the legislature that completes a long journey to transform an idea into draft legislation and the draft into law. The United Nations (UN) General Assembly has photographed one of those magic moments—the decision to pass the UN Convention on the Prevention of Discrimination against Persons with Disabilities, 2006 (Fig. 9.1). The snapshot marks the end of 30 years of struggle for civil rights for people with disabilities, beginning with rehabilitation laws in 1973, the Americans with Disabilities Act (ADA) in 1990 that became the blueprint for international law, and, subsequently, passage of the UN convention preventing discrimination against people with disabilities.

The snapshot also marks the beginning of an amazing journey: implementation means operationalization of an idea, codified in legal principles and transformed into law.

Will civil society create this photograph for nanotechnology law?

One of the startling aspects of nanotechnology is its ability to bring science fiction into real daily life. Using nanotechnology to create new medicines, and to anticipate illness that can be treated before it starts, is a charismatic promise of nanotechnology. The

Global Health Impacts of Nanotechnology Law
Ilise L. Feitshans

ISBN 978-981-4774-84-0 (Hardcover), 978-1-351-13447-7 (eBook)
www.panstanford.com

promise of nanotechnology fits squarely into the international legal system's conceptual matrix for precautions protecting health and operationalizing health rights. US Supreme Court Associate Justice William O. Douglas wrote about inevitable but inexplicable moments of social change, when previously frozen social forces give way to "the dynamic component of history" [1], times when irresistible opportunities exist for rethinking and consciously changing harmful, intransient policies that hurt people, while advancing an evolutionary leap forward in science, technology and the well-being of the human species. Perhaps nanotechnology's arrival in commerce is the marker for one of those dynamic moments in history that Associate Justice Douglas described.

Figure 9.1 Decision to pass the UN Convention on the Prevention of Discrimination against Persons with Disabilities, 2006, United Nations General Assembly. Source: United Nations.

One facet of this irresistible dynamic component of history is the elasticity of nanotechnology markets and uses, which can provide a turning point for bringing together new commerce and sustainable development that promotes enhanced corporate responsibility. *Small but Important Things,* the Swiss Nanotech Report 2010 issued by the Swiss Federation Department of Home Affairs, State Secretariat for Education and Research (SER), offers the appealing concept that nanotechnology can become a key tool toward conserving natural resources and thus creating sustainable development, because small quantities of materials already in use will be needed when engineering and manufacturing nanoparticles, noting "sustainability will be a key merit of nano. Small in size intrinsically translates to sustainable in terms of material use, energy consumption, and excellence of performance.... In computing, for example, the energy consumption per arithmetic operation has decreased a trillion times over the past 60 years. But in the same time span, the total energy consumption of computing has increased by orders of magnitude," quoting Heinrich Rohrer, IBM Fellow Emeritus, Nobel Prize in Physics 1986, jointly with Gerd Binnig, for the design of the scanning tunneling microscope (STM) [2]. In the same report, Rohrer adds, "Microelectronics was an unintentional, but huge and extremely successful step towards a sustainable world. Nano-systems, which will replace many computer-controlled macro-systems or perform new tasks in a new way, are expected to add also new dimensions to sustainability. However, when talking about sustainability we should always keep in mind that it is not just a matter of solving a single task most efficiently and effectively, but that it is the sum of all the tasks" [3].

Applying these principles with forethought should create a flexible framework for nanotechnology laws that will be attractive to stakeholders so that people will want to comply with the laws—just like people will usually stop at red lights in traffic, without first reading the statute. The success of the World Health Organization's (WHO) constitutional language proves that when civil society desires, definitions can be crafted that create a workable legal framework for solving problems that could not have been anticipated when writing laws. The promise offered in the WHO *Guidelines for Workplace Exposure to Manufactured Nanomaterials* suggests that the partnership between science, commerce, and regulation can continue to advance mutually beneficial opportunities, even before legislatures and international parliamentary assemblies have codified existing best practices into law. These goals, although requiring careful stakeholder input and forethought by guideline authors can have a positive impact upon the programs that emerge and the tasks to be achieved. Stating a clear purpose and offering the remedies for failure to meet that purpose has enabled legislatures to tackle important but difficult occupational safety and health questions, consumer health issues, and environmental health during the life cycle of use for engineered nanoparticles. Achieving the goal of implementing new technologies for stated societal goals, despite scientific uncertainty by creating flexible frameworks for risk governance of nanotechnology, is not, therefore, an impossible dream.

9.1.1 Legislative Drafting Solutions: Benefits of Using Criteria That Avoid Definitions, Numbers, or Labels

The eloquent words of nanoparticle researcher Jiayuan Zhao, formerly at the University of Lausanne, Switzerland, capture the essence of the nanotechnologies policy dilemma. Speaking at the public forum "Law and Science of Nanotechnology: Perfect together?", she said, "What is so fascinating about nanoparticles is that they are so very small. If we scale down to the nanoscale, properties change. We all have aluminum foil at home. At the nanoparticle level, aluminum is easy to explode. So it is good to put in a rocket, but not your lunchbox. The point is to make good use of the special properties of nanoparticles" [4]. For legislative drafting and policymakers, the innate polar opposite features of ordinary products used daily creates a troublesome dilemma: the same products used for destruction can be used for good, and the same products that can heal can cause harm. As in the case of asbestos, both aspects of any product may be indispensable for society.

When a high-rise fire in an apartment building in London claimed the lives of small children in 2017, public outcry about the premature death of society's posterity renewed public interest in using asbestos in residences because of its ability to contain and prevent fires, once it was revealed that new equipment that was used to refurbish the building enabled the fire to leap to the highest floors. There is no clear measure of how to weigh the death today of young children against the potential death of insulation workers who may manifest illness from today's exposure to asbestos after 20 or 30 years. Surely some parent may feel that the decision not to use asbestos robbed their child of at least as many life-years as the latency period for the workers. It is unclear, too, when social values will change the calculus of risk that once deemed a given risk acceptable but no longer accepts that risk after a tragic unforeseen outcome. Despite all the evidence of risk that brought asbestos litigation and regulation, someone has had the unfortunate experience that the absence of asbestos resulted in a greater irreparable harm. No one knew how many children might die when compared to workers; not every worker manifests the disease, not every worker will live long enough to realize the impacts of their exposure, and not every child will be caught in a high-rise fire. Legislators must tell crying mothers and grieving fathers and their families that a hard choice was made to discourage or forbid the use of asbestos, and reflect upon the consequences of the legislative choices once a time arrives when people are unhappy about the results of the legislative choice that was made.

Many applications of nanotechnologies that appear promising or even magical also have unknown or unquantified risks and are therefore fraught with uncertainties. In the 20th century, gambling about these huge risks to all humanity was the hallmark of global science policy and it paid off with major benefits that touch everyone's daily life. When nuclear international research efforts started, no one knew whether the planned products would work or not, and everyone on earth feared the harms that could end humanity in the event of an accident. Nonetheless, society funded and developed large-scale industries using the research from these efforts until such technologies became outdated; now governments are prepared to put aside nuclear weapons and face the issues of fate and hazardous waste at the end of the product life cycle, without bankrupting the military industry or harming large populations who were previously at risk. At the same time, a positive health impact of many technologies designed for the military, now applied for civilian use in in daily life, cannot be overlooked: the internet, cell phone communications, and electronic books are just a few examples.

Political will surrounding nanotechnologies therefore creates an important open question regarding the way that society will view the logical extremes of risk management dilemmas in daily life. The same substance in one context can heal and in a different context can cause undesirable harm. Nanosilver toxicity is not desirable for infants and toddlers according to the court of appeals that heard litigation on the subject for almost three years, but it is desirable when confronted with rodents and bacteria in shipping goods or transporting food. Thus, the policy dilemma: the very same properties of nanotoxicity that presented a dilemma when making the risk assessment for toddler exposure is used in microbicides in condoms to fight sexually transmitted diseases. It is the role of science and legislature to promote discovery that advances progress for the greater public good, despite uncertainties and without blowing up the world or unleashing new monsters from fractured genetic materials in the human genome, and to anticipate the remedies that society will need in the event of negative consequences.

Figure 9.2 (Left) Red- and (right) orange-colored juices at a Society for Risk Analysis meeting. Photo by Dr. Ilise L. Feitshans.

This real-life example from the reception at a Society for Risk Analysis meeting demonstrates very clearly the dangers of using labels and numbers instead of functional criteria for legislative drafting. By happenstance, participants at the meeting were told they could have orange juice at breakfast, and people lined up for the orange-colored juice in the dispenser on the right in Fig. 9.2, only to be shocked by the taste because the orange-colored juice was not a citrus juice but several juices mixed with mango. By contrast, the red-colored juice in the dispenser on the left in Fig. 9.2 was made from blood oranges, which are red and therefore had the flavor and vitamin C that the people drinking the orange-colored juice expected. Furthermore, the red-colored juice was labeled correctly as orange juice because it was made from fruits of an orange tree. And, the orange-colored juice was labeled correctly as mixed. But a toxicologist with degrees in science tasted the orange-colored juice and said, "Eww, this tastes weird. I'm not drinking this orange juice," when, in fact, it was mixed juice and there was nothing wrong with the juice. Instead, the red-colored juice in the dispenser next to it had all the desired properties and taste of proper fresh orange fruit juice, but few people knew it was orange juice or felt adventurous enough to taste it. There was waste and spoilage when people refused the red-colored juice, too. This small object lesson teaches important points for constructing draft legislation that will be useful to people it governs, while addressing difficult policy dilemmas.

First, the limitations of labeling are illustrated in this example, because even smart and sophisticated people may overlook easily accessible information such as the labels on a juice dispenser. Second, they may respond inappropriately when they react on the basis of bad information. If the legislature, confronting a similar problem of popular confusion between two products or two sets of risks, tries to label one of them with just a few words, then similar confusion may follow. But to be fair, it is equally likely that a long explanation stating the characteristics of the juice would also easily be ignored. If people will not pay attention to one-word labels such as "Orange" and "Mixed," it is hard to imagine they will read a long description such as "made with several wonderful fruits" or "although red, this juice has orange fruit juice chemical composition, taste, and vitamin C."

Any legislature wedded to the proposition that labels are necessary must therefore be prepared to categorize the labels and train the end users about the information on the labels, as required by the law for global harmonization of chemical safety in the examples from the Globally Harmonized System for the Classification and Labeling of Chemicals (GHS) (see Fig. 9.3).

This has two implications for labels. The nature of the target recipient for the information on the label is key. First, it is deceptively simple to offer consumers a label, but at the same time, it is not clear whether the label will be appropriately understood. Labels for use by the general public require striking a different balance in the calculus of risk management and must take into account the limited time and resources of populations of consumers for whom detailed safety data training may not be practical. Second, detailed labels can be very useful if training is also required. Therefore, within the stream of commerce for transport or in a workplace, detailed training about how to read labels makes perfect sense when the products they label are especially dangerous and when the end user is required to apply the information correctly. So, too, numerical criteria, such as lists with cut-offs for a level of exposure are useful tools but limited when data changes about the impact of exposure or when new problems emerge that were not on the list. Internationally, lists are expensive to maintain because a globally respected panel of experts must be convened at government expense in order to credibly revise lists, even if the enabling legislation requires that list be revised in a fixed period of time (annually, every five years, once a decade, etc.).

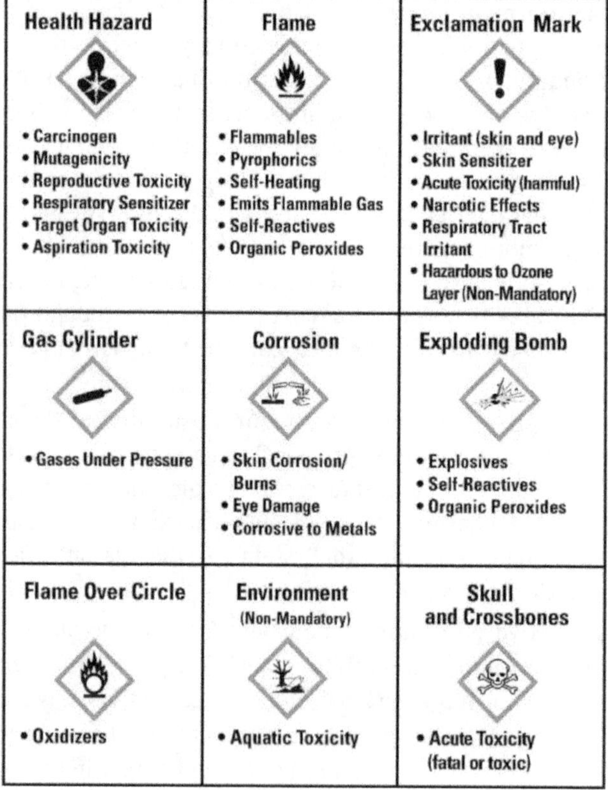

Figure 9.3 Examples of labels from the GHS.

Numerical cut-offs are easy to game and provide full employment for eager lawyers whose clients may not be motivated by the purposes and intent of the law and would like to try evading the law without punishment.

The European Union (EU) questions the applicability of numbers-based criteria in its recommendation. Criteria based on a functional analysis of anticipated problems therefore may not sound popular and require much forethought by the drafters and the legislature that is making compromises and voting on the draft law. But flexible criteria in the long term offer the most likely path for actually creating a workable set of laws that will endure the test of time, and work for both flavors of orange juice: orange the fruit and orange the color.

9.1.2 Chart: Key Points for Legislation Harmonizing Nanotechnology Standards

Table 9.1 offers a unique law-writing toolkit designed to stimulate forethought and conversation about key issues before codifying best practices. This toolkit is useful regardless of whether the drafters are preparing an international or a multilateral treaty, statutory law, a regulatory framework, municipal rules, guidelines, or a code of practice. Although checklists are merely guides and not a master, and therefore can inevitably miss key points or dwell upon unimportant points, this table starts with areas of consensus and asks what is not examined that might prove to be important. Since legislation and all rulemaking writing involves conscious choices by the authors that are later tempered by political judgments of a larger group, this table offers suggestions for preparing a sharp clear draft, bearing in mind the potential opposition and possible compromises that will lay ahead for the draft. Once a draft leaves its initial committee that has articulated its conception, the new draft takes on a life of its own. The fate of the draft as it morphs through other committees and winds its way through legislative process can be predicted but not fully controlled. This table reflects this realpolitik.

Table 9.1 Law-writing toolkit

Points requiring consensus	Approach	Key question
1. Cyclical review with notice to parties who are the subject of review Taking into account changes in political will and new discoveries that will reshape the policy response to these issues requires a flexible regulatory framework. The framework cannot simply be a one-shot firecracker approach that looks at a situation, arguably finds problems, but then ceases to monitor the situation for long-term effects.	Regulatory efforts must be reviewed periodically to refresh the program, and the terms of the regular audit must be clearly stated in the regulation. Scope of the audit, new scientific evidence to be considered, types of impacts of interest, and an inventory of new or emerging topics of concern must be evaluated on a regularly scheduled agenda and records kept of the deliberations and findings. A legislative judgment should be made concerning whether to codify the hierarchy of controls for safety engineering as it applies to some nanomaterials. A team of experts within an administrative agency can be delegated to determine the appropriate criteria for determining compliance with each step of the hierarchy In the alternative, the law may adopt by reference a specific code of practice such as but not limited to WHO *Guidelines for Workplace Exposure to Manufactured Nanomaterials*, American Conference of Governmental Industrial Hygienists (ACGIH) standards, International Organization for Standardization (ISO) standards, or Organisation for Economic Co-operation and Development (OECD) working papers.	What is the question that we are not asking ourselves, and when we finally ask it, are we confronting it properly? How does the problem material or substance that is targeted by this legislation change in a different context? **Dose** Does the applicability or requirements stated in the text change with a greater or smaller dose? In the alternative, is this a "one size fits all" regulation? Synergy with other materials To what extent does the rule take into account the context, and are there special provisions for anticipated problems based on synergy with specific materials, or exclusions (such as smoking as a confounding variable for asbestos cancer liability or interactions of certain drugs that may be harmful even though either drug is very effective alone)? Does the rule take into account specific populations who may have special exposures, such as women exposed to cosmetics or patients exposed to specific treatments or drugs? Is there an environmental life cycle concern regarding the disposal or biodegradation of the material at the end of its use? **Social context** The law may address or choose to ignore bioethical questions, appoint a panel to address bioethical questions, or institute a procedure for hearing complaints from the general public. If so, who has standing to complain as a matter of right under this law? Must harm have happened already, or can the harm be anticipated? Can a person or group complain on behalf of other people? Are the stakeholders limited to individuals and nongovernmental organizations (NGOs), or can they include corporations and governments? **Quality of material** Is there a source for determining the quality of material to be measured against a common standard, either set forth by law or set forth in the scientific community? Is it a fixed standard that sets forth specific terms for compliance or a performance standard that sets forth regulatory goals that must be achieved? Are these standards indefinite in duration or set to be updated on a regular basis, such as annually, biannually, or on a fixed regulatory cycle?

Points requiring consensus	Approach	Key question
2. Key functional statement tied to the stated purpose, instead of definition of the most important terms Such laws address risks, while balancing the development of commerce. The hallmark of 20th-century scientific precedents, the emerging laws for nanotechnology, can ensure that every stakeholder—manufacturers, researchers, consumers, or workers—will have the very best information at their fingertips when making decisions about nanotechnology use and its implications for the future.	The most sweeping legislation eschews precise definitions, defining circumstances, and desired outcomes instead of the exact conditions that trigger regulation.	What is the harm we are trying to prevent? What is the purpose of this law? What is the remedy for the consequence? Are there similar situations that are unimportant under this law? Will there be liability? For whom? How will liability be penalized?
3. Limited use of lists, numbered limits, and specification-based criteria Lists are nice to have but a dangerous master.	Think carefully about possible scenarios where data will change rapidly. Some data can be included, but provide what to do when it becomes outdated. Nanotechnology "at 100 nm or less" poses questions about aggregates and clusters, may address harmless uses because they fit the criteria for size, and may miss larger-sized applications that pose a risk or danger.	Is there a specific list that is trustworthy that can be incorporated by reference? Is there a group of experts to include? How will the revision of the list be justified under law (is it periodic or data driven)? How will revision be financed? What happens if there is an important change but the experts do not meet or the list is not revised?
4. Inputs for stakeholder comments that will be applied by regulators	Stakeholder comments must be solicited in open, transparent calls for stakeholder input and then reflected in the text and implementation strategy to achieve regulatory goals. To make sure that the rules are flexible, these comments can be requested periodically as well as in the event of modification or change in the law	Who are the constituents? Who is the primary stakeholder? Do the stakeholders have the ability to speak for themselves? Are there restrictions on stakeholder free speech?
5. Private right of action to ensure compliance	In the event that the law is ineffective or cannot reach new issues of personal harm, populations or corporations that have been harmed can be granted a right to sue after demonstrating evidence connecting the actions complained about and the harm complained about Who can sue? Who sues if the person or people who have been harmed are dead?	What harm is acceptable? Must there be actual harm or can there be a strong likelihood of harm? Can an injunction be granted to prevent harm? Must the causation be direct, or can there be a presumption that some activities caused harm and the defense must rebut the presumption of harm?
6. Positive incentives for compliance	Often the tax code is used as a tool to incentivize people or corporations to obey the law by rewarding some types of deeds with favorable tax situations or tax credit. Positive incentives refers to benefits created by legislative design, an approach that provides a positive alternative to the claim that laws merely tell people what NOT to do.	What are the special benefits for compliance: tax credits, preferences in bidding for contracts, or preferential treatment under law? Who has the right to such benefits? Is here a limit on how long the benefits can be applied or how large they can be? How should one apply for such benefits?
7. Policy judgments regarding illegal acts and penalties	When all else fails there may be fines, civil penalties, the rights for people who were harmed to collect damages, or even criminal penalties for failure to obey the law.	Does the law have civil or criminal penalties or both? Are the penalties framed fairly? Are the penalties a last resort or upfront deterrence tactics?

9.2 Stakeholder Tools for Translating Scientific Consensus into Governance for Nanomaterials

WHO's remarkably well-crafted old language from the midtwentieth century has proven to be prescient: the text is flexible to embrace new problems that could not have been foreseen at the time of its writing. And yet, it is not viewed as outdated. WHO's Constitution provides an early example of a flexible framework that does work. Authors of the WHO Constitution did not know about every disease that could impact human health, but they clearly understood that simply listing the known illnesses of their era was an unwise approach for drafting a vibrant constitution. The authors also understood that definitions of key terms concerning health and ill-health are fluid, cross several categories, and are dynamic and ever-changing: science is capable of changing the very meaning of the terms "illness," "health," and "disease" or "infirmity." WHO's activities have spawned expert working groups in bioethics, delivery of health care, training, social determinants of disease, and women's health disparities. The WHO constitutional language is a paragon of model text for flexible frameworks employing criteria clearly enabling the agency to examine questions involving the right to life and security of the person, addressing new problems without squashing industry or new technology.

Countless nanotechnology guides and sets of best practices are available on the web and in textbooks. Their words confuse further the question of unknown risks and what is the right course of action for a well-intended employer or an instructor giving training, but they can provide evidence for a standard of care expected by the general public, whether as consumers merely reading labels or as members of a jury weighing evidence about due diligence before the courts. The inevitable solution to this knotty problem of nanolaw proliferation is an international code with unified obligations and harmonized terms of art, especially definitions regarding the use and limits of applied nanotechnology. Along with consensus, there must be clear language in order to draft precise and predictable legislative text. Definitions must synthesize key scientific "nano" concepts but should use a "functional analysis" that scrutinizes the use of nanomaterials in their context. To do so requires a flexible framework to embrace new data and new developments in technology systematically.

And the system must anticipate that it is geared to address hot or burning policy questions that may not be easily answered empirically but must nonetheless be addressed from a political standpoint despite weak data at the so-called frontiers of science [5]. For this reason, there has been an emerging trend in the early 21st century toward a new form of regulatory governance by international scientific consensus, favoring Memoranda of Understanding (MoU). Typically, the parties signing the MoU have some regulatory authority or hold moral sway over a large constituency of stakeholders but do not have the time or power to go to a legislature in order to convert their wish to work together into statutory law. For example, the photograph in Fig. 9.4 shows Dr. John Howard and Dr. Sameera Al-Tuwaijri (National Institute for Occupational Safety and Health [NIOSH] and SafeWork, respectively) signing an MoU in Washington, DC (April 23, 2008). Although each signatory is the head of a major administrative agency, there is no specific legal basis for the informal agreement to exchange data, scholars, and information about programming. Instead, the MoU fills a necessary void in the infrastructure of each agency, clearly stating the directorate's desire to work with the other members of the MoU . This type of agreement is crucial for creating exchanges of expertise and allowing agency resources to be spent on expert time required for having effective conferences and standard-setting discussions, creating guidelines and recommended best practices, and, of course, ensuring that staff will cooperate with the cosignatory agency.

In the international arena of nongovernmental organizations (NGOs), MoUs provide an ideal vehicle for encouraging capacity building without a commitment of government funds and, more importantly, without committing governmental policy to informal activities. For example, in a host of MoUs, such as its agreements with NIOSH, the ILO, and a host of national safety and health organizations, the American Society of Safety Engineers (ASSE) can tap the resources of the MoU partners without changing the formal policy agenda of either party to the MoU. Typically, text of the MoU format allows for exchange of expertise, use of library or archival facilities, free registration or participation in conferences and meetings, and an agreement to mutually publicize each other's activities through their regular communications with the general public and members in newsletters, periodicals, and social media. The MoU states that the ASSE and the ILO will work together toward the common objective of preventing illness and injuries in the workplace across all industry sectors through advocacy, awareness promotion, knowledge development, information dissemination, and application of relevant standards and industry best practices in the community and workplace. "As there are no global marketplace boundaries today, and with a large number of our 32,000 occupational safety, health and environmental professional members continuing to work in countries and on projects around the world, this agreement will help us move forward in preventing injuries and illnesses worldwide," said ASSE President Warren K. Brown, CSP, ARM, CSHM. "This agreement also reflects the value of the SH&E profession and ASSE's growth" [6]. Another feature of most MoUs found in the ASSE–ILO agreement is that neither party commits to paying anything, thus rendering the agreement's budget neutral and more likely to provide added value that will be welcomed by the organizational infrastructure and its constituents.

Figure 9.4 Signing ceremony. Dr. John Howard and Dr. Sameera Al-Tuwaijri (NIOSH and SafeWork, respectively) sign a Memorandum of Understanding in Washington, DC (April 23, 2008). Source: US Government.

9.2.1 Syllabus of "Nanotechnology Law and Health Policy: Science Serving Society"

Nanotechnology Law and Health Policy: Science Serving Society

Dr. Ilise L. Feitshans, JD and ScM and DIR

Lectures 1–12

Grading methods: Course requirements for assessment

Reading assignments: Required reading

Suggested supplemental text

Proposed paper topics

9.2.1.1 Course objectives

1. Provide an overview of current health laws under international law that regulate or attempt to outline the emerging risks from application of nanotechnology in order to maximize the benefits and minimize the risks from industrial applications of nanotechnology.
2. Provide students with an opportunity to have a deeper understanding of the real challenge for civil society when implementing nanotechnology: whether a revolution for commerce and industry also holds the momentum for revolutionary innovations in public policy for consumers' "right to know," workplace exposures impacting personal health, and the environment.
3. Provide students with an opportunity to broaden their knowledge in related health law questions so that they can understand when law helps and when law must change to meet new social needs, using nanotechnology as the vehicle for understanding key policy analysis techniques for addressing new risks and unknown dangers to public health.
4. Provide students with an opportunity to have a deeper understanding of (i) access issues in public health and (ii) economic issues of health care as intellectual property, including but not limited to pharmaceuticals and equipment.

Upon completion of this course, students will have an understanding of several key aspects of the types of regulatory frameworks that can be used in health policy and law in the private sector, by national governments, and internationally and a reasonable grasp of the importance of nanotechnology in society as found in products in daily life.

9.2.1.2 Proposed syllabus: lectures 1–12

Lecture 1: What is health?

1. Icebreaker: What is your favorite UN agency and your future favorite job? Name three things you like about it and three risks it presents to your personal health. What will the "revolution" in nanotechnology mean to the things you just described?
2. What is the WHO definition/related analytical constructs/ other legal definitions under international laws? What is the limit? Does this include preventive programs inoculations vaccines research testing, etc.?

Legal definitions of health

A survey of national laws that apply the WHO definition and national laws that function to protect health despite the absence of a constitutional requirement to protect health or the absence of a coherent national system to protect health. What does a rights-based analysis of health offer a country that has no right to health under law?

Readings

WHO Constitution; comments excerpted from several books, including Ilise L. Feitshans, *Bringing Health to Work.*

Lecture 2: What is nonhealth, disability, and are there healthy disabled people?

The future of nanomedicine and the ethical questions that new medicines raise, Neurodiversity Issues, United Nations Convention on the Rights of People with Disabilities, ILO *Guidelines for Managing Disability in the Workplace,* WHO *World Report on Disability 2009 and Updates,* related documents

Lecture 3: What is law?

- What is the state obligation regarding health?

  (*Parens patriae* and concepts from national laws)

- What makes the incantation of words into law?

  (Comparison of custom, usage, and "real" law from a legislative or executive body of a nation, international organization, regional association of nations, or municipal government)

Does law triumph over tyranny?

- The debate regarding "hard law" and "soft law"
  - Concepts of law: as a means of social control, as an ethical or moral code governing society
  - Extralegal systems of rules: terrorism, fascism, corporate compliance, and their implications for international governance: the difference between laws made by legitimate governmental bodies compared to corporate internal governance, guidelines, best practices, voluntary compliance, and adherence to standards by NGOs such as the International Organization for Standardization (ISO)
  - Does the text matter more than the act of implementation or vice versa?
  - Are there illegal laws or times when the law goes too far?

Lecture 4: Emerging risks: What we know, what we don't know and what policymakers need to know in face of scientific uncertainty

In-class quiz: 10 questions and guest lecture by a policymaker

Lecture 5: Gender-specific issues: Nanomedicine to address disability among women

Respected public health literature suggests a developing understanding of health outcome differences based on age, sex, class, ethnicity, and race. How should we glean the appropriate policy perspective that applies this knowledge? Can nanomedicine apply it without creating a new institutional basis for discrimination?

Experiences and law are brought together to examine the following questions:

- Are there gender specific illnesses, injuries, or disease?
- How does epidemiology explain gender-based disparities in prevalence and incidence of major illnesses among women?
- Are there disproportionate negative impacts among women associated with specific cancers, specific illnesses, or vulnerability to specific injuries or disease?
- What difference does paying work make in this equation?

Lecture 6: Environmental laws that impact health

Life cycle analysis of nanomaterials and the environment

- Where nanopackages for medicine go (life cycle and waste disposal issues)
- Nanomaterials that clean brownfields (emerging research in the use of nanomaterials to break down toxic waste)
- Regulations/overlap with the private sector/rule making and standard setting; «orphan» drug patenting, human life, and related genetic pharmaceuticals; leading cases; special equipment and related medical devices for disabled populations

Lecture 7: Food and the role of nutrition

- Nanotechnology and the role of the government in preserving, sanitizing, and shipping food
- The European Food Agency, the Food and Agriculture Organization (FAO), and WHO report the role of the government in producing and protecting food quality and ensuring equitable distribution of food
- Genetically modified foods as a precedent in Europe, the United States, and Brazil
- Role of nonconsumable nanoparticles in shipping, packaging, storing, and sanitizing food
- Folding emerging research about migration of nanoparticles in food into the regulatory framework: Will the "generally

recognized as safe" (GRAS) notion in administrative law survive under the new nanotechnology regulatory governance?

Lecture 8: Health in the workplace and nanoparticle exposure

Laws around the world address the right to information about workplace exposures to materials that are considered to be toxic or hazardous under international law. The GHS and several international instruments provide a framework for discussing the risks and proper handling of nanomaterials in the workplace.

- Key aspects of the GHS
- Applicably to nanomaterials in commerce
- Design of Safety Data Sheets (SDS)
- Use and training requirements
- The hierarchy of controls: NIOSH recommended exposure limits (RELs) for nanomaterials
- KOSHA (Korean OSHA) standards and recommendation and WHO *Guidelines for Workplace Exposure to Manufactured Nanomaterials*: areas of consensus and areas of conflict. How should the conflicts be resolved?
- Role of precautionary principles and universality in shaping global nanotechnology regulations in practice and under law

Lecture 9: Genetics and overview of possible applications of new genetic technologies

- Impact on prenatal and neonatal care
- Impact on health care delivery
- New directions for genetic testing in the workplace
- Silk jeans or spider silk genes? The future of genetic testing in the workplace

Lecture 10: The next generation: Reproductive health and nanomedicine

Should all people with disabilities be allowed to reproduce? Who decides? What incentives are there for increased assistance? What priority should be given to Maternal and Child Health (MCH) programs in light of new definitions of parenting that extends reproductive health to lesbian, gay, bisexual, transgender, and intersex (LGBTI) communities and new approaches to fertility that do not require women to bear their own children?

- Ilise L. Feitshans, Is there a human right to reproductive health?, *Texas Journal of Women and the Law*, **Fall 1998**
- International Commission on Occupational Health (ICOH), Brazil, February 2003, Fp 40.6, Working for Reproductive Health at Work: Towards Maternalistic Views of Health and Survival, free paper session, Reproductive Hazards in the Workplace (FPS 40)
- American Society of Law Medicine and Ethics, October 1, 2001, Working for Reproductive Health at Work: Health and Survival of the Next Generation (also presented at Yale Medical School, Spring 2002)
- Invited article for the State of the Art Reviews (STAR) on Ethics and Occupational Health Practice, *Occupational Physicians Ethical Obligations to Pregnant Workers*, September 2002, report from CLINAM, European medicine agency, regarding the future of nanomedicine and presymptomatic treatments

Lecture 11: The political process and the future of the law

Will Rogers, the sage comic from 19th-century United States, is credited with saying, "When they make a law, it's a joke, and when they make a joke, it's a law."

Is this still true?

This lecture examines the legislative process in several contexts, discussing the nuts and bolts of how an idea for draft legislation can become a law, with examples from real-life ideas that became law and also a short discussion of lobbying and how to be sure of having your say before a legislative body.

Nanotechnology Perceptions article: "Forecasting Nano Law: Defining Nanotechnology," Ilise L. Feitshans

What is legislation?

- What does law do (rules of the game; moral, ethical, or religious values: right and wrong translates to legal and illegal under written laws)?
- What are the characteristics of laws: Realistic goals? Affordable? Fair? Enforceable? Ideals? Social change?

What happens when there is no law? There ought to be a law.

- Outside the legislature: hoopa, hype, and silence
- Legislative activity: looking to the future to correct a problem

Legislative activity in response to public outcry

- Sponsor: the gleam in the eye becomes a draft, introduction of the bill to the legislature as a whole
- Committee meetings and hearings: testimony by constituents and interested groups that influence the views and votes of legislators
- Amendments, compromises, and modifications of the draft legislation
- Introduction of the draft legislation on the floor: Debate?

  Note: In the case of many legislatures, the real fight is in committee, not on the floor.
- Passage of the bill, awaiting executive signature

Outside the legislature

- Newspaper articles "planted"
- Editorials by partisans, media hype
- Letters from constituents to legislators
- Active controversy?

Voila: The bill is signed into law by the executive (mayor, governor, president)

Role-playing for legislative drafting

Review the written materials and decide who you want to be—legislator, constituent, lobbyist, or the media.

Do you favor the proposed law or oppose it? Why? What are you willing to do so that your views win the battle over the law?

The students take a turn!

Work sheet for the in-class debates:

- Think about your views on this bill. Are you for it or against it?
- What are the arguments for it?
- What are the arguments against it?
- Which interest groups are negatively impacted by the problem?
- How would they be helped by the proposed solutions?
- Which groups are not affected by the problem?
- How would their life change if the bill becomes law?
- What changes would you make in the proposed legislation?
- Would you vote for it or against it?
- What other topics are more important than this bill to you so that you might not even want it to go out of committee to the floor?
- What would you trade off to get this bill passed or to stop this bill from becoming law?

(Rule of thumb: Never watch how sausages or laws are made!)

Lecture 12: Having your say!

Student presentation of final papers on corporate compliance policy to grapple with an issue of nanotechnology risk management under public health law (Fig. 9.5)

Grading methods: Course requirements for assessment

Quiz during the fourth class: Students will have a 10-question in-class handwritten quiz based on the readings and discussions from the first through third lectures. This will count for 25% of the final grade.

Papers

Students will write a paper on any topic of nanotechnology that they wish and one corporate compliance policy to grapple with any aspect of implementing the public health protections concerned in the paper, with a strategy for including the views of stakeholders (10–15 pages). This will count for 50% of their final grade. The professor offers a "free ride policy" that a first draft of the paper may be submitted for comment and revision without any penalty up to three weeks before the end of the semester. This will be given a grade that *does not count*, and students will be expected to improve this paper, taking into account comments. The paper should have the target audience of a policymaker with a budget, using the model of a topic from one of the lectures and one corporate compliance policy to grapple with any aspect of health law in the workplace that they wish (10–15 pages). It is expected that comments made by the professor under the free-ride policy will be incorporated into the final paper. The remaining 25% of the final grade will be based on class participation.

Figure 9.5 Conference pictures. Photos by Dominique Charoy.

Reading assignments

Readings for the course are the basis for in-class discussion. Only materials from the first three lectures will be the basis for the in-class quiz in the fourth session. Readings are required for following the discussion. The readings are designed to form a basis for student research on topics of their own choice for the corporate compliance policy that will be their final paper. Students are urged to locate new resources, such as UN publications, on their own as references for their papers.

9.2.2 City of Cambridge Nanomaterial Use Survey

Instructions: Please answer the questions below to the best of your ability using information already on hand or readily available to you or others in your organization. All information provided on the survey will be considered proprietary in nature and will be protected by the city of Cambridge under the state public records law. Please use additional sheets of paper, if necessary. If you have questions or comments about this survey, please contact Sam Lipson, Director of Environmental Health, Cambridge Public Health Department, by telephone at 617-665-3838 or by email at slipson@challiance.org. Alternatively, you are also welcome to contact Deputy Chief Gerard Mahoney or Lieutenant Michael Hughes, Cambridge Fire Department, by telephone at 617-349-4944 or by email at LEPC@cambridgefire.org.

1. Please complete and submit this survey within 30 business days of your receipt or as soon as possible.

Definition: In this survey *engineered nanoscale materials* include materials composed of particles or structures with at least one dimension between 1 and 100 nm (1 nm = one billionth of a meter). These materials are referred to as "engineered" because they are manufactured and used purposefully to make use of size-related properties.

Questions

1. Does your organization presently manufacture, handle, or process engineered nanoscale materials within the city of Cambridge (please refer to the Appendix for examples)?

   ______ **No**. Please sign and date this survey and return to the Cambridge Fire Department at the address listed on page 1.

   ______ **Yes**. Please provide answers to Questions 2 through 6 below.

   ______ **Decline to respond**. If you would prefer not to disclose this information, please indicate this and please let us know what concerns you may have about sharing this detail. We may contact you to determine whether we can meet your concerns, while ensuring that we gather the basic information we have been asked to collect. Please sign, date, and return this survey to Sam Lipson, Cambridge Public Health Department, at the address listed on page 1.

2. Please provide the complete street address for all facilities within the city of Cambridge where engineered nanoscale materials are fabricated, handled, processed, or stored.

Facility A	Facility B	Facility C
______________	______________	____________
______________	______________	____________
______________	______________	____________

(Please use additional sheets of paper, if necessary, to list all facilities.)

3. Please provide a contact name, telephone number, and email address for a contact person at each facility listed in Question 2. In some cases, this may be the same person.

   Facility A:

   Facility B:

   Facility C:

   Note: If you intend to provide information for more than 25 discrete engineered nanoscale materials, please contact the Cambridge Public Health Department for further direction before completing Questions 4, 5, and 6.

   - Please provide the information listed below for each engineered nanoscale material that is fabricated, handled, processed, or stored by your organization. Please refer to the list in the Appendix but do not limit yourself to this list if any material you work with is not listed.
   - Chemical name, formula, and Chemical Abstracts Service (CAS) number (if specific to the nanoscale material).
   - General description of the material (e.g., organic or inorganic, any known physical and chemical properties, including size, phase, and flashpoint).
   - General description of the physical state or conditions in which the material is manipulated (e.g., handled dry, wet, in solution, powder, granules, threads, sheets, etc.).

- Approximate mass of each type of nanoscale material on-site at each facility where such material is fabricated, handled, or processed. Please *estimate* the mass or *give a range* (e.g., approximately 40 oz. or < 1 oz., 1 oz.–1 lb., 1 lb.– 5 lb., >5 lb.).
- Location. Include sufficient information to allow an emergency responder to quickly locate the materials, such as floor, room number, shelf or area, and 24–7 contact name and phone number for that area.

5. Please list any occupational health and safety protocols, practices, or assessments for the nanoscale materials listed in Question 4. If you use nanoscale materials that become or can become free (aerosolized, airborne, waterborne) at any point during production, processing, handling, transfer, or storage while they are present in your facility, please try to answer the following questions in as much detail as you can:
   - Does your facility/lab use personal protective equipment (PPE) or other protective equipment (fume hoods or biosafety cabinets, negative airflow, sticky mats, high-rate ventilation, fine-particle filters) to prevent exposure?
   - Does your hazard response staff training include information about potential hazards of nanoscale materials?
   - Does this staff training include specific clean-up and recapture procedures for nanoscale materials?
   - Does your facility have a written emergency response plan for the release of nanoscale materials?
   - Do the material SDS available for the nanomaterials handled at your facility include information about risks posed by these materials that are specific to the nanoscale (vs. parent compound or material)?
   - Does your facility label all waste streams containing nanoscale materials?
   - Have you or your staff performed a risk or exposure assessment?
   - Does your facility conduct area exposure monitoring or personnel exposure monitoring? Please describe the

monitoring regime. Can nanoscale materials become airborne under any conditions anticipated?

- Does your facility have a medical surveillance plan or arrangement for occupational health services that addresses potential health risks posed by nanoscale materials?

Alternatively, you may submit your relevant environmental health and safety plan or chemical hygiene plan in lieu of a response to these questions. If some question is not fully addressed within the safety plan, please include necessary additional details along with the document.

6. Does your facility have a chemical hygiene plan on file with the Cambridge Local Emergency Planning Committee Office, and if so, does this plan include specific information about the engineered nanoscale materials listed in Question 4?

If your facility does **not** have such a plan on file, please contact the Local Emergency Planning Committee at 617-349-4944 or via e-mail at LEPC@Cambridgefire.org to determine whether such a requirement exists.

Please provide information for the person responsible for completing this survey:

Name: ______________________________ **Title:** ________________

Signature: ______________________________ **Date:** __________

Telephone: ______________________ **Email:** ____________________

Thank you for your cooperation! Cambridge strives to be an exceptional setting for exploration into the frontiers and applications of science and technology. By establishing a transparent, responsible relationship with emerging technology firms to address reasonable public safety and occupational concerns, we hope to build on good health and safety practices already in place and to encourage adoption of those practices universally.

9.3 There "Oughtta Be" a Law: A Stakeholder Simulation of the "Nano-Law" Debate

Sometimes, talking about a proposed legislation is less effective than actually trying to draft text and then hear what other people believe is good, bad, important, or silly in the proposed law. This role- playing simulation of a legislature could be anywhere and could address any topic. Indeed some legislatures drift far away from their proposed topic once they start to hold hearings about draft laws. As shown in Fig. 9.6 many laws are written in places that are quite remote from the places where the laws will be implemented. Is this distance from the trenches of the actual real-life problem good for creating clear thinking about complex problems or bad because the players in the political games that form these laws are distanced from the actual daily-life problems? Will the UN be the home of an international nanotechnology regulatory regime by treaty or convention under international law? The roles sketched here can be expanded or condensed on the basis of the committee's time limits or the political views of the actors. Here is an opportunity for life to imitate art, if rehearsing for legislative activity about nanotechnology. Try to identify a group or societal goal that can be embraced with passion; think about the consequences, potential supporters, and potential opposition to your goals; and, above all, enjoy having your say!

Figure 9.6 Entrance to the Palais des Nations, home of the United Nations in Geneva, Switzerland, where many international laws are made. Photo by Dr. Ilise L. Feitshans.

The first version of this simulation was hosted by the University College Dublin (UCD), Ireland, in collaboration with the Institute for Work and Health, Lausanne, Switzerland, program as part of the 4th NanoImpactNet Integrating Conference and the 1st QNano

Integrating Conference "From Theory to Practice: Development, Training and Enabling Nanosafety and Health Research,"

February 27 to March 2, 2012, Symposium (E): There "Oughtta Be" a Law: A Stakeholder Simulation Session on the "Nano-Law," Debate Chair Ilise L. Feitshans.

Role-playing: Participants "You Can't Tell the Players without a Score Card"

Discussion of the Definition of "Nanoparticle" for the Purposes of Regulation

Your Legislature, Four Hundred Thousandth Session

Subcommittee on the Future of Prospective and Possible Nanotechnology Research,

in the Committee on Fashion Economics

in the Your Legislature.

Agenda:

1. Presentation of nanotechnology report by **Dr. Nobel Prize Winner**.
2. Public comments by **Dr. X Tremereaction**, author of "No Known Risks," Executive Director, Foundation for the Advancement of Nanotechnology (FAN)
3. Public comments **Dr. Stan Dupp**, professor emeritus of physical, chemical, and biological sciences and philosophy and ethics at Large Major University and executive director of Retired Scientists Against Progress (RSAP)
4. Informational questions by legislators and staff
5. Questions and comments from the general public in attendance

Your legislature has convened this special Subcommittee of the Committee on Fashion Economics in order to hear the expert reports and accept public comments from stakeholders and voters regarding the following:

Policy question: How can the benefits of nanotechnology be realized, while minimizing the risks to public health?

Subcommittee hearings on the question "Should this subcommittee recommend adoption of the draft definition in the proposed bill text?"

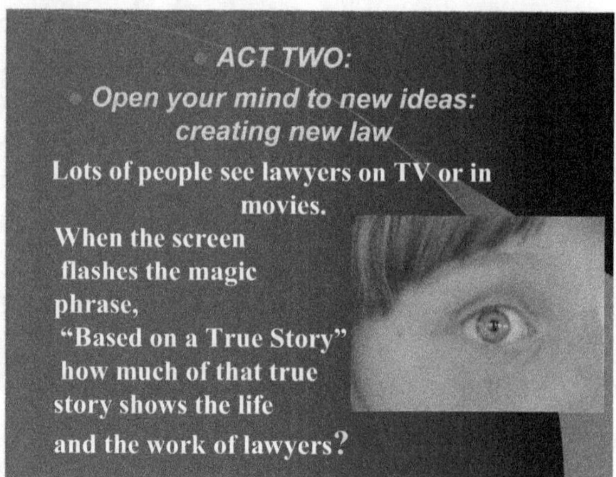

Figure 9.7 An eye-opening concept: much of the work of good lawyering is outside the courtroom. Photo by Dr. Ilise L. Feitshans.

Nanotechnology is "the understanding and control of matter at dimensions between approximately 1 and 100 nanometers, where unique phenomena enable novel applications."

9.3.1 The Players

Dr. Nobel Prize Winner, expert who presents testimony that remains unquestioned, although appreciated

Dr. Nobel Prize Winner has been researching nanoparticles since 1951. Dr. Winner taught Andrew Maynard the periodic table of elements and discovered three new elements in the same year. The electron movement in these three elements combined to form a new synthetic element and eventually when bonded to titanium formed the strongest molecules on earth ever known to humanity. In popular culture, these molecules were the prototype for the material in the red cape worn by Superman. Dr. Winner has lectured at every major university on earth and plans to do video conferencing from the skylab and the moon, but unfortunately, when visiting Dr. Winner's own child in *Parents Describe Your Work Day*, neither the teacher nor the students had heard of any of this work. Dr. Winner is respected internationally for brilliance, integrity, and objectivity and for writing large sentences with newly invented words, especially verbs, without ever once using a subject or predicate in the same sentence. Dr. Winner is the lead author of the major nanotechnology report before the Your Legislature's Subcommittee on the Future of Prospective and Possible or Strategically Important Nanotechnology Research Advisory and Action Plan, in the Committee on Fashion Economics in the Legislature.

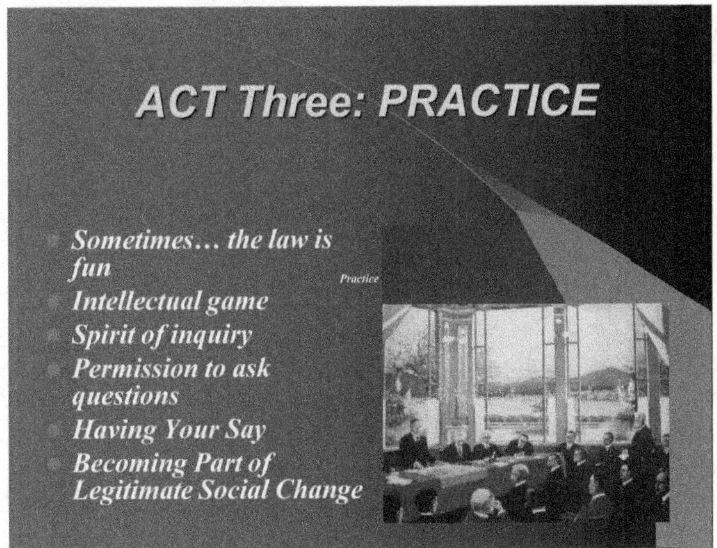

Figure 9.8 View of legislative drafters. In this painting, now owned by the United Nations, inside the Palais des Nations in Geneva, Switzerland, men from dozens of nations convened to write the Treaty of Versailles that ended World War I in Europe and established the ill-fated League of Nations. Not seen in these chairs are women from the United States and other nations representing the Quakers, including Dr. Alice Hamilton and Mrs. Edith Wilson, wife of US President Woodrow Wilson (for whom the road in Geneva, Quai President Wilson, is named). Those ladies, who did not have the right to vote in the United States at the time that they were lobbying to end the war, lobbied in the corridors, demanding peace and ongoing negotiations but were not seated at the negotiating tables. Some historians believe that this disconnect between the gender-specific roles of lobbyists and negotiators had an indirect role in shaping policies that failed. Consequently, draft texts written by men for the needs of men that ultimately were written into treaties would have been more successful if women's voices as stakeholders had been more formally and clearly heard. Photo by Dr. Ilise L. Feitshans.

Dr. X Tremereaction, author of "No Known Risks," Executive Director of FAN

Dr. Tremereaction is not particularly famous but is extremely well connected, gracious, and very charming. Dr. Tremereaction has been principal science and technology advisor to three US presidents, to President Sarkozy of France when Sarkozy was the chief executive for Europe, economic advisor to the Greek Parliament for the EU, and former personal tutor for the current king of Morocco when the king was a child. A resident of Alexandria, Egypt, Dr. Tremereaction is mutlilingual (speaks five languages and reads seven others fluently). He is a major advocate in favor of nanotechnologies and insists that there are no known risks from nanomaterials.

A. Vocat, legal advisor to the legislator, explaining procedures, keeping order at the meeting, and discussing these points in light of the unrelated policy issues before the legislature, whose real opinion remains unseen

Vocat is qualified to practice in the United States, Europe, Asia, and Dakar. A native of Ireland, Vocat worked in the legislature throughout times of great turmoil, briefing one legislator about particularly difficult policy issues. Vocat knows nothing of science but is world respected for knowledge of procedure in several legislatures. Vocat also lectures at the British Parliament once a year in continuing education seminars for parliamentarians but has no political aspirations of one's own. Vocat is particularly astute at measuring the pulse of the legislative temperament about a proposed bill. A keen strategist, Vocat has an excellent track record persuading opposition to get their buy-in for legislative compromise following orders from Vocat's boss, the committee chair. Vocat also has the unusual ability to memorize the entire legislative agenda for a session, knows the bill numbers by heart, and therefore can instantly prioritize the importance of any policy question in a given legislative session or committee hearing. Nanotechnology, however, does not rank among Vocat's list of vital matters.

Dr. I. M. Smartbutunappreciated, research assistant serving as rapporteur for the legislator

Dr. Smartbutunappreciated graduated from the finest medical school in the world as an honors student pulmonologist but then went to public health school because caring for people with lung cancer from cigarette smoking made Dr. Smartbutunappreciated realize that even lifestyle choices about destructive behaviors have an important public health dimension. One day during a required course on the law of public health at Harvard School of Public Health, Dr. Smartbutunappreciated heard for the first time about the major cases that changed US society forever, such as *Jacobsen v. Mass* (1896), which required a private citizen to have a vaccine even against their personal fear that the vaccine would harm them and, despite the statistical studies that showed some people would get sick, perhaps even die, because of the publicly administered vaccines. Dr. Smartbutunappreciated became fascinated by the dance between public health and social control, which on the one hand promotes individual health and sound choices but on the other hand enforces measures that take away private rights and civil liberties in areas such as quarantine, forced vaccination, distribution of publicly funded medicine, and cigarettes and alcohol. Is nanotechnology yet another area for this dance of social control? As rapporteur, Dr. Smartbutunappreciated has no right to express opinions but can ask endless procedural questions in order to clarify the record and organize the written comments submitted by the general public.

Dr. Stan Dupp, professor emeritus of physical, chemical, and biological sciences and philosophy and ethics at Large Major University and executive director of RSAP

Dr. Dupp wants the world to be better educated but is skeptical of "big science" funding that does not have transparency for stakeholders. Dr. Dupp became obscenely wealthy as the inventor of the world-famous Dupptometer casting system that was commonly used in video games to simulate trout fishing and fly fishing 20 years before video games available to the general public could simulate sports such as baseball, golf, and skiing to improve the athlete's skills.

Dr. Dupp became disillusioned with fame, fortune, and glory and the hunt for vast funding for the next project (an adaption to trap big fish such as sharks) when the military bought up all of the Dupptometer intellectual property rights under the name of a private corporation and applied the technology to nuclear weaponry. Dr. Dupp vowed never to allow undisclosed applications of new inventions to occur again and instead used fame, fortune, and academic prestige to begin a campaign for transparency in corporate purchasing around the globe.

Legislator Hon. Francine Jacob, France, whose sole question is, "How does nanotechnology, as defined for the purposes of this proposed law, improve the life of the Roma women who are my constituents?"

Hon. Jacob is skeptical about the importance of nanotechnology. She is convinced that too much funding goes to wasteful projects that most people can't understand, such as "big science," fashion design, governmental infrastructures filled with overpaid and underworked functionaries, and the military. From a political standpoint, she views these institutions as the same thing. She has several key corporate donors who back her election campaigns, but those companies are owned in large part by members of her constituency. She spent an entire lunch with A. Vocat, listening to the information before Your Legislature prior to this subcommittee meeting. She is convinced that the primary problem impacting negatively on the lives of her constituents is early forced marriage of Roma women but is open to the possibility that regardless of their age they require an excellent education but that unfortunately their own ethnic culture and national government structures have conspired together to prevent them from obtaining education in the past. She is open to the possibility that nanotechnology can benefit all humanity and is therefore worthy of funding if that benefit includes specific programs to educate Roma women by welcoming them to career paths that have high-paying jobs, especially but not limited to entrepreneurial ownership, business school and management training, executive-level private sector work in nanotechnology industries, nanomedicine, and engineering. *Therefore, she is willing to vote in favor of any definition that the committee chair wants, if she can secure a guarantee that the bill that leaves the committee will have specific requirements for nanotechnology training and higher education for Roma women.*

References

1. Vern Countryman, Preface, in *The Douglas Opinions*, Random House, 1977.
2. Heinrich Rohrer, (6 June, 1933–16 May, 2013) was a Swiss physicist who shared half of the 1986 Nobel Prize in Physics with Gerd Binnig

for the design of the scanning tunneling microscope (STM). The other half of the prize was awarded to Ernst Ruska.

3. *Small but Important Things*, Swiss Nanotech Report 2010, Swiss Federation Department of Home Affairs, State Secretariat for Education and Research (SER), p. 3.
4. Available on DVD and on YouTube: http://youtu.be/vEmorvYOJz8; *Lois et sciences des nanotechnologies: un ensemble parfait?* Débat public sur des questions émergentes de notre époque Samedi 17 août Law and Science of Nanotechnology: Perfect Together? A Public Discourse about the Emerging Issues of Our Times, Musée d'histoire des sciences Parc La Perle du Lac Genève Suisse Nanotechnologie: pour la protection des droits humains de la santé *Nanotechnology: Science Protecting the Human Right to Health*, Ilise Feitshans, Avantages et risques de la Nanotechnologie: la position de l'Assemblée parlementaire du Conseil de l'Europe (Benefits and Risks of Nanotechnology: The Parliamentary Assembly's Position, Council of Europe), Tanja E. J. Kleinsorge, Cheffe du Secrétariat, Commission des questions sociales, de la santé et du développement durable, Assemblée parlementaire du Conseil de l'Europe, Strasbourg, France (Head of the Secretariat, Committee on Social Affairs, Health and Sustainable Development, Parliamentary Assembly of the Council of Europe), Sciences et société civile: est-ce que les gouvernements peuvent en faire plus? (The Role of Science in Our Society: Can't Our Governments Do More?) Nicola Furey, Vice President de Earth Focus Foundation, moderator Nastassja Lewinski.
5. Ilise Feitshans, *Forecasting Nano Law: Presentation for ASSE*, American Society of Safety Engineers Annual Meeting, Chicago, IL, June 2011.
6. Chris Sanford, Shaping destiny: ASSE signs MOU with ILO, *Facilites Safety Magazine*, June 2009, reprinted in the International Occupational Hygienists Association (IOHA) newsletter, **7**(3):15–17, December 2009.

Chapter 10

There Oughtta Be a Law

10.1 Synthesizing Nanotechnology Advances into Harmonized Legislative Text

Traditional lawmakers who made difficult policy decisions with remarkably little empirical evidence worked hard to create regulations so that in future generations decision making would be easier. The 20th century's legacy of codified norms and standards can was designed to be applied for ensuring the health of all people. The legacy includes a richness of international law and norms that discuss human dignity and freedom; universal, fundamental needs of humans in society in order to survive; and the right to life. Universal norms, such as those codified in international instruments, outlast lifetimes and reach across geographic borders and ethnicities and national jurisdictions to enrich all human society. Their essential character for sustaining human existence transcends all job descriptions. Since illness and health are inherent in the human condition, occupational health systems and, in turn, the laws that institute and govern them are essential to preserving health (however health may be defined in a given context). Of all such programs, those that impact occupational health are inevitably the most important tools for achieving the goals of programs that protect life and public health, because the microcosm of ecosystem health impacts can be found in the workplace. From this perspective, the

Global Health Impacts of Nanotechnology Law
Ilise L. Feitshans

ISBN 978-981-4774-84-0 (Hardcover), 978-1-351-13447-7 (eBook)
www.panstanford.com

link between work, health, and survival is inescapable: perpetuating civilization depends on having healthy new generations who enjoy the fruits and replenish the labors of their ancestors.

Regulatory frameworks at the international level are merely the tip of the iceberg. Every nation and several regional organizations offer a legislated nanotechnology program whose underlying enabling legislation authorizes appropriations, following a strategic agenda. Creating adaptable regulations that will apply robust science into regulatory frameworks is crucial, regardless of whether those frameworks were modified by or created for nanotechnologies. Harmonizing emerging nanotechnology regulation offers opportunities for stakeholder involvement, in partnership with government and business, to fix old social problems when replacing old industries with applications of nanotechnology. Harmonization remains essential because all it takes is one nation or one state or municipality to insist on its own methods and approaches in its jurisdiction and then poof—everyone has two conflicting sets of standards to obey. It is no secret that one of the most expensive obstacles to commerce is the waiting period, sometimes years, to sort out which jurisdiction's laws apply in case there is a conflict of laws. Expensive and avoidable!

A harmonized framework can at the same time achieve key social goals. Removing embedded gender discrimination that creates health disparities between men and women in the workplace and in society in general is a key development that will remain after nanotechnology's tidal wave of economic and social change.

10.2 Nanotechnology Risk Management and Public Health Protection Act Model

10.2.1 Section 1: Legislative Findings

10.2.1.1 Impact of nanotechnology to galvanize new commerce despite potential risks to public health

1. In the next five years, trillions of dollars will be spent on research and development funding for the application of nanotechnology, touching the economy globally, across almost

every industry: food processing for retail markets, cosmetics, paintings and coatings, agriculture, equipment, and packaging.

2. Nanotechnology was expected to represent about three trillion dollars of the US gross domestic product (GDP) by 2015. Heralded as a "revolution," comparable to the Industrial Revolution of the 19th century, capable of changing daily life in a manner compared to the invention of the car, the sheer economic importance of nanotechnology will change several antiquated systems regarding industrial processes, scientific understanding and categorization of chemical informatics, and, ultimately, the health care delivery systems that must use or correct the end products of these changes.
3. Application of nanotechnology in industry outstrips assessment of the risks from exposure. Therefore, the science of protecting public health lags behind the science of research and development that creates new elements and new intellectual properties.
4. Risks associated with the application of such technologies are much slower to emerge than the many new vistas of prosperity and efficiency that nanotechnology promises to humanity throughout the world.
5. The specter of new economic frontiers with wider horizons for new products and the attendant commerce from their trade has caused many opinion leaders in science, law, and health policy to herald nanotechnology as an unprecedented opportunity for human development and growth.
6. Yet, known dangers of many of the substances whose molecular structures are changed using nanotechnology has caused alarm among scientists and policymakers who fear that unfettered use of such new technologies can unleash a public health crisis in the event of explosion or spill of engineered or manufactured nanomaterials (MNMs).
7. Thus, the law and policy question arises, and its answer remains unclear until new legislation is created to fill this void: *What law, if any, applies to protect the general public, nanotechnology workers, and their corporate social partners from both liability and preventable harms?* Risk management,

employing best practices and exposure assessment and tools for risk communication, provides the best answer to the problem of avoiding liability and foreseeable harms and for anticipating problems that cannot be predicted at the time of this writing.

10.2.1.2 Risk management based on due diligence to promote innovation and protect public health

1. Historically, law and science have successfully partnered together throughout the 20th century to take on "big science" projects that were fraught with risk, and succeeded in harnessing atomic energy and decoding the human genome without blowing up the world or unleashing mutant monsters.
2. Therefore, even though qualitative data to protect exposed people and the greater ecological system that surrounds the human environment lags behind industrial use, research, and application of nanotechnology to consumer products, scientists and public health officials have an obligation to the general public to institute scientific oversight employing precautionary principles that will enable products to enter commerce, while:
   a. Safeguarding the public health for consumers
   b. Protecting people exposed to nanotechnology applications in their workplace
   c. Protecting workers' children who are indirectly exposed through passive secondary exposure to nanoparticles in workplace products brought home on cars and in clothes or through direct transplacental exposure during pregnancy
   d. Protecting the delicate balance of the ecosystem of the greater planetary environment
3. Therefore, a commission is needed to ensure the free flow of data throughout the process of risk management for the purposes of exposure assessment, risk assessment, risk mitigation through education and training, and stakeholder feedback and to ensure the ongoing flow of data in a cyclical process.

4. The commission shall be empowered with the oversight of communication but shall not ensure the quality of data, which shall remain the responsibility of those who send the information. The commission shall also have power and oversight for bioethical issues in the applications of nanotechnology in medicines; in the event of explosions, spills, or other emergencies; and in daily consumer exposures.

10.2.1.3 Precedent justifying regulation of nanotechnology

1. Although application of nanotechnology may be unprecedented, the concept of creating preventive policies that contain or manage risk, while incubating new industries, is not new—it is the hallmark of "big science" in the 20th century that fostered nuclear research, genetics, and astrophysics, just to name a few, and industries that fail to accept the harness of regulation run the risk of failing: asbestos use is one example.
2. Work health and survival are inextricably linked; therefore, innovation must be balanced by regulation of consumer choices and workplace exposures in order to protect posterity. As noted by the American Society of Safety Engineers (ASSE) on its website, "A safe and healthy place is a fundamental right. We believe that sound SH&E practices are both socially responsible and good business."
3. The small quantities of materials involved in nanotechnology call into question the application of scientific constructs applied under law, such as threshold values and safe levels of exposure, because nanoparticles that may cause harm fall below the baseline of existing regulatory limits.

10.2.1.4 Scientific uncertainty regarding unquantified risks is the sole point of scientific consensus

1. Testimony from Safe Implementation of Innovative Nanoscience and Nanotechnology (SIINN) of the European Union, juridical experts, and the Council of Europe consistently established that "the precautionary principle is directly applicable to emerging nanotechnologies . . . inadequate information, until the results from research studies can fully elucidate the characteristics of MNMs that may potentially pose a health risk (warrants) precautionary measures."

2. Scientists and governments agree that the application of nanotechnology to commerce poses important potential risks to human health and the environment, but the risks are unknown.
3. Several examples of high-level, respected reports that express this concern have been brought to the attention of this legislature. The authors of such reports include the Swiss Federation (Precautionary Matrix 2008), the Royal Society on Environmental Pollution (UK2008), the German Governmental Science Commission ("SRU"), public testimony sought by US National Institute for Occupational Safety and Health (NIOSH; February 2011), the Organisation for Economic Co-operation and Development (OECD) working group (since 2007), the World Health Organization (WHO), the Council of Europe in its report *Nanotechnology: Balancing Benefits and Risks to Public Health and the Environment* (//assembly.coe.int/ASP/NewsManager/EMB_NewsManagerView.asp?ID=8693&L=2), plus the discussions among several industrial groups, the consortium headed by the Center for International Environmental Law and various nongovernmental organizations (NGOs).

10.2.1.5 Impact of nanotechnology on constituents

1. A response to this problem requires risk communication within the context of risk management using information that has been gathered through exposure assessment, risk assessment, hazard analysis, and data collection and using the tools of risk mitigation and best practices so that manufacture, use, handling, transport, and safe disposal of nanotechnology applications is feasible and so that risks are understood by people exposed to these risks.
2. Workers, their families and dependents, consumers in the general public, and industrial purchasers of nanotechnology applications in various products have the right to know of these contents, their hazards, and the steps for preventing harms, with special attention to the needs of vulnerable populations such as but not limited to aged persons, persons with disabilities, women who are pregnant or nursing, and

children who face secondary passive exposures to these products.

3. The state of the art of nanotechnology risk assessment and risk management is undeveloped, and therefore the legislature cannot draw bright lines regarding the size, function, character, or end use of nanotechnology applications that would either limit or promote ongoing use.
4. In the face of undefined, unquantified risk, industrial purchasers and users in the general public need a regulatory framework that is dynamic in order for meaningful risk communication to be feasible and to change as new circumstances dictate.
5. These issues have been successfully addressed in the workplace context only by the risk mitigation techniques that form part of sound risk management that constitute the internationally accepted and widely praised Globally Harmonized System for the Classification and Labeling of Chemicals (GHS).
    a. The rationale behind the GHS provides consistent and predictable data from manufacturers and suppliers to downstream users. Each link in the supply chain must operationalize the right to know by providing training about materials and attaching that information to products using Safety Data Sheets (SDS). The list of chemicals covered is impressive. It is a long list, and there is widespread consensus about the dangers of these substances.
    b. The GHS has been viewed by many companies as a sigh of relief due to the previous confusion regarding labeling and the right of companies to know the accurate information from their suppliers regarding chemicals that are shipped around the world and used in thousands of different applications in international trade. Until the GHS there were so many laws governing this topic, no one set of rules applied everywhere and many rules seem to apply needlessly.
    c. The success of the GHS proves there can be voluntary compliance and that the government can regulate in

partnership with industry when global trade so dictates. Suppliers, distributers, and end users need predictable laws and consistency, and they need reliable quality when purchasing goods from abroad or shipping their goods from their home country.

   d. The GHS has a three-pillar approach: (i) labeling, (ii) SDS, and (iii) regularly required training with risk communication about safe handling, transport, and possible health impacts.

6. Filling the substance of these pillars meets current civil society needs for risk mitigation, risk management, and risk communication about nanotechnology.

10.2.1.6 Responsible development of nanotechnology is not a spectator sport

1. Positive incentives for voluntary compliance form a vital alternative model to "punishment and deterrence" models for government regulation of organizational activity, while preserving limited administrative resources for oversight and enforcement and also while encouraging responsible development of nanotechnology through compliance with law.
2. Positive incentives for voluntary compliance are offered herein:
   a. To prevent organizational misconduct, especially within corporations but not limited to corporations as defined by existing laws
   b. To implement these voluntary compliance by recognizing positive achievements toward compliance using performance standards rather than bright line numbers for minimum and maximum exposures and related enforcement
3. Every effort should be made to encourage the development of positive incentives for compliance with this law and to reduce administrative costs by granting limited waiver of fines or penalties in the event of self-reporting of violations, harms, or

spills discovered through routine internal audits or reported due to emergency.

10.2.1.7 Bioethical concerns and benefits of nanomedicine and nanotechnology applications must be addressed by a permanent commission

1. On the basis of evidence presented in expert reports and testimony from stakeholders, the legislature has therefore determined that nanotechnology's revolution for commerce can also revolutionize public health.
2. Nanomedicine, a successful application of nanotechnology, will require society to rethink ancient notions that are the building blocks of social constructs regarding the nature of disease and its treatment, and the prejudices encountered by people who suffer from illness, forcing collective rethinking about ill-health.
3. Legal terms such as "disabled" and "healthy" will take on a new meaning once treatments may be required or commonplace using nanotechnology. Therefore, an unprecedented opportunity exists to benefit from both the nanotechnology revolution and the revolutionary social change that recognizes individual human potential under international laws by preventing discrimination against people with disabilities, permanently reshaping civil society for the better. The commission shall report on such changing paradigms and shall make recommendations for improving access to care, delivery of clear and competent instructions for informed consent, and related impacts of nanomedicine on society.
4. Key public health policy benefits of nanomedicine include:
    a. Rethinking the distribution of public health care and delivery of health services
    b. Rethinking the role of public health compared to private insurance
5. Bioethical issues associated with nanotechnology applications therefore are wide ranging and will change as new techniques are developed and population experience provides empirical

evidence to guide the future direction of policy that is best resolved by an ongoing permanent advisory body in the Department of Health comprising experts, industrial leaders, consumers, and individuals from the general public.

6. The bioethical issues regarding the application of standards for protecting consumers, directly exposed workers, indirectly exposed children of consumers and workers, and the ecosystem of the planet's environment loom large in the case of possible accidents such as explosions from combustible materials the public uses every day, such as aluminum and gold, and from spills of nanoparticles whose fate and transport remains poorly understood or unknown when they cross previously impermeable barriers at the protein corona between cells, across the blood–brain barrier, and inside cells.
7. These developments must be monitored and controlled while being encouraged and therefore must be subject to ongoing scrutiny and review of data through a multidisciplinary commission comprising experts, industry leaders representative of watchdog NGOs, and private members of the general public from all walks of life, who shall examine the flow of data, goods, and research activities in commerce, reported annually by the authority created under this act.

10.2.1.8 Call for harmonization of nanotechnology regulations worldwide via collaboration and coordination with parallel regulatory bodies

1. A vibrant corpus of international law justifies governmental actions to address concerns about the operationalizing of the precautionary principle in order to protect workers, consumer exposures, and the public health. Accidental explosions and spills also threaten the environment and large populations.
2. Nonetheless, in the field of nanomaterials, many different institutions are working on different focal points: standardization organizations, regulators, scientists, economic or health organizations, industries, and others. Countless nanotechnology guides and sets of best practices are available on the web and in textbooks. These words confuse further the question of unknown risks and what is

the right course of action for a well-intended employer or an instructor giving training, the inevitable solution to nanolaw proliferation is an international code with unified obligations and harmonized terms of art, especially definitions regarding the use and limits of applied nanotechnology. Quite often, a given aspect of nanotechnology has more than one definition among respected references.

3. A desire to regulate nanotechnology has been clearly articulated by hundreds of draft laws and several key statutes at the national, multinational, and United Nations level. Every nation has a statutory basis for its nanotechnology program; plus there are several key emerging laws at the regional and international levels. The desire to regulate is clear, but too many laws exist or are emerging for any one rule of law to reign sovereign.
4. Therefore there is a call for harmonization of the hundreds of existing and emerging nanotechnology laws:
   a. To meet commerce's need for predictability in trade
   b. To meet the private and public need for data collection regarding risks
   c. To answer the universal need for consistent requirements regarding risk communication within industry and among industrial users and consumers of nanotechnology applications
   d. To provide risk commutation to directly impacted and secondary passive exposed populations, to protect the environment, and to promote the legal viability of nanotechnology industrial users from future risk of liability from unregulated harms, while encouraging a culture of innovation for new industry

WHEREFORE it is hereby enacted the "Nanotechnology Risk Management and Public Health Protection Act."

10.2.2 Section 2: Purposes

1. To construct a regulatory framework to anticipate, recognize, evaluate, control, and confirm the absence or presence of

risks to human health and the environment in each context on a case-by-case basis and to apply precautionary principles recognized under law to control and manage the risks associated with the presence of nanotechnology applications in the environment impacting human health or impacting the ecosystem, in the human body, and in workplaces by using the tools of risk management and risk communication when applying nanotechnology

2. To require and develop ongoing regulatory mechanisms for oversight and monitoring of communication about risk, safe handling, and projected long-term health effects of nanotechnology products, their biological fate, and transport in systems, including but not limited to the ecosystem, nonhuman species of organisms, and human beings
3. To provide ongoing contact with the directly impacted workforce, their families and dependents, and the general public regarding risk communication for primary and secondary exposures to nanotechnology applications in society
4. To protect consumers, including industrial users, as purchasers of nanotechnology applications in goods by requiring and implementing universal risk communication that is clear and understandable by ordinary voting citizens who are not part of the scientific community
5. To provide positive incentives for compliance with requirements for risk communication about applications of nanotechnology, mindful of the practical reality that such partnerships are needed because of the limited resources available for watchdogs and enforcement of important laws that embed precautionary principles into the daily work of public health

10.2.3 Section 3: Definitions

"Commission" means the "Commission for Bioethical Oversight of Nanotechnology Risk Communication in the Department of Health" established under this act.

"Corporate criminal sanctions" refers to penalties imposed on the corporate entity itself and not individuals, as punishment for proven harms in violation of law. Such penalties include but are not limited to forbidding specific activities, suspending licences or revoking permits, or barring the corporation from engaging in specific commercial activities, such as obtaining government contracts for goods or services.

"Cyclical approach to risk management" means that compliance activities such as audits, risk mitigation programs, and risk communication components of risk management are not one-shot firecracker approaches to evaluating and reducing risk. Each step in compliance must be regularly repeated on a cyclical schedule.

"Due diligence" is the fundamental concept for crafting and implementing effective in-house management systems for occupational health compliance that avoids liability. In many legal systems, the law requires that every effort be made in advance to protect the public health and manage risk, regardless of whether that risk can be quantified. The ability to prove that such efforts exist on a systematic basis throughout the employer's company, the so-called paper trail is dependent entirely upon proving this concept of due diligence. Compliance programs offer concrete proof of due diligence and thus allow an employer to enjoy a presumption of compliance in areas of the law where the limits are unknown.

"GHS" means the UN-based multinational system for Globally Harmonized System for the Classification and Labeling of Chemicals, involving hundreds of nations and hundreds of UN agencies working in collaboration with NGOs.

"Positive incentives for compliance" refers to tax credits, prizes, awards, and other incentives to good citizen corporations.

"Risk communication" for the purposes of this act embraces the GHS's three pillars of labeling; transfer of data sheets regarding the safe handling, transport, or use of materials and their potential adverse health effects; and disclosure of such information to workers, consumers, end users, and the general public.

"Risk management" is a process that embraces exposure assessment, risk assessment, risk mitigation, and risk communication. Risk management is proven by documented due diligence.

"Risk mitigation" refers to a result-oriented process designed to prevent, detect, report, and correct potentially dangerous conditions that can result in harm to human health or the global environment by involving in-house compliance programs.

"In-house compliance infrastructure" features but need not be limited to (i) managerial statements in writing that demonstrate the enterprise commitment to workplace safety and health and to protection of the global environment in order to reduce or stabilize the global disease burden; (ii) documentation of the use of due diligence to steer the components of the compliance infrastructure using internal audits on a cyclical basis that can capture health disparities, isolate particular exposures that have heightened hazards, and provide documentation of the best practices that were applied in response to potential harm; (iii) in-house communication to staff including but not limited to interactive video training and web-based eLearning regarding the safe response to problematic conditions in the workplace (regardless of whether chemical or circumstantial, and embracing emergency response); (iv) two-way vertical communication that enables complaints about problems to be recorded with response in a timely manner using hotlines and in-house newsletters and intranet; (v) documented ongoing interaction with regulators, insurers, consumers, suppliers, end users, and the general public in advance of developments and in the case of emergency; and (vi) any tools that withstand judicial or scientific scrutiny that meet and maintain lawful occupational safety and health goals.

"NSDS" refers to Nanomaterial Safety Data Sheets, the basic tool of risk communication under this act.

"Stakeholder" refers to any party that may be indirectly or directly harmed by the failure to disclose risk communication as required by this act.

10.2.4 Section 4: Scope

1. This act shall apply to end-use impacts of engineered or MNMs on the basis of the context in which their nanoparticles function and their character and fate once functionalized,

without regard to size, aggregation, or agglomeration with additional particles in clusters or the reason they were used.

2. This act shall apply to any or all materials in commerce that self-characterize as "nanomaterials" "using nanotechnology" or listing "nano" among the product features in advertisements, labeling, or related publicity. For the purposes of this statute, jurisdiction requiring compliance with the provisions of this law is triggered by the use of "nano" in trade names.
3. Failure to disclose the risk communication required by this act in order to avoid jurisdiction of this act shall result in fine and penalties enumerated herein, including but not limited to possible criminal prosecution of organizations.
4. Fines and penalties may be waived despite failure to disclose upon presentation of evidence of systemic efforts to capture the best-possible data and to use due diligence in risk management of applications of nanotechnology.
5. Voluntary compliance is encouraged through awards that will be established by the commission created herein.
6. Nothing in this act shall create liability on the part of administrators in the commission or enforcement authorities for the use of NSDS risk communication or manufacturer-generated data that is inaccurate, misleading, or unreliable. The sender of the nanotechnology risk disclosure is solely responsible for ensuring its accuracy and for any or all liability arising from errors in the data transmitted.

10.2.5 Section 5: Commission Established

10.2.5.1 Commission for Bioethical Oversight of Nanotechnology Risk Communication in the Department of Health

1. Delegation of authority to propose risk communication for nanosafety: For the purposes of evaluating fairness, informed consent, social justice, and human health parameters of nanomedicine and any applications of nanotechnology subject to the provisions of this act, this act hereby establishes the

Commission for Bioethical Oversight of Nanotechnology Risk Communication in the Department of Health.

2. The purpose of the commission is to act as a conduit with oversight to confirm the ongoing free flow of information from the manufacturers, industrial consumers, and product distributors of applications of nanotechnology to workers, consumers, the general public, and end users.
3. The commission shall hear complaints from organizations and individuals regarding the failure to disclose risk communication required by this act:
   a. The commission shall have the power to authorize remedies to correct said failures.
   b. The commission shall have the power to set penalties and fines consistent with the purposes of this Act, and to refer matters to criminal investigators if necessary.
   c. The Commission shall have the power to create positive incentives for voluntary compliance.
   d. The commission shall have the power to waive fines and penalties upon clear showing of due diligence in risk management leading to risk communication for workers, consumers, and end users of the nanotechnology applications involved.
4. The commission shall review and synthesize best practices into a biannual report, maximizing benefits of nanotechnology, while minimizing risk.

   Report contents:

   a. The thematic scope of the report shall examine the changing social meaning of terms such as "disabled" and "healthy" that will take on a new meaning once treatments using nanotechnology may be required or commonplace, mindful of the unprecedented opportunity that exists to benefit from both the nanotechnology revolution and the revolutionary social change that recognizes individual human potential under international laws by preventing discrimination against people with disabilities.

b. The subject of the report shall examine key public health policy benefits of nanomedicine, including:

   i. Rethinking the distribution of public health care and delivery of health services

   ii. Rethinking the role of public health compared to private insurance

c. Bioethical issues associated with nanotechnology applications.

d. Bioethical issues regarding the application of standards for protecting consumers, directly exposed workers, indirectly exposed children of consumers and workers, and the ecosystem.

e. Emergency planning and first responder protections in the case of possible accidents such as explosions.

f. Long-term impact on the global disease burden (GDB) from nanomaterials the public uses every day, such as aluminum and gold, and from spills of nanoparticles whose fate and transport remain poorly understood or unknown when they cross previously impermeable barriers at the protein corona between cells, across the blood–brain barrier, and inside cells.

5. Commission proceedings:

   a. Commission deliberations shall be open to the public.

   b. Commission findings and reports shall be posted on the web.

   c. Commission reports and materials for risk communication shall be downloadable free of charge.

   d. The commission shall provide and maintain web-based blogs with discussions of best practices, NSDS, and related aspects of risk management.

   Exception:

   e. The commission and its members shall make no warranty of the validity of data transmitted, and responsibility for the quality of blog or message content shall remain solely with the sender of the information.

6. Commission composition:
   a. The commission shall comprise no less than 27 members and not exceed 99, with an odd number divisible by 3 for its voting membership.
   b. The commission shall provide an opportunity for a democratic voice for stakeholders in the process and for the monitoring and follow-up of nanotechnology applications in commerce and related risk mitigation and risk management activities.
   c.
      i. There shall be established a panel of commissioners representing a balanced cross section of civil society, with duly authorized representatives of stakeholders from:
         - Leading corporations
         - Governments: national, local, and municipal
         - Nonprofit organizations and international NGOs
         - Scientific academies
         - International organizations
         - Consumer groups
         - Public health research institutes
         - Professional associations
         - Private individuals
      ii. Members may represent also a collaborative or collective group of organizations.
7. Voting:
   a. Voting strength shall be equal among all members.
   b. At least two-thirds of the members shall be present to create a quorum for voting.
   c. Sessions shall be transparent and open to the public, except for voting, where ballots shall be secret.
   d.
      i. Voting commission members shall each provide no less than four sessions of focus groups regarding the

understandability of labels, NSDS, and training for risk communication per year.

ii. Each voting commission member shall report the results of focus groups to the commission as a whole to determine whether the methods used are readable and clear and, when necessary, alert the recipients to scientific uncertainty that requires further research.

10.2.6 Section 6: Rewards for Proven Due Diligence in Risk Management

10.2.6.1 Positive incentives for compliance with risk disclosure provisions, risk management to enhance the quality of risk communication, and risk mitigation techniques to reduce avoidable exposures

1. Within the first six months of its establishment, the commission shall seek stakeholder comments and vote on a list of innovative positive incentives for compliance with the provisions of this act.
2. Positive incentives for compliance shall include but not be limited to competitive prizes, awards for the broad-reaching and high level of risk communication achieved through interactive programming with vulnerable populations or high-risk groups, and recognition of achievements in research and development of safer products.

10.2.7 Section 7: Risk Management to Promote Responsible Development of Nanotechnology

1. Components of risk management systems:
    a. Data for hazard analysis: The degree of acceptable risk, the methods of risk assessment, and the measures of effectiveness for the same or similar hazards shall be determined by the context in which they arise, and therefore may differ depending upon circumstances.

b. Databases: To achieve the purposes of this act, acceptable types of databases include but are not limited to:
   i. Quality-controlled inventory of products incorporating nanomaterials or resulting from nanoscience- and/or nanotechnology-based food or feed processes based on substantiated, statistically tested claims and random samples of new products likely to stem from nanoscience or the nanotechnologies
   ii. Quality-controlled, remotely accessible, searchable archives of comparable characterization, toxicological, and exposure information
   iii. Quality-controlled, remotely accessible, searchable archives of risk assessment and test methods
   iv. Quality-controlled, remotely accessible, searchable archives of safety equipment and equipment characteristics

c. Exposure assessment:
   i. Analytical methods and instruments required to assess the (external) exposure of populations and the (internal) exposure of organs in the body, favoring noninvasive approaches.
   ii. Analytical methods and instruments required to characterize, detect, and trace inorganic and organic nanomaterials in food and feed matrices, preferably in a high-throughput mode; pathways for exposure assessment, risk assessment, risk mitigation, and risk management.

2. The components listed here must be refined into risk communication:
   a. Flow if data is multidimensional, but must include views of stakeholders.
   b. Inside the entity: This requires soliciting the views and experiences of the staff.
   c. For persons entitled to this information outside the entity, focus groups shall be convened to determine readability

and comprehension of written materials such as but not limited to labels, NSDS, and materials generated for risk communication training and coursework.

d. Risk communication requires ongoing discourse between the commission, regulated entiites under this act, and stakeholders in order to confirm that risk management activities are undertaken, that those activities are visible and open to the commission's oversight, and that the disclosures made through risk communication are understandable to stakeholders.

e. The commission has the discretion to require an entity governed by this act to rewrite its risk management plan in the event of system failure leading to emergency incidents that threaten the public health with harm.

3. Apply the data using the hierarchy of controls within the context of an in-house compliance system that has been documented and tested by the governed entity. The components of the four key steps to reduce risks include but are not limited to:
    a. Eliminate or minimize risks at their source.
    b. Reduce risks through engineering controls or other physical safeguards.
    c. Provide safe working procedures to reduce risks further.
    d. Provide, wear, and maintain personal protective equipment.
4. Due diligence in auditing and periodic reviews shall be required to be demonstrated as proof of compliance with the requirements of this act in the event of violations, in the determination of fines and penalties, and in order to determine the rewards for positive incentives for compliance.
5. NSDS:
    a. For the purposes of this act, governed entities must produce, develop research to support, and transmit NSDS to the next user in the chain of distribution.
    b. NSDS shall follow the basic form of SDS generated in compliance with the GHS, with the exception, however, that NSDS will pay particular attention to the known and

unquantified risks of exposure to the nanoparticles or nanomaterials involved.

6. Cyclical approach to risk management:
   a. Each entity is required, to warrant the validity of information transmitted in the NSDS, that it has followed the key steps of risk management: *anticipate, recognize, evaluate, control,* and *confirm* the absence or presence of risks to human health and the environment in each context on a case-by-case basis and to apply precautionary principles recognized under law to control and manage the risks associated with the presence of nanotechnology applications in the environment impacting human health or impacting the ecosystem, in the human body, and in workplaces.
   b. Each governed entity must repeat the anticipate, recognize, evaluate, control, and confirm cycle annually or as new processes so dictate but in no case less often than once a year.
7. Stakeholder views to be incorporated in procedures and measures of effectiveness, with focus groups of stakeholders for labels and risk communication seminars ("train the trainer").

10.2.8 Section 8: Integration with the GHS

The GHS is hereby incorporated by reference to the extent that it may apply to applications of nanotechnology in the workplace and to workplace exposure to nanomaterials.

1. Adoption and incorporation by reference of the justification and modes of practice in the GHS model means that the three pillars of the GHS are also the fundamental three prongs of enforcement for this act.
2. NSDS
3. Risk communication through training and education regarding the safe handling, transport, and potential health and safety hazards of the materials.

4.

   i. To the extent that GHS data conflicts with data transmitted under the terms of this act, the commission shall review both sets of data and make a determination regarding which rule is most respectful of scientific precautionary principles in the context.

   ii. In such cases, entities using products involved shall have the right to apply for and be granted a temporary waiver of the requirements of this act, provided that:

      - The entity is not already subject to fines and penalties under this act.
      - The entity demonstrates due diligence in the development and implementation of its in-house risk management programs for compliance with this act.

10.2.9 Section 9: Enforcement Provisions

1. Individual citizens can trigger enforcement by petitioning, with clear and convincing evidence, to the commission created under this act.
2. The procedures for petitions shall be set forth by the commission on or before the second session of its meetings, consistent with administrative procedures and constitutional requirements of the jurisdiction in which the commission shall reside.

10.2.10 Section 10: Penalties

1. Failure to disclose risk communications for labels, NSDS, or training and education by a commercial or government entity that is involved in the manufacture or direct purchase and use of nanomaterials in order to avoid risk management and risk communication required in compliance with this act shall result in fines and penalties enumerated herein, including possible criminal prosecution of organizations.

a. Individuals
    i. Wilful refusal to engage in risk management and risk communication pursuant to the purposes of this act by individuals who are found to have wilfully violated this act shall be subject to fines and penalties.
    ii. Such fines and penalties can be waived in the discretion of the oversight authority after a showing of good-faith use of due diligence to disclose and risk management system maintenance that use approved and established methods not specifically set forth by the oversight authority in this act.
b. Corporate persons who have violated this act by failing to disclose use of nanomaterials to persons exposed or failing to institute risk management procedures with proven due diligence shall be subject to fines as set forth by the oversight authority and penalties including criminal sanctions, including but not limited to:
    i. Prohibition from participating in major contracts initiated or administered by this jurisdiction
    ii. Liability in a court of law
    iii. Publicity regarding their wrongdoing to be made available by the media using the information published by the oversight authority in its official annual report
c. In the event of a proven causal relation between failure to disclose nanomaterials use and death of one or more individuals, the corporation that failed to disclose shall be held criminally liable for harm in the form of:
    i. Prohibition from participating in major contracts initiated or administered by this jurisdiction
    ii. Absolute liability in a court of law without immunity based on the nature of the business, cost of compliance with this act, or benefits of the application of nanotechnology involved

iii. Publicity regarding its wrongdoing to be made available by the media using the information published by the oversight authority in its official annual report

2. Waiver of penalties, fines, or liability
   a. In the event of self-discovery of violations of the risk communication provisions of this act, a commercial or government entity that is involved in the manufacture or direct purchase and use of nanomaterials followed by prompt efforts to report, detect, and correct the violation may face reduced fines or penalties
   b. Upon demonstrated documented due diligence to comply with this act, fines may be reduced or waived.
   c. There shall be no presumption of validity for the due diligence of risk management, but established proof of due diligence shall weigh heavily in the commission's decisions regarding penalties and fines.

10.2.11 Section 11: Saving Clause

In the event that any of the terms or provisions of this act are declared invalid or unenforceable by any court of competent jurisdiction or any federal or state government agency having jurisdiction over the subject matter of this act, the remaining terms and provisions that are not effected thereby shall remain in full force and effect.

10.2.12 Section 12: Effective Date

1. This act shall take effect 30 days after passage.
2. Start-up grace period:
   a. This act shall take effect 30 days after passage with a 6-month start-up period for the preparation of risk communication materials for implementation and dissemination.
   b. No fines or penalties shall be assessed during the 6-month start-up period.

10.3 Keeping the Canary Alive with Safe and Healthful Uses of Nanotechnology: Birthright or Gift to Humanity?

A long time ago, before electricity was used in homes, people mined coal by crawling around the inner bowels of the earth using axes or other implements or using their bare hands. Ivy Pinchbeck, in a brilliant, unsung treatise about the life and work of female miners in 18th- and 19th-century England and France (yes, two or three hundred years ago, women mined coal and other ores in England and France) described how women went into parts of the mines where men who were too highly paid to be risked and horses who were too large and who cost too dear to replace would not go [1]. She described women working naked, on their knees, and crawling in the veins of coal, working long hours, descending before daybreak, and emerging when it was again night, without ever seeing the sun. Pinchbeck described their working conditions and how often women were buried alive from mine cave-ins or escaping gas. Engravings of such women working, which were used to illustrate a point about the contributions of a doctor in England at a conference on the history of occupational health in 1998, support Pinchbeck's statements. Many women died working in the coal mines in these very difficult working conditions. And people work and die in coal mines two or three centuries later.

The 19th-century French political novelist Emile Zola [2] in *L'Assemoir* also paints a bleak landscape of the coal mining villages in southern France, similar to Pinchbeck. Zola described the life of coal miners in the south of France in the early 19th century. Where there was famine and little to eat and threats of strikes and threats on the person of the mine owners, it was the outraged female workers who had the hardest, dirtiest, and least paid jobs. Women also were the leaders in strikes and civil unrest, and eventually, they won fairer contracts with better wages for better work but never attained economic parity or social equality.

In those days, too, there were crude emergency rescue systems in the coal mines. If there was a cave-in or other disaster, few, if any, workers got out alive. The coal that warmed houses in Paris and London and warmed food for urban families of wealth came at a high price in human life. Sometimes there were emergencies when people could get out if leaking gas also emerged from the innermost layers of the earth. And to protect human life in the event of gas escape, as feeble or inadequate as that attempt may have seemed to be with 20th-century hindsight, there were canaries in the mine. Among cautious mine owners there were always the sensitive, the vulnerable, canaries in the mines. Especially selected for a mysterious genetic trait that was observed but that remains not well understood, canaries were posted at key locations in the mines because they were known to be more sensitive to a change in air quality that could mean the loss of precious life.

The canaries in the mines who sang and chirped sang of life; their chirping meant all was fine. But their silence meant death of their hypersusceptible pulmonary system, and it meant that it was time to sound the warning to evacuate the mines. While not everyone could get out in time, the death of the canaries meant "get out of the mines." And accident investigators debated the time lag between the death of the canaries and the actual command to evacuate the mines and how much time was really needed to get out, especially if there were so many people working in the mines. Was there time to do more work before announcing the canary had died, before evacuating the mines?

Should every dead canary give cause for evacuation? Couldn't a canary just die of old age on the job or of some other intervening cause?

Could a vengeful worker willfully deprive the canary of oxygen in order to call a strike, to cause a disruption, or to prevent some new technology from being used in the mines? For revenge or terrorism? Or by accident?

How many canaries died before a foreman was certain it was necessary to evacuate the mines?

Stopping work costs money. It was an era when coal production was reaching unprecedented demand and when there was unprecedented economic competition between two international superpowers of the time: France and England. And most importantly and most troublesome of all, the unanswered eternal question remained: Once you believed the canary is dead, truly dead from gas and not dead from sabotage or other intervening causes, *how much time*?

How much time does the mine owner have between the death of the canaries—the genetically selected population—prized for their vulnerability, until the inevitable destruction of everyone who cannot get out of the mine?

No one really knows. Gas fluctuates in its ability to travel, depending on many factors: weather, wind, production quotas, size of the working population inside the caverns, and size and level of development of the work spaces in the mines. There is not much decision time, and any mistaken evacuation is expensive and economically costly. But even when there is consensus that the canary is dead, no one really knows the exact evacuation time that remains.

For centuries there has been an ongoing battle to make better, more sensitive surrogates for canaries—sensitive to gasses and other potential harms. Alarms make noise rather than falling silent when we expect disaster. Ongoing monitoring charts and graphs the peaks and valleys in exposure to various substances and railway cars and cameras to speed people through and record evacuations, and nanosensors detect small quantities in places where humans cannot go. And we don't have many women crawling around naked in the depths of the mines, even in third-world nations, where there may still be small children in the mines.

Civil society remains plagued by many silent killers that emit from the workplace. These harms negatively impact vulnerable populations, our canaries, in our mines. Preconceptual exposures for a parent may bring effects in the offspring from exposure to toxic substances; fetal exposures to noise, teratogens, contagious or infectious disease, or perhaps even beryllium may cause unreportable or undetected harm. Remote and far-fetched? Abstract and insignificant? The link is a fine thread of posterity that demands attention and investigation all the same.

Nanotechnology in the workplace, in consumer products such as food and cosmetics, refrigerators, and cars, and as an increased environmental burden, impacts not merely so-called workers but people from all classes in professions and all walks of life.

Nanotechnology's potential health hazards have been observed but not yet studied in detail in a whole host of vulnerable populations that are not expected to have such experience, even though the impacts appear in public health databases around the world in unprecedented numbers in previously considered "safe" populations. Questions of lung impairment, litigated to extremes about preventable asbestos exposures, re-emerge as carbon nanotubes (CNTs) and nanosilver pose new threats of migration into the cavernous depths of human lungs and courts and regulators have responded with concern about the potential long-term negative health impact upon particularly sensitive populations. These vulnerable populations (small children, disabled people, and people who live at the margins of society without adequate social protection) are the contemporary canaries in the mines. Protecting their good health through responsible development of nanotechnology can save many lives in the world outside.

Uncertainty in science is a wildcard that sharpens the edge of the dilemma faced by all potential stakeholders, regulators, or legislative drafters. As noted by the Royal Commission on Environmental Pollution of the United Kingdom (2008, paragraph 1.39), ". . . in the early stages of a technology we don't know enough to establish the most appropriate controls for managing it. But by the time problems emerge, the technology is too entrenched to be changed without major disruptions . . . ," and (2008, paragraph 1.40), "The solution to this dilemma is not simply to impose a moratorium that stops development, but to be vigilant with regard to inflexible technologies that are harder to abandon or modify than more flexible ones" [3].

When the canary died in the mines, there were no epidemiologists or toxicologists around to examine the body, its DNA, the surrounding air quality, or the quality of its food intake. No one could ever really know whether the canary died because of sabotage or old age or inherent genetic weakness or happenstance or actual exposure to toxins in that workplace, standing alone as an independent variable or in combination with a host of other toxins in the surrounding environment. People understood only that the canary had died.

This work has been designed to awaken public awareness to the notion that some canaries may have died but that many more canaries are still alive and working; some are actually thriving! Somehow civil society has not thought clearly about how to harmonize the zillions of existing laws before writing new laws when confronted with risks that seem new, and not exhibited forethought about whether there is a safe way out. There are no zero-risk scenarios, no undisputable facts, and no inescapable scientific certainties, but there is time to think carefully about the issues as emerging nanotechnologies, nanomedicines, and regulatory frameworks try to diligently make the best effort when planning the future for civil society. Law already exists, so drafting law is not the immediate need. Even though a harmonized regime similar to the GHS and possibly under the auspices of the United Nations or another international governing body created by treaty could improve the questions of standardization, these tools are only useful if applied with forethought. Understanding law in order to manage the risks and implementing law consciously addressing the right to know is a genuine universal need.

Figure 10.1 Discussion of "Law and Science of Nanotechnology: Perfect Together? Nanotechnology under International Law Protecting Public Health," Museum of the History of Science, Geneva, Switzerland, public forum available on YouTube, August 13. 2013. Dr. Ilise L. Feitshans speaks. Photo by Dominique Charoy.

The blueprint for saving civilization, protecting posterity, and ensuring an excellent quality of health and life for all is already embedded in the DNA of commerce and innovation; society need only trigger the force of making these traits in our system dominant and then operationalizing our intuitively best practices.

Dear Reader,

Thank you for reading all of this text to the very end. Remember, the power to make these dreams happen is in you. Read, think, and act locally and globally to move forward a safe and healthful future applying nanotechnology.

References

1. Ivy Pinchbeck, *Women Workers and the Industrial Revolution 1750-1850*, London School of Economics Studies in Economic and Social History, George Routledge, London, 1930.
2. Emile Zola, *L'Assemoir*, 1876.
3. Royal Commission on Environmental Pollution, Sir John Lawton CBE, FRS, Twenty-Seventh Report: *Novel Materials in the Environment: The Case of Nanotechnology*, presented to Parliament by Command of Her Majesty, November 2008. See also UK Department for Environment, Food, and Rural Affairs, 2010, research into the likelihood and possible pathways of human exposure via inhalation arising throughout the life cycle of a selection of commercially available articles containing carbon nanotubes – CB0423. http://www.defra.gov.uk/.

Acknowledgments with Gratitude!

My deceased parents, Sylvia Feelus Levy and Jack Levy, Esq., are the beacon for all the principles articulated in my writing.

My children, Jay Levy Feitshans and Emalyn Levy Feitshans, remain the moral compass for all my work.

Like the fine threads within a silken tapestry, writing this book gathers together principles and experience from all walks of life. These strands of reasoning provide the basis for the next generation to ultimately amass political will with the momentum to address the politics of human health, because it is a universal right that society must protect in order to survive. The work to be done in these times touches destiny.

This book is the natural outgrowth of my undergraduate masters and doctoral work in the jurisprudence of health.

My research and writing about nanotechnology is unabashedly inspired by the stellar work of the United States' national treasure and my personal professional role model, Dr. John Howard, MD MPH JD LLM. Dr. Howard honored me by visiting Geneva, Switzerland, for a full week while I was the coordinatrice of the 5th edition of the *ILO Encyclopaedia of Occupational Health and Safety*. His clear thinking and his vast knowledge of so many important disciplines in science and the humanities (as exemplified by his then-37-page résumé) has permanently set the bar very high for personal professionalism. His unforgettable prescient lecture at the International Labour Organization (ILO), "Nanotechnology: The Newest Slice of Daily Economic Life," on Thanksgiving Day, November 28, 2008, was a real holiday treat! He delivered his urgent and exciting message so calmly and clearly, despite its cutting-edge, exciting knowledge and undeniable complexity. His lecture has influenced my work for the rest of my life. I learned more from that man in a week than I had learned in schools for many years, and we are lucky that he has the power that comes with two presidential appointments by two different US presidents—Bush and Obama—as different as black and white.

I speak of these emerging global health and safety issues from an academic perspective. I have never worked for either labor or management sides and have always worked in government and academia. My government experience truly spans the gamut: from the lofty international memoranda drafting at the United Nations and for multinational nonprofits in several nations to the dusty local level of municipal laws for the state of New Jersey (concerning special education for children with disabilities) and the Board of Health in my beloved native city of New York, regarding confidential testing of minors for HIV/AIDS in the 1980s. I am a woman, and that influences my professional history and thus my writing. Unlocking the death grip of global gender apartheid is, in my opinion, the key to creating global policies that will ensure survival of civil society.

Risks associated with the application of any new technology are much slower to emerge than the many new vistas of prosperity and efficiency that it promises to humanity throughout the world. In particular, the *Precautionary Matrix for Synthetic Nanomaterials: Version 1.0* (Federal Office of Public Health [FOPH], Federal Office for the Environment [FOEN], Berne, 2008) is a clear document whose authors truly care about the convergence of economic development and science that advances human progress and yet is mindful of the public health. Such consensus is an ideal that a huge country like mine could only hope to achieve, but it is a real-life working vehicle for robust assessment of applying nanotechnology under law. I commend my colleague Dr. Michael Riediker, Institute for Work and Health, University of Lausanne, for his work on this remarkable, brief, but effective document.

There are many people in Geneva, Switzerland, to thank, and many more around the world.

Everything that can be accomplished using the research from this book was made possible because of the amazing faith and encouragement of Dr. Mark Hoover, PhD, CHP, CIH, Senior Research

Scientist, Division of Respiratory Disease Studies, National Institute for Occupational Safety and Health (NIOSH), Centers for Disease Control and Prevention (CDC), Morgantown, WV, USA. Dr. Hoover has the enthusiasm for my work found amongst the best cheerleaders, combined with the belief that the project is doable and well worth his considerable time and effort to see it through. He has worked hard to convince leaders of academia, publishing, and the scientific community to look closely at the policy analysis and conclusions in my writing.

The blueprint for operationalizing the precautionary principle of the so-called right to know, discussed in this book in detail, owes a huge debt of professional gratitude to the principles articulated by Dr. Alice Hamilton (1869–1970; yes, she lived to be 101 years old) in her pathbreaking autobiography, *Exploring the Dangerous Trades* (1943). Dr. Hamilton lived half a century before she had the right to vote, and had no children. Both facts are a great loss to humanity because of the influence her vote and her offspring would have had on all civil society. The fundamental importance to civil society of exercising the right to know about information pertaining to public health rings clearly like a well-tuned bell throughout her written work.

At the Geneva School of Diplomacy, three men of unique vision about education and the role of international human rights law for providing better governance in the future made possible my book: Dr. Alfred de Zayas, who shepherded my book proposal from start to end and allowed me to research US ratification of the global harmonization for chemical safety as part of my research for his course (my text became comments for the official public docket); Dr. Jon-Hans Coetzer, who immediately grasped the importance of this topic as an early bridge across the trans-science legal landscape; and Dr. Colum De Sales Murphy, who once compared my work to Rostopovich playing cello in the Berlin metro and gave me intellectual space for my book.

Raymond Azoulay, the reluctant but beloved benefactor of the old doctoral effort, provided a place to study and write, when no one else in Switzerland would do so for over a year in Geneva, Switzerland. Raymond is a remarkable man—quatrilingual: native speaking Arabic; scholarly Hebrew, which he chants magically in rare Sephardic sounds whenever he leads a religious service; mesmerizing Moroccan French (that I rarely understand); and fluent English where he earns his daily living as a chemist at the University of Geneva.

Vera De Ruvo has been sitting in her house in White Plains New York, waiting for me to be recognized and prestigious *for decades*. Because Vera knows me since I was a child and was a close friend of my parents, she recalls when I first started writing about European diplomatic behavior in international relations for the Columbia University School of International Affairs while I was an undergraduate in Barnard College. She knows how proud my father was when he convinced me to leave the international policy analysis program there in order to follow my deeper passion of studying law, even though to me personally the two subjects were inextricably linked. I love each subject as differently and completely as one does one's own child, and I had a vision of the mutual interdependence of these subjects, even when the school administrations did not quite think so. Vera recalls also how I started motherhood with the ambitious goal of having a baby and a doctorate at the same time, only to find that I was obliged to abandon my studies when my ex-husband abandoned our family and then my parents died, all within three years' time.

Consequently, Vera can tell you how I struggled to finish my masters of science in public health at Johns Hopkins University (instead of a doctorate) when none of the faculty cared about occupational health law and policy and when public health curricula had not yet spread its wings to encompass extensive programs in international health that are based on the jurisprudential underpinnings of health as a human right. Therefore, Vera knows better than anyone how profound a healing and closure it is for me at last to finish both: studies in occupational health at the doctoral level and studies in international relations concerning health as a human right.

Vera's fidelity and dedication to me are a remarkable gift from a dying mother to her children, her grandchildren, and her best friend. When my mother, Sylvia Feelus Levy, understood that cancer would soon end her life, as it did just weeks after this promise, she exacted from Vera a very unusual promise. Noting that Vera has sons but no grandchildren of her own, my mother asked Vera to serve as grandmother to my children, who very soon had no grandmother of their own. It was a very odd request, a very unusual gift, and a remarkable legacy. Vera has faithfully attended every birthday party, class play, graduation, and family celebration and treated me as if I were her daughter ever since, including telephoning me every week to wish me a Shabbat shalom, even when I am in Switzerland!

Similarly, Franklin Illfelder has been a friend of our family since my daughter was in nursery school. He has called me every week regardless of where I live and worries about my problems more than my own brother. I could not plan such a friendship, and therefore I cannot thank him enough for his tireless kindness. So, too, Robert Niemic, Esq., a recently retired official of the US federal government and a close friend of my ex-husband since law school, appeared to me one day on Facebook, asking about my life. He then proceeded to take a keen interest in my writings, and I can only hope that as a result, my ex-husband actually does know something of my recent work.

Don Brown, CEO of Digital 2000 Productions, and his entire staff have been, in a word, amazing in their support for this research. If I had a dollar for every time I have relied on Don and his team, I would give it to him and he would be very rich! Together we have created over a dozen videos spanning important policy issues, such as "Lessons Learned from Three Centuries of Occupational Health Laws," "Is There Global Gender Apartheid?" "Nanotechnology and the Law," and "OSHA 35, Still Alive!" (the last video features an interview with Dr. John Howard), and a documentary about legislative history that answers the question "What would the world be like if there never had been an Occupational Safety and Health Act in the United States?"

Herb Siegel, former head of the office in the ILO in Geneva, Switzerland, that arranges a clearinghouse for information exchange called CIS (the title changed but the acronym has remained from Centre pour l'information scientificqiue) greeted me when I was just a child in college. I came to the ILO for the first time, thanks to the visionary faith of my faculty advisor Prof. Peter Juviler of Barnard College, Columbia University. In hand, I had only a letter of introduction from Barnard College to Ms. Betty Hamnet, who was a

Barnard alumna and the UN librarian at the time. I found the ILO by accident, and the then head of the CIS, Siegel, met me with leering curiosity—who was this woman from New York City that stumbled upon the ILO? He gave me a paper application to fill out by hand, which I did. I stayed until 7:00 p.m. that night that I met him for the first time. I poured my heart and soul into that handwritten form that probably was promptly discarded, but the notions I wrote on that paper organized my thinking into a set of ideas that has stayed with me the rest of my life. The things I wrote about for the ILO became part of my senior scholar honors thesis for Barnard College: "Occupational Health from an International Perspective." Then nearly 20 years later, this text became the blueprint for the article "Occupational Health as a Human Right," written for the 4th edition of the *ILO Encyclopaedia of Occupational Health and Safety* while I was teaching legislation at the Columbia Law School, Columbia University.

I learned first-hand about making a difference using the weapons and tools of international human rights law from my former boss in the Application of Standards Branch of a United Nations agency, ILO, in Geneva, Switzerland. My boss was K. T. Samson, who is credited with single-handedly founding the Human Rights Division of the legal staff, which existed in the ILO for one brief, shining moment. After a particularly difficult negotiation in which he tried to convince a diplomat to enforce international conventions protecting the freedom of association, he looked out the window toward the captivating beauty of Mont Blanc and said, "We don't move those mountains, but we push, we push."

Many years later, Dr. Sameera Al-Tuwaijri brought me back to that magical office, like a character in the Broadway play *Brigadoon*. Nothing in the ILO had changed while I raised my family, and my new office held the very same view of Mont Blanc. Thanks to her, my work as coordinatrice of the ILO encyclopedia for years rekindled my desire to learn, write, and bring a new vision of safety and health to our field.

My newlywed husband, Dominique Charoy, has not had much time to enjoy this journey, but he is patient and kind to indulge my interest and the massive time required for this work and a great travel companion when I present talks in Greece and Japan. Our chapter in my quest to make the world safer and healthier is yet to be written, but I know it will be full of meaningful adventures that will make a difference for the work, health, and survival of our society.

I know my work; it's writing. Reading is not a passive or natural task. I write, you read. Thank you one and all for reading this work.

Ilise L. Feitshans

2018

Index

CPSIA information can be obtained
at www.ICGtesting.com
Printed in the USA
BVHW072115071218
535054BV00010B/193/P